Lecture Notes in Physics

The Lecture Notes in Physics

The series Lecture Notes in Physics (LNP), founded in 1969, reports new developments in physics research and teaching – quickly and informally, but with a high quality and the explicit aim to summarize and communicate current knowledge in an accessible way. Books published in this series are conceived as bridging material between advanced graduate textbooks and the forefront of research and to serve three purposes:

- to be a compact and modern up-to-date source of reference on a well-defined topic
- to serve as an accessible introduction to the field to postgraduate students and nonspecialist researchers from related areas
- to be a source of advanced teaching material for specialized seminars, courses and schools

Both monographs and multi-author volumes will be considered for publication. Edited volumes should, however, consist of a very limited number of contributions only. Proceedings will not be considered for LNP.

Volumes published in LNP are disseminated both in print and in electronic formats, the electronic archive being available at springerlink.com. The series content is indexed, abstracted and referenced by many abstracting and information services, bibliographic networks, subscription agencies, library networks, and consortia.

Proposals should be sent to a member of the Editorial Board, or directly to the managing editor at Springer:

Christian Caron
Springer Heidelberg
Physics Editorial Department I
Tiergartenstrasse 17
69121 Heidelberg / Germany
christian.caron@springer.com

G. Gallavotti

The Fermi-Pasta-Ulam Problem

A Status Report

Springer

Editor

Giovanni Gallavotti
Università Roma La Sapienza
Dipartimento di Matematica
Piazzale Aldo Moro, 2
00185 Roma, Italy
Giovanni.Gallavotti@roma1.infn.it

G. Gallavotti (Ed.), *The Fermi-Pasta-Ulam Problem: A Status Report*, Lect. Notes Phys. 728 (Springer, Berlin Heidelberg 2008), DOI 10.1007/ 978-3-540-72995-2

ISSN 0075-8450
ISBN 978-3-642-09209-1 e-ISBN 978-3-540-72995-2

Springer is a part of Springer Science+Business Media
springer.com

Softcover reprint of the hardcover 1st edition 2008

Cover design: eStudio Calamar S.L., F. Steinen-Broo, Pau/Girona, Spain

Contents

1

Introduction to FPU

G. Gallavotti

Università Roma La Sapienza, Dipto. di Matematica, Piazzale Aldo Moro, 200185 Roma, Italy
giovanni.gallavotti@roma1.infn.it

1.1 The FPU Experiment and Its Ramifications

The FPU experiment opened the way to molecular dynamics simulations as well as to the use of computers to study fundamental questions on the foundations of Statistical Mechanics.

In the 1960s–1970s most efforts were concentrated on equilibrium statistical mechanics (the theory of phase transitions and of the critical point with the important successes of the Renormalization Group), but at the same time a strong attention was dedicated to the original scope of Fermi, namely to understand the ergodic hypothesis and its apparent contradiction with the results of the FPU experiment.

It is well known that Fermi started his scientific publications with a study of the ergodic hypothesis and with a "proof" of generic ergodicity [1]. The proof was incomplete because it only proved, extending an argument by Poincaré, that no smooth surface could divide phase space into two regions containing open sets. Even though he probably realized that there was a problem with the proof of his far more general claim, he was certainly not convinced and, around 1953–54, he started the experiment to check the idea that essentially any nonlinearity would lead to a system satisfying the ergodic hypothesis. The Hamiltonian of the systems of N oscillators $\mathbf{r} = (p_x, r_x)$, $x = 1, ..., N$, considered by FPU was $(N = 64, r_0 = r_N = 0)$

$$\sum_{x=0}^{N-1} \frac{p_x^2}{2m} + \sum_{x=1}^{N} \frac{1}{2}(r_{x+1} - r_x)^2 + \frac{\alpha}{3}(r_{x+1} - r_x)^3 + \frac{\beta}{4}(r_{x+1} - r_x)^4 .$$

The outcome was quite against the conjectured conclusion and a new era was opened to understand why the tempting explanation of ergodicity as due to "any" (reasonable) nonlinearity failed. Several developments followed as consequences of the attempts to clarify the above "FPU phenomenon"; perhaps the main one was the realization that the FPU dynamics could be

G. Gallavotti: *Introduction to FPU*, Lect. Notes Phys. **728**, 1–19 (2008)
DOI 10.1007/978-3-540-72995-2_1

approximated, at least if the initial conditions were close to certain special ones ("long waves"), by an integrable partial differential equation whose integrals of motion would only be approximate integrals of the FPU equations, but which could remain almost constant for long times. Several aspects of this development, started by M. Kruskal and N. Zabusky [2], are represented in the reviews in this collection.

Another attempt was to appeal to the KAM theorem, discovered at about the same time of the FPU experiment; the theorem seemed to offer a natural explanation to the phenomenon of the apparent lack of ergodicity: for low enough energy the FPU model would show quasi periodic motions whose initial data would fill most of phase space [3].

The explanation was soon realized to be untenable: the energy had to be too close to a minimum if the KAM estimates (known since Kolmogorov's work and, for chains or lattices of oscillators, not greatly improved since) had to guarantee quasi periodic motions. And worse, the closeness to a minimum value had a strong dependence on the number N of oscillators (i.e. it decreased to 0 exponentially in N at least). Not to mention that a proof of applicability of the KAM theorem to the FPU model was not general enough to cover all the interesting cases; in this respect the proof that at the very low energy most motions were quasi periodic has been eventually completed in all details [4], but it has not added new direct information about the explanation of the FPU results.

The question of why the FPU motions did not show ergodic behavior underwent a long period of research, and it was soon suggested that at small energy equipartition could be essentially true only for what concerned the part of the energy that was located in the normal modes with longest wavelength [5]. However the meaning of "essentially" gradually evolved into a problem of time scales: the energy could not stay confined to the longest wavelength (if initially it was concentrated there) forever [5]. After a transient time equipartition would be reached; but the transient time had a stretched exponential dependence on the energy above a minimal one.

1.2 A Guided Tour of This Volume

The initial idea for this volume emerged quite some time ago, at the conference "FPU fifty years later" ("FPU cinquanta anni dopo") which took place in Rome in June of 2004, at the Physics Department of "La Sapienza." All the questions briefly outlined in the previous section were addressed and debated and are to some extent reflected by the various contributions collected here as substantial and carefully edited extensions of the lectures delivered.

The order of the contributions reflects the aim of preparing the interested but unexperienced reader through a gradual understanding starting from general analysis and proceeding towards more specialized topics.

In the contribution by Lichtenberg et al. [6] a study of one-dimensional chains is performed: it presents an approach different with respect to the one in the subsequent reviews [5, 7]. More attention is dedicated to the equipartition following initial data that have short wavelength and other fundamental problems are studied; like theory of Lyapunov exponents and heat transport in FPU or more general lattices (like Klein–Gordon lattices). The authors often make the attempt to present heuristic theories. The approach of this work is also interesting for its personal touch: it shows, if compared to the other reviews, that the subject is so rapidly developing that the historic perspective of different groups can be remarkably different even when dealing with the same physical phenomena. I have decided not to try to invite the authors of this and of the other reviews to make the various viewpoints agree, a possibly unrealistic aim, but to leave the papers as the authors decided to write them so that readers could see by themselves how different the perspectives can be.

The general introduction ends with a description of numerical methods, like the definition and use of equipartition indicators to study the problems of low energy and long wave length or Lyapunov exponents. For the latter a heuristic theory is presented: it appears to give extremely good agreement with simulations and therefore it leaves us with a challenge to understand why.

Such aspects are stressed and developed also in the review by Cencini et al. [8] where the question of the "long standing and controversial problem of distinguishing chaos from noise in signal analysis" is first discussed. And it is interesting as it would seem that this is an impossible task, if understood in its literal meaning. Then the authors discuss the importance of Chaos for the theory of macroscopic transport properties (in systems not necessarily close to a ground state). The idea is presented that, after all, Chaos is not as important as many, including the writer, think it is; it is not necessary even if one limits oneself to understanding the simplest macroscopic transport properties and nonequilibrium phenomena.

In reality by Chaos the authors mean a signal emitted by a dynamical system which has positive finite Kolmogorov–Sinai (or "KS") entropy (i.e. essentially a system with bounded phase space) and by noise—a signal emitted by a dynamical system which has infinite KS-entropy. The second question is the relevance of Chaos for diffusion and heat conduction; the conclusion reached is that it is not necessarily relevant as shown by way of a few examples. The discussion, supported by simulations, may not be convincing for every reader but it is certainly stimulating as the given examples are witty and have theoretical interest, being different from the trivial remark that all numerical simulations are, strictly speaking, necessarily nonchaotic although they provide effective means to exhibit chaos properties.

Metastability, i.e. the approach to equipartition within an "exponential" time scale, is discussed by [4] in great detail for one-dimensional chains, the classical FPU case, together with results that are by now classical, thanks mainly to the work of the authors. It clearly emerges that even at small energy, equipartition appears to be eventually reached; however, the time scale increases as a stretched exponential in the inverse of the energy in excess over its minimum. Thus if the initial data have a specific energy below a suitable threshold the motion is metastable in the above sense, while for energies above the threshold the motion approaches equipartition on a time scale of the order of an inverse power of the specific excess energy over the ground state. This is a result on which there seem to be few doubts since the 1980s. It is an aspect of the FPU motions which goes far beyond the analysis in the subsequent [9]; but there is no mathematical result that allows realistic estimates (i.e. close, conceptually, to the results suggested by the numerical simulations).

Unfortunately the authors have chosen not to discuss one of the most interesting developments that were presented at the conference; namely, the fact that this picture might be strongly dimension dependent. In the FPU lattices of higher dimension (i.e. with x located on a lattice Z^d of dimension $d \geq 2$ rather than on Z as above) the time scale for the equipartition at low energy seems to be very fast, i.e. to increase, as the energy E approaches its minimum, as a power in E^{-1} (rather than a stretched exponential appearing in the $d = 1$ case). In this sense Fermi's idea that any perturbation would turn a system of harmonic oscillators into an ergodic system would be vindicated. It seems that we shall have to wait a little more time to be sure of this new scenario, which would make the dimension play a key role in the problem of equipartition, making the case of dimension 2 a "marginal case" and the dimension 3 (or more) cases as cases in which the behavior follows the most naive scenario about equipartition, i.e. an approach on a time scale of the order of an inverse power of E, even at low energies.

The authors present the problems from their personal historical perspective: which is very interesting because most of the work described was developed by them, and achieved in spite of a sometimes strong criticism.

It should be stressed that so far the metastability scenario is not proved in any case, at least not as rigorously as such an important issue would deserve.

The relation of the FPU motions $t \to r_x(t)$ with the integrable equations is examined by Bambusi and Ponno [9] from the point of view of recent developments in perturbation theory and the results of Kruskal and Zabusky are interpreted (and extended to more general equations) determining conditions for the possibility of explaining the FPU "recurrences" in terms of an approximating the FPU motions with solutions of integrable PDE equations. If $\varepsilon \stackrel{\text{def}}{=} \frac{E}{N}$ is the value of the energy of the lowest momentum mode divided by the number of particles then $\mu = \varepsilon^{\frac{n-2}{4}}$ is a good scaling parameter for solutions of the form $r_x(t) = \mu^{\frac{2}{n-2}} u(\mu x, \mu t)$ with a nonlinearity $U(r) = \frac{1}{2}r^2 + \frac{r^n}{n}$, n being an integer parameter $n \geq 3$. The review studies the FPU lattice with periodic boundary conditions ($r_0 = r_N$ rather than $r_0 = r_N = 0$) and distinguishes between

(a) long-time concentration of the solution on low lying modes: i.e. closeness to the solution of the PDE that $u(\xi, \tau)$ satisfies to lowest order in μ in the formal limit $\mu \to 0$. The latter approximates the solution $r_x(t)$ for a time of the order of an inverse power of ε; or the stronger property:

(b) metastability: i.e. closeness to the solution of the PDE that $u(\xi, \tau)$ satisfies to lowest order in μ in the formal limit $\mu \to 0$; it approximates the solution $r_x(t)$ for a time of the order of a stretched exponential of ε, $O(e^{+b\varepsilon^{-a}})$ for some $a, b > 0$.

The reported (new) result is the validity of (a) as an application of the so called averaging method: (a) is proved for the case $n = 3$. In this case also metastability is conjectured (but not proved) to be a consequence of the integrability of the PDE associated with the leading order as $\tau \to 0$; and if so metastability would also imply quasi periodicity of the motion, hence possibly the "recurrence" phenomena discovered in the FPU experiment.

For the higher n cases, property (a) should hold (but this is not proved) for $n = 4, 5$ (and if $n = 4$ then it should also imply metastability and recurrence because the PDE is still integrable). The general conjecture is that in fact metastability (i.e. property (b)) occurs as long as the PDE formally satisfied by u does not lead to a blow up of the solution in a finite time. The authors conjecture therefore that for $n \leq 5$ there is metastability, while for $n \geq 7$ (where blow up is established) it holds at most as long as there is no blow up. The case $n = 6$ is a critical case and the results could depend on the initial data in a stronger way.

The next review is by James and Sire [10] with the analysis of the recurrent motions and of the approximability of solutions of FPU type of systems by simpler systems can also be studied, without trying to connect them with integrable PDEs by the methods of perturbation theory, in alternative to [9]. The method links the simplicity of certain motions of FPU-like systems to the possibility of describing them by simple ODEs with few degrees of freedom via the methods of bifurcation theory; such motions develop on small dimension invariant manifolds. The theory leads to a systematic analysis of special kinds of motions, "breathers and travelling breathers" [10]. This is an approach which follows classical paths of bifurcation theory and is an interesting development parallel and complementary to the approach in the review [9].

It explains some of the FPU phenomena and stands on the use of extensions to infinite dimension of the center manifold theorems. It is important to stress that the motions that are found by the authors are quite different from the ones that are associated with long wavelength initial data (i.e. of the kind considered in the papers [5, 9]) because the energy is not concentrated on long wavelengths but should be regarded as distributed in packets involving modes of all wavelengths and producing wave forms with strong localization properties (breathers are localized in space and therefore delocalized in their momentum components).

In the contribution by Paleari and Penati [7] the methods of analysis of the one-dimensional chains are described in great detail and the chaotic motions are described in quantitative detail. In the frame of the research on the threshold for metastability several chaotic motions properties that develop and eventually lead to equipartition have been studied; for instance, the energy can be shared by various collections of modes which can be divided into energy sharing groups, whose energies become equalized on the same time scale. The time scales are systematically studied via an accurate analysis of the Lyapunov exponents of the motions. The authors review the numerical algorithms used for the analysis and the criteria for quantitative estimates of the "equipartition degree" of the energy in the waves on a given group. The studies are not confined to the initial data of "FPU type," i.e. with energy concentrated on long wavelengths, and they discuss the energy flow from many initial data thus providing results that at least in principle may relate the regimes considered in [5, 9, 10]. The question of how to measure the degree of energy sharing within "clusters of modes" requires the use of "sharing indicators" and leads to evidence that the low energy evolutions show metastability phenomena not only if the initial data have long wavelength but also if they have, instead, short wavelength. However the exponentials time scales are not described by stretched exponentials with the same stretching exponent, but the exponent is smaller for the short wavelengths.

The series of reviews is completed by B. Rink [11] with a description of the recent work [4], proving the applicability of KAM theory in FPU chains with energy very close to a minimum: this leads to a full realization of a program initiated by Nishida [12] who solved it under a condition that could be checked only exceptionally. The problem was completely solved only recently with the paper reproduced here in extended version and covering the fixed extremes boundary conditions (the periodic boundary conditions had been solved earlier). It is important to keep in mind that most of the results that have been obtained on FPU, including in particular the ones mentioned in the collection edited here, are not mathematically completely proved; therefore, it is of interest to present one of the few complete results on the subject, i.e. the applicability of the KAM theory to the actual FPU system at energy extremely close to the ground state energy. Even though the closeness to the minimum energy is so tiny (and infinitesimal as $N \to \infty$) to be of little physical interest in the interpretation of the FPU experiment the result is not obvious, and is nontrivial, because the FPU system has symmetries, which generate degeneracies in the frequency spectra of the motions of very low energy (not only in their harmonic approximation but also in their simplest averaged versions): and degeneracies correspond to "resonances" and resonances are the key difficulty (as well as source of interesting behavior) in the KAM theory.

Given the importance of the FPU system it is surprising that the conceptual problem of KAM applicability, even if only in principle, was left with several open questions for so many years: the reason has to be found, possibly, in the distrust that physicists have consistently shown towards the KAM results, considered not only esoteric but also quite irrelevant given the orders of magnitude of the time scales involved.

The original work of FPU is easy to find it in the libraries: but the readers may find it useful and stimulating to have it ready at hand while getting the glimpse, offered here, of the vast literature that it has influenced. For this reason the paper is appended to this introduction.

Roma: April 2007 *Giovanni Gallavotti*

References

1. Fermi, E.: Beweis, daß ein mechanisches normalsystem im algemeinen quasi-ergodisch ist. Physikalische Zeitschrift, 24, 261–265, 1923. Reprinted in paper n. 11a.
   Fermi, E.: Note e Memorie (Collected papers), Accademia dei Lincei and University of Chicago Press, vol. I, 1961, e vol. II, 1965.
2. N.J. Zabusky, M.D. Kruskal Interaction of "Solitons" in a Collisionless Plasma and the Recurrence of Initial States, Physical Review Letters, 15, 240–243, 1965.
3. F.M. Izrailiev, B.V. Chirikov, Statistical properties of a nonlinear string, Soviet Physics Doklady, 11, 30–34, 1966.
4. B. Rink, Proof of Nishida's conjecture on anharmonic lattices, Communications Mathematical Physics, 261, 613–627, 2006.
5. G. Benettin, A. Carati, L. Galgani and A. Giorgiui: The Fermi–Pasta–Ulam problem and the metastability perspective, Lect. Note Phys. 728 (2007).
6. A.J. Lichtenberg, R. Livi, M. Pettini and S. Ruffo, Dynamics of oscillator chains, Lecture Notes in Physics **728** (2007).
7. S. Paleari and T. Penati: Numerical methods and results in the FPU problem, Lecture Notes in Physics **728** (2007).
8. M. Cencini, F. Cecconi, M. Falcioni and A. Vulpiani: Role of chaos for the validity of statistical mechanics laws: Diffusion and conductance, Lecture Notes in Physics **728** (2007).
9. D. Bambusi and A. Ponno: Resonance, metastability and blow up in FPU, Lecture Notes in Physics **728** (2007).
10. G. James and Y. Sire: Center manifold theory in the context of infinite one-dimensional lattices, Lecture Notes in Physics **728** (2007).
11. B. Rink: An integrable approximation for the Fermi–Pasta–Ulam Lattice, Lecture Notes in Physics **728** (2007).
12. Nishida, T., A note on an existence of conditionally periodic oscillation in a one-dimensional anharmonic lattice, Memoirs of the Faculty of Engineering of Kyoto University, 33, 27–34, 1971.

1.3 Appendix: Reprint of Fermi, Pasta, Ulam – Studies of Nonlinear Problems (1955)

N° 266.

After the war, during one of his frequent summer visits to Los Alamos, Fermi became interested in the development and potentialities of the electronic computing machines. He held many discussions with me on the kind of future problems which could be studied through the use of such machines. We decided to try a selection of problems for heuristic work where in absence of closed analytic solutions experimental work on a computing machine would perhaps contribute to the understanding of properties of solutions. This could be particularly fruitful for problems involving the asymptotic—long time or "in the large" behavior of non-linear physical systems. In addition, such experiments on computing machines would have at least the virtue of having the postulates clearly stated. This is not always the case in an actual physical object or model where all the assumptions are not perhaps explicitly recognized.

Fermi expressed often a belief that future fundamental theories in physics may involve non-linear operators and equations, and that it would be useful to attempt practice in the mathematics needed for the understanding of non-linear systems. The plan was then to start with the possibly simplest such physical model and to study the results of the calculation of its long-time behavior. Then one would gradually increase the generality and the complexity of the problem calculated on the machine. The Los Alamos report LA–1940 (paper N° 266) presents the results of the very first such attempt. We had planned the work in the summer of 1952 and performed the calculations the following summer. In the discussions preceding the setting up and running of the problem on the machine we had envisaged as the next problem a two-dimensional version of the first one. Then perhaps problems of pure kinematics e.g., the motion of a chain of points subject only to constraints but no external forces, moving on a smooth plane convoluting and knotting itself indefinitely. These were to be studied preliminary to setting up ultimate models for motions of system where "mixing" and "turbulence" would be observed. The motivation then was to observe the *rates* of mixing and "thermalization" with the hope that the calculational results would provide hints for a future theory. One could venture a guess that one motive in the selection of problems could be traced to Fermi's early interest in the ergodic theory. In fact, his early paper (N° 11 *a*) presents an important contribution to this theory.

It should be stated here that during one summer Fermi learned very rapidly how to *program* problems for the electronic computers and he not only could plan the general outline and construct the so-called flow diagram but would work out himself the actual *coding* of the whole problem in detail.

The results of the calculations (performed on the old MANIAC machine) were interesting and quite surprising to Fermi. He expressed to me the opinion that they really constituted a little discovery in providing intimations that the prevalent beliefs in the universality of "mixing and thermalization" in non-linear systems may not be always justified.

A few words about the subsequent history of this non-linear problem. A number of other examples of such physical systems were examined by calculations on the electronic computing machines in 1956 and 1957. I presented the results of the original paper on several occasions at scientific meetings; they seemed to have aroused considerable interest among mathematicians and physicists and there is by now a small literature dealing with this problem. The most recent results are due to N. J. Zabusky.[(1)] His analytical work shows, by the way, a good agreement of the numerical computations with the continuous solution up to a point where a discontinuity developed in the derivatives and the analytical work had to be modified. One obtains from it another indication that the phenomenon discovered

(1) Exact Solutions for the Vibrations of a non-linear continuous string. A. E. C. Research and Development Report. MATT–102, Plasma Physics Laboratory, Princeton University, October 1961.

is not due to numerical accidents of the algorithm of the computing machine, but seems to constitute a real property of the dynamical system.

In 1961, on more modern and faster machines, the original problem was considered for still longer periods of time. It was found by J. Tuck and M. Menzel that after one continues the calculations from the first " return " of the system to its original condition the return is not complete. The total energy is concentrated again essentially in the first Fourier mode, but the remaining one or two percent of the total energy is in higher modes. If one continues the calculation, at the end of the next great cycle the error (deviation from the original initial condition) is greater and amounts to perhaps three percent. Continuing again one finds the deviation increasing—after eight great cycles the deviation amounts to some eight percent; but from that time on an opposite development takes place! After eight more i.e., sixteen great cycles altogether, the system gets very close—better than within one percent to the original state! This supercycle constitutes another surprising property of our non-linear system.

Paper N° 266 is not the only work that Fermi and I did together. In the summer of 1950 we made a study of the behavior of the thermonuclear reaction in a mass of deuterium and wrote a report, LA-1158, which is still classified. The problem is of enormous mathematical complexity, involving the hydrodynamics of the motion of the material, the hydrodynamics of radiation energy, all interwoven with the processes of the various reactions between the nuclei whose probabilities and properties depend i.a., on temperature, density, and the changing geometry of the materials. The aim of this work was to obtain, by a schematized but still elaborate picture of the evolution of all these physical processes, an idea of the propagation of such a reaction. This was to complement a previous work by Everett and myself, dealing with the problem of ignition of a mass deuterium. Assuming an ignition somehow started in a large volume, one wanted to evaluate the prospects of propagation of the reactions already started. Many ingenious schematizations and simplifications had to be introduced in order to describe the process, without the possibility of calculating in exact detail the innumerable geometrical and thermodynamical factors. The results of our computations on the chances of propagation were negative and the report played an important role in channeling imagination and energies towards a search for a different scheme for a successful hydrogen reaction. This was indeed found later on on a different basis. All the calculations on which the work of the report is based were performed on desk computers and slide rules. The subsequent massive and lengthy work on the electronic computer machines (organized and performed by von Neumann, F. and C. Evans and others) confirmed in large lines, qualitatively and to a good degree quantitatively the behavior of the system as estimated and predicted in our report—with its combination of intuitive evaluations, schematized equations and hand calculations.

S. M. Ulam.

266.

STUDIES OF NON LINEAR PROBLEMS

E. Fermi, J. Pasta, and S. Ulam

Document LA-1940 (May 1955).

Abstract.

A one-dimensional dynamical system of 64 particles with forces between neighbors containing nonlinear terms has been studied on the Los Alamos computer Maniac I. The nonlinear terms considered are quadratic, cubic, and broken linear types. The results are analyzed into Fourier components and plotted as a function of time.

The results show very little, if any, tendency toward equipartition of energy among the degrees of freedom.

The last few examples were calculated in 1955. After the untimely death of Professor E. Fermi in November, 1954, the calculations were continued in Los Alamos.

This report is intended to be the first one of a series dealing with the behavior of certain nonlinear physical systems where the non-linearity is introduced as a perturbation to a primarily linear problem. The behavior of the systems is to be studied for times which are long compared to the characteristic periods of the corresponding linear problems.

The problems in question do not seem to admit of analytic solutions in closed form, and heuristic work was performed numerically on a fast electronic computing machine (MANIAC I at Los Alamos).(1) The ergodic behavior of such systems was studied with the primary aim of establishing, experimentally, the rate of approach to the equipartition of energy among the various degrees of freedom of the system. Several problems will be considered in order of increasing complexity. This paper is devoted to the first one only.

We imagine a one-dimensional continuum with the ends kept fixed and with forces acting on the elements of this string. In addition to the usual linear term expressing the dependence of the force on the displacement of the element, this force contains higher order terms. For the purposes of numerical work this continuum is replaced by a finite number of points (at most 64 in our actual computation) so that the partial differential equation defining the motion of this string is replaced by a finite number of total differential equations. We have, therefore, a dynamical system of 64 particles with forces acting between neighbors with fixed end points. If x_i denotes the displacement of the i-th point from its original position, and α denotes the coefficient of the quadratic term in the force between the neighboring mass points and β that of the cubic term, the equations were either

$$x_i = (x_{i+1} + x_{i-1} - 2\,x_i) + \alpha\,[(x_{i+1} - x_i)^2 - (x_i - x_{i-1})^2] \qquad (i = 1\,,2\,,\cdots,64), \tag{1}$$

or

$$x_i = (x_{i+1} + x_{i-1} - 2\,x_i) + \beta\,[(x_{i+1} - x_i)^3 - (x_i - x_{i-1})^3] \qquad (i = 1\,,2\,,\cdots,64). \tag{2}$$

α and β were chosen so that at the maximum displacement the nonlinear term was small, e.g., of the order of one-tenth of the linear term. The corresponding partial differential equation obtained by letting the number of particles become infinite is the usual wave equation plus non-linear terms of a complicated nature.

Another case studied recently was

$$\ddot{x}_i = \delta_1\,(x_{i+1} - x_i) - \delta_2\,(x_i - x_{i-1}) + c \tag{3}$$

(1) We thank Miss Mary Tsingou for efficient coding of the problems and for running the computations on the Los Alamos MANIAC machine.

where the parameters δ_1, δ_2, c were not constant but assumed different values depending on whether or not the quantities in parentheses were less than or greater than a certain value fixed in advance. This prescription amounts to assuming the force as a broken linear function of the displacement. This broken linear function imitates to some extent a cubic dependence. We show the graphs representing the force as a function of displacement in three cases.

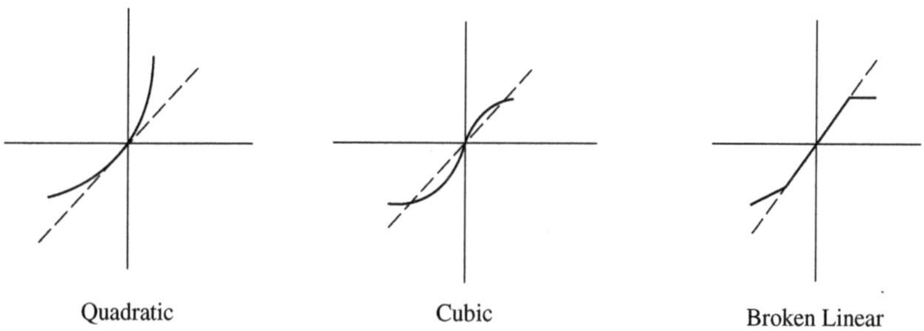

The solution to the corresponding linear problem is a periodic vibration of the string. If the initial position of the string is, say, a single sine wave, the string will oscillate in this mode indefinitely. Starting with the string in a simple configuration, for example in the first mode (or in other problems, starting with a combination of a few low modes), the purpose of our computations was to see how, due to nonlinear forces perturbing the periodic linear solution, the string would assume more and more complicated shapes, and, for t tending to infinity, would get into states where all the Fourier modes acquire increasing importance. In order to see this, the shape of the string, that is to say, x as a function of i and the kinetic energy as a function i were analyzed periodically in Fourier series. Since the problem can be considered one of dynamics, this analysis amounts to a Lagrangian change of variables: instead of the original $\dot{x}_i$ and x_i, $i = 1, 2, \cdots, 64$, we may introduce a and $\dot{a}_k$, $k = 1, 2, \cdots, 64$, where

(4) $$a_k = \Sigma x_i \sin \frac{ik\pi}{64}.$$

The sum of kinetic and potential energies in the problem with a quadratic force is

(5 a) $$E^{\text{kin}}_{x_i} + E^{\text{pot}}_{x_i} = \frac{1}{2}\dot{x}_i^2 + \frac{(x_{i+1} - x_i)^2 + (x_i - x_{i-1})^2}{2}$$

(5 a) $$E^{\text{kin}}_{a_k} + E^{\text{pot}}_{a_k} = \frac{1}{2}\dot{a}_k^2 + 2a_k^2 \sin^2 \frac{\pi k}{128}$$

if we neglect the contributions to potential energy from the quadratic or higher terms in the force. This amounts in our case to at most a few percent.

The calculation of the motion was performed in the x variables, and every few hundred cycles the quantities referring to the a variables were computed by the above formulas. It should be noted here that the calculation of the motion could be performed directly in a_k and $\dot{a}_k$. The formulas, however become unwieldy and the computation, even on an electronic computer, would take a long time. The computation in the a_k variables could have been more instructive for the purpose of observing directly the interaction between the a_k's. It is proposed to do a few such calculations in the near future to observe more directly the properties of the equations for $\ddot{a}_k$.

Let us say here that the results of our computations show features which were, from the beginning, surprising to us. Instead of a gradual, continuous flow of energy from the first mode to the higher modes, all of the problems show an entirely different behavior. Starting in one problem with a quadratic force and a pure sine wave as the initial position of the string, we indeed observe initially a gradual increase of energy in the higher modes as predicted (e.g., by Rayleigh in an infinitesimal analysis). Mode 2 starts increasing first, followed by mode 3, and so on. Later on, however, this gradual sharing of energy among successive modes ceases. Instead, it is one or the other mode that predominates. For example, mode 2 decides, as it were, to increase rather rapidly at the cost of all other modes and becomes predominant. At one time, it has more energy than all the others put together! Then mode 3 undertakes this role. It is only the first few modes which exchange energy among themselves and they do this in a rather regular fashion. Finally, at a later time mode 1 comes back to within one percent of its initial value so that the system seems to be almost periodic. All our problems have at least this one feature in common. Instead of gradual increase of all the higher modes, the energy is exchanged, essentially, among only a certain few. It is, therefore, very hard to observe the rate of " thermalization " or mixing in our problem, and this was the initial purpose of the calculation.

If one should look at the problem from the point of view of statistical mechanics, the situation could be described as follows: the phase space of a point representing our entire system has a great number of dimensions. Only a very small part of its volume is represented by the regions where only one or a few out of all possible Fourier modes have divided among themselves almost all the available energy. If our system with nonlinear forces acting between the neighboring points should serve as a good example of a transformation of the phase space which is ergodic or metrically transitive, then the trajectory of almost every point should be everywhere dense in the whole phase space. With overwhelming probability this should also be true of the point which at time $t = 0$ represents our initial configuration, and this point should spend most of its time in regions corresponding to the equipartition of energy among various degrees of freedom. As will be seen from the results this seems hardly the case. We have plotted (figs. 1 to 7) the ergodic sojourn times in certain subsets of our phase space. These may show a tendency to approach limits as guaranteed by the ergodic theorem. These limits,

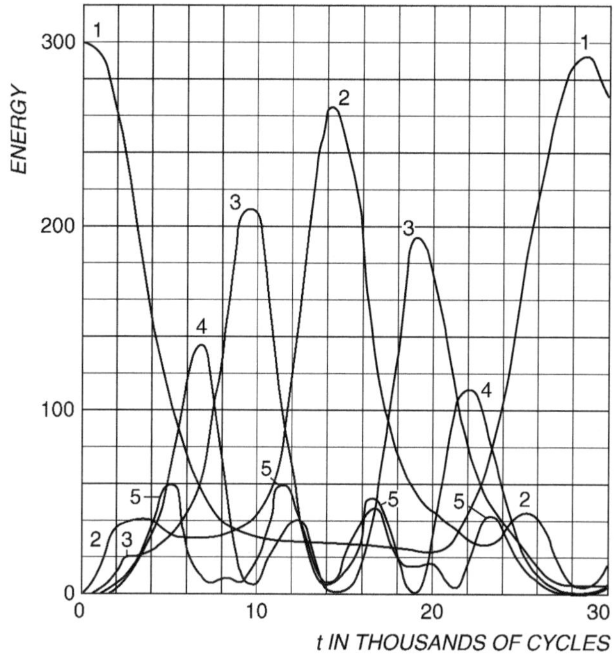

Fig. 1. - The quantity plotted is the energy (kinetic plus potential in each of the first five modes). The units for energy are arbitrary. N = 32; α = 1/4; δt^2 = 1/8. The initial form of the string was a single sine wave. The higher modes never exceeded in energy 20 of our units. About 30,000 computation cycles were calculated.

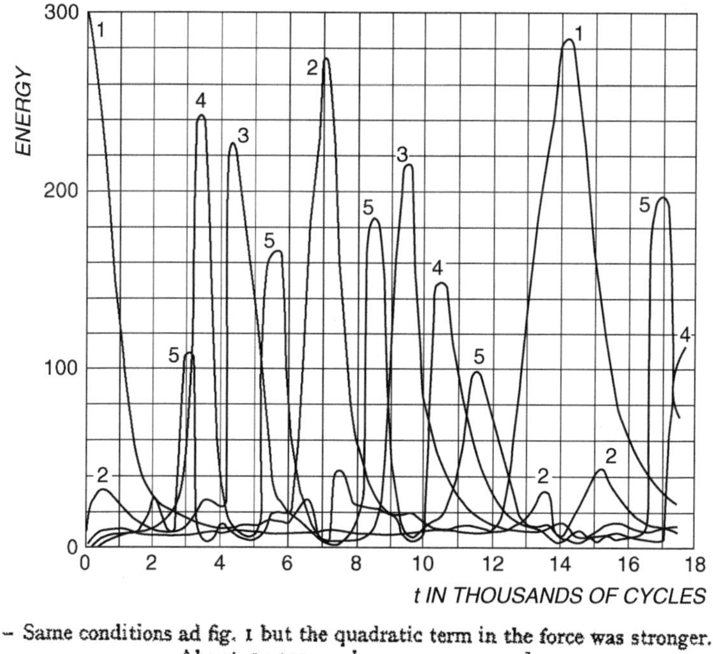

Fig. 2. - Same conditions ad fig. 1 but the quadratic term in the force was stronger. α = 1. About 14,000 cycles were computed.

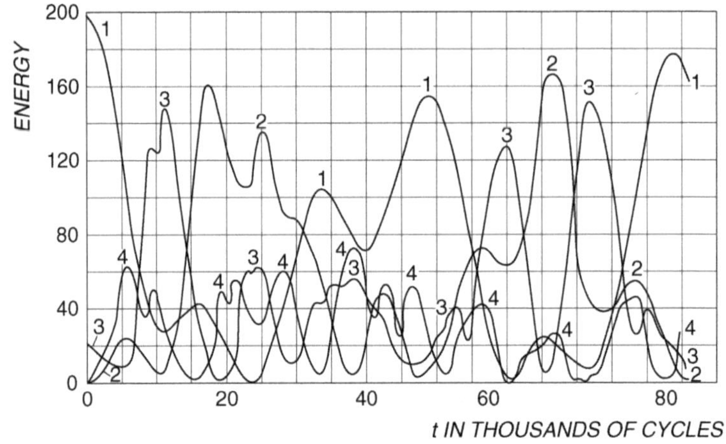

Fig. 3. – Same conditions as in fig. 1, but the initial configuration of the string was a "saw-tooth" triangular-shaped wave. Already at $t = 0$, therefore, energy was present in some modes other than 1. However, modes 5 and higher never exceeded 40 of our units.

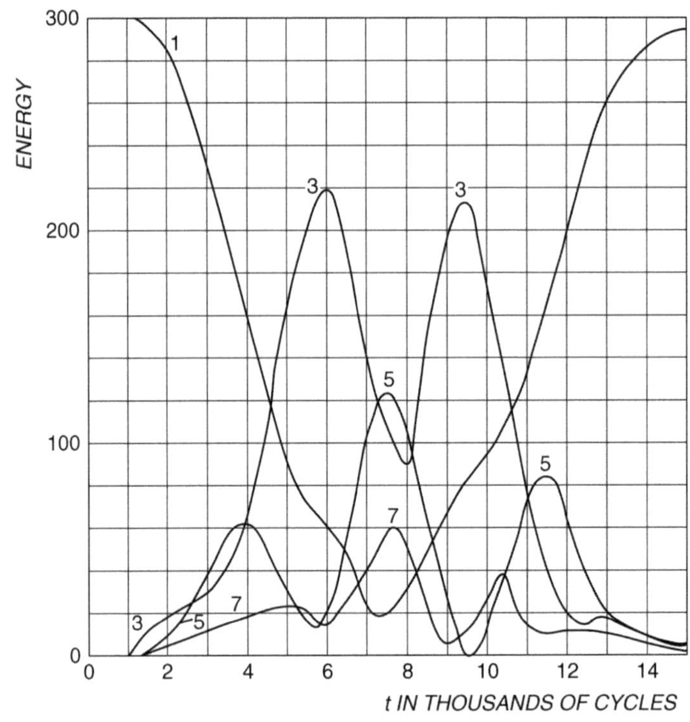

Fig. 4. – The initial configuration assumed was a single sine wave; the force had a cubic term with $\beta = 8$ and $\delta t^2 = 1/8$. Since a cubic force acts symmetrically (in contrast to a quadratic force), the string will forever keep its symmetry and the effective number of particles for the computation is $N = 16$. The even modes will have energy 0.

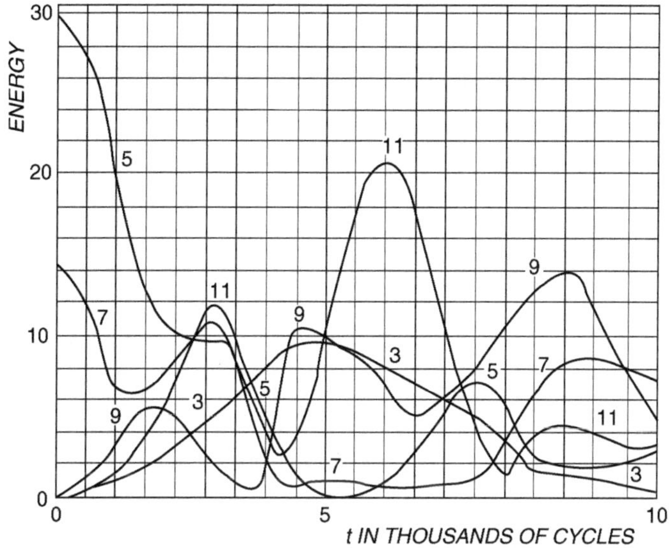

Fig. 5. – $N = 32$; $\delta t^2 = 1/64$; $\beta = 1/16$. The initial configuration was a combination of 2 modes. The initial energy was chosen to be 2/3 in mode 5 and 1/3 in mode 7.

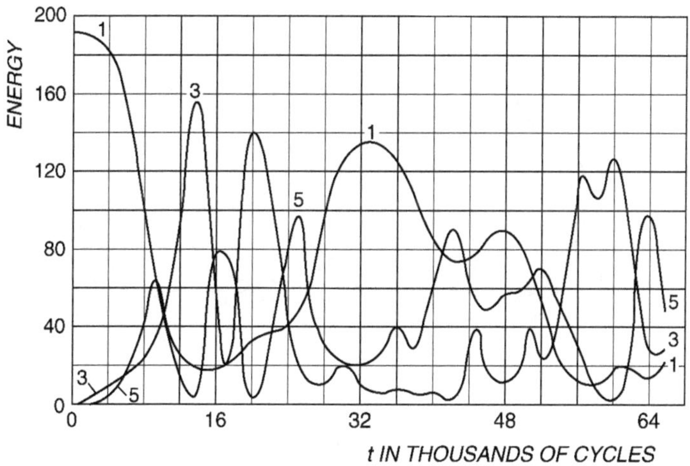

Fig. 6. – $\delta t^2 = 2^{-6}$. The force was taken as a broken linear function of displacement. The amplitude at which the slope changes was taken as $2^{-5} + 2^{-7}$ of the maximum amplitude. After this cut-off value, the force was assumed still linear but the slope increased by 25 percent. The effective $N = 16$.

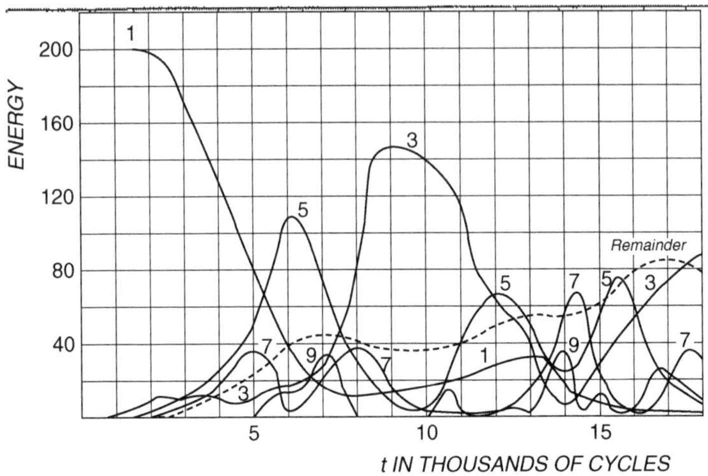

Fig. 7. – $\delta t^2 = 2^{-6}$. Force is again broken linear function with the same cut-off, but the slopes after that increased by 50 percent instead of the 25 percent charge as in problem 6. The effective N = 16.

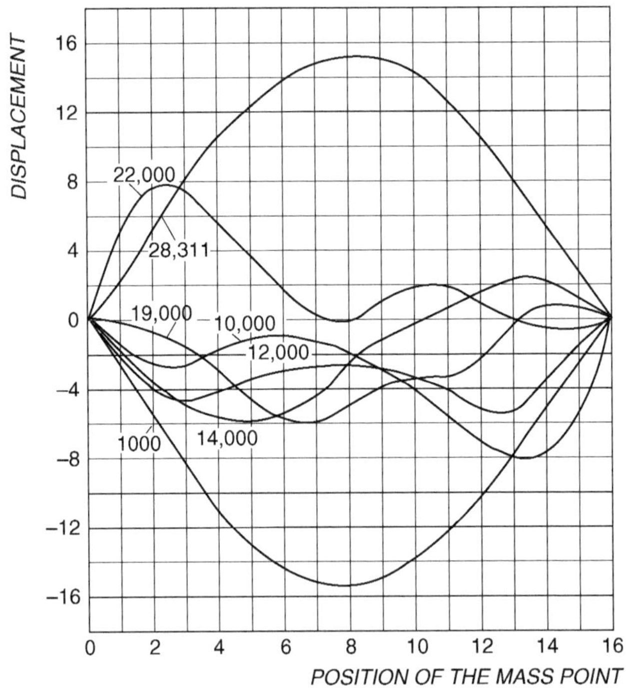

Fig. 8. – This drawing shows not the energy but the actual *shapes*, i.e., the displacement of the string at various times (in cycles) indicated on each curve. The problem is that of fig. 1.

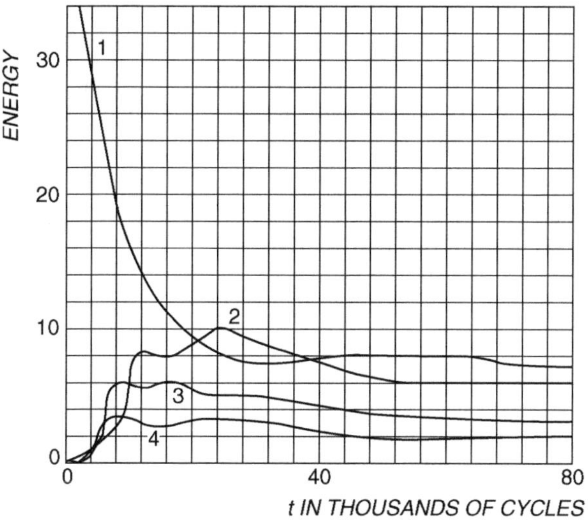

Fig. 9. This graph refers to the problem of fig. 6. The curves, numbered 1, 2, 3, 4, show the time averages of the kinetic energy contained in the first 4 modes as a function of time. In other words, the quantity is $\frac{1}{\nu}\sum_{i=1}^{\nu} T^{i}_{a_k}$. ν is the cycle no., $k = 1, 3, 5, 7$.

however, do not seem to correspond to equipartition even in the time average. Certainly, there seems to be very little, if any, tendency towards equipartition of energy among all degrees of freedom at a given time. In other words, the systems certainly do not show mixing. [2]

The general features of our computation are these: in each problem, the system was started from rest at time $t = 0$. The derivatives in time, of course, were replaced for the purpose of numerical work by difference expressions. The length of time cycle used varied somewhat from problem to problem. What corresponded in the linear problem to a full period of the motion was divided into a large number of time cycles (up to 500) in the computation. Each problem ran through many "would-be periods" of the linear problem, so the number of time cycles in each computation ran to many thousands. That is to say, the number of swings of the string was of the order of several hundred, if by a swing we understand the period of the initial configuration in the corresponding linear problem. The distribution of energy in the Fourier modes was noted every few hundred of the computation cycles. The accuracy of the numerical work was checked by the constancy of the quantity representing the total energy. In some cases, for checking purposes, the corresponding linear problems were run and these behaved correctly within one percent or so, even after 10,000 or more cycles.

It is not easy to summarize the results of the various special cases. One feature which they have in common is familiar from certain problems in me-

(2) One should distinguish between metric transitivity or ergodic behavior and the stronger property of mixing.

chanics of systems with a few degrees of freedom. In the compound pendulum problem one has a transformation of energy from one degree of freedom to another and back again, and not a continually increasing sharing of energy between the two. What is perhaps surprising in our problem is that this kind of behavior still appears in systems with, say, 16 or more degrees of freedom.

What is suggested by these special results is that in certain problems which are approximately linear, the existence of quasi-states may be conjectured.

In a linear problem the tendency of the system to approach a fixed "state" amounts, mathematically, to convergence of iterates of a transformation in accordance with an algebraic theorem due to Frobenius and Perron. This theorem may be stated roughly in the following way. Let A be a matrix with positive elements. Consider the linear transformation of the n–dimensional space defined by this matrix. One can assert that if $\bar{x}$ is any vector with all of its components positive, and if A is applied repeatedly to this vector, the directions of the vectors $\bar{x}$, $A(\bar{x}), \cdots, A^i(\bar{x}), \cdots,$ will approach that of a fixed vector $\bar{x}_0$ in such a way that $A(\bar{x}_0) = \lambda(\bar{x}_0)$. This eigenvector is unique among all vectors with all their components non-negative. If we consider a linear problem and apply this theorem, we shall expect the system to approach a steady state described by the invariant vector. Such behavior is in a sense diametrically opposite to an ergodic motion and is due to a very special character, linearity of the transformations of the phase space. The results of our calculation on the nonlinear vibrating string suggest that in the case of transformations which are approximately linear, differing from linear ones by terms which are very simple in the algebraic sense (quadratic or cubic in our case), something analogous to the convergence to eigenstates may obtain.

One could perhaps conjecture a corresponding theorem. Let Q be a transformation of a n–dimensional space which is nonlinear but is still rather simple algebraically (let us say, quadratic in all the coordinates). Consider any vector $\bar{x}$ and the iterates of the transformation Q acting on the vector $\bar{x}$. In general, there will be no question of convergence of these vectors $Q^n(\bar{x})$ to a fixed direction.

But a weaker statement is perhaps true. The directions of the vectors $Q^n(\bar{x})$ sweep out certain cones C_α or solid angles in space in such a fashion that the time averages, i.e., the time spent by $Q^n(\bar{x})$ in C_α, exist for $n \to \infty$. These time averages may depend on the initial $\bar{x}$ but are able to assume only a finite number of different values, given C_α. In other words, the space of all direction divides into a finite number of regions R_i, $i = 1, \cdots, k$, such that for vectors $\bar{x}$ taken from any one of these regions the percentage of time spent by images of $\bar{x}$ under the Q^n are the same in any C_α.

The graphs fig. 1–9 show the behavior of the energy residing in various modes as a function of time; for example, in fig. 1 the energy content of each of the first 5 modes is plotted. The abscissa is time measured in computational cycles, δt, although figure captions give δt^2 since this is the term involved directly in the computation of the acceleration of each point. In all problems the mass of each point is assumed to be unity; the amplitude of the displacement of each point is normalized to a maximum of 1. N denotes the number of points and therefore the number of modes present in the calculation. α denotes the coefficient of the quadratic term and β that of the cubic term in the force between neighboring mass points.

We repeat that in all our problems we started the calculation from the string at rest at $t = 0$. The ends of the string are kept fixed.

2

Dynamics of Oscillator Chains

Allan J. Lichtenberg[1], Roberto Livi[2], Marco Pettini[3] and Stefano Ruffo[4]

[1] Electrical Engineering and Computer Science Department University of California, Berkeley, CA 94720-1770, USA
ajl@eecs.berkeley.edu
[2] Dipartimento di Fisica and CSDC, Università di Firenze via G. Sansone 1, 50019 Firenze, Italy
livi@fi.infn.it
[3] INAF – Osservatorio Astrofisico di Arcetri Largo Enrico Fermi 5, 50125 Firenze, Italy
pettini@arcetri.astro.it
[4] Dipartimento di Energetica "S. Stecco" and CSDC, Università di Firenze, and INFN, via s. Marta 3, 50139 Firenze, Italy
stefano.ruffo@unifi.it

Abstract. The Fermi–Pasta–Ulam (FPU) nonlinear oscillator chain has proved to be a seminal system for investigating problems in nonlinear dynamics. First proposed as a nonlinear system to elucidate the foundations of statistical mechanics, the initial lack of confirmation of the researchers expectations eventually led to a number of profound insights into the behavior of high-dimensional nonlinear systems. The initial numerical studies, proposed to demonstrate that energy placed in a single mode of the linearized chain would approach equipartition through nonlinear interactions, surprisingly showed recurrences. Although subsequent work showed that the origin of the recurrences is nonlinear resonance, the question of lack of equipartition remained. The attempt to understand the regularity bore fruit in a profound development in nonlinear dynamics: the birth of soliton theory. A parallel development, related to numerical observations that, at higher energies, equipartition among modes could be approached, was the understanding that the transition with increasing energy is due to resonance overlap. Further numerical investigations showed that time-scales were also important, with a transition between faster and slower evolution. This was explained in terms of mode overlap at higher energy and resonance overlap at lower energy. Numerical limitations to observing a very slow approach to equipartition and the problem of connecting high-dimensional Hamiltonian systems to lower dimensional studies of Arnold diffusion, which indicate transitions from exponentially slow diffusion along resonances to power-law diffusion across resonances, have been considered. Most of the work, both numerical and theoretical, started from low frequency (long wavelength) initial conditions.

Coincident with developments to understand equipartition was another program to connect a statistical phenomenon to nonlinear dynamics, that of understanding classical heat conduction. The numerical studies were quite different, involving the excitation of a boundary oscillator with chaotic motion, rather than the excitation of

A.J. Lichtenberg et al.: *Dynamics of Oscillator Chains*, Lect. Notes Phys. **728**, 21–121 (2008)
DOI 10.1007/978-3-540-72995-2_2

the entire chain with regular motion. Although energy transitions are still important, the inability to reproduce exactly the law of classical heat conduction led to concern for the genericity of the FPU chain and exploration of other force laws. Important concepts of unequal masses, and "anti-integrability," i.e. isolation of some oscillators, were considered, as well as separated optical and acoustic modes that could only communicate through very weak interactions. The importance of chains that do not allow nonlinear wave propagation in producing the Fourier heat conduction law is now recognized.

A more recent development has been the exploration of energy placed on the FPU or related oscillator chains in high-frequency (short wavelength) modes and the existence of isolated structures (breathers). Breathers are found as solutions to partial differential equations, analogous to solitons at lower frequency. On oscillator chains, such as the FPU, energy initially in a single high-frequency mode is found, at higher energies, to self-organize in oscillator space to form compact structures. These structures are "chaotic breathers," i.e. not completely stable, and disintegrate on longer time-scales. With the significant progress in understanding this evolution, we now have a rather complete picture of the nonlinear dynamics of the FPU and related oscillator chains, and their relation to a wide range of concepts in nonlinear dynamics.

This chapter's purpose is to explicate these many concepts. After a historical perspective the basic chaos theory background is reviewed. Types of oscillators, numerical methods, and some analytical results are considered. Numerical results of studies of equipartition, both from low-frequency and high-frequency modes, are presented, together with numerical studies of heat conduction. These numerical studies are related to analytical calculations and estimates of energy transitions and time-scales to equipartition.

2.1 Historical Perspective and Background Theory

2.1.1 Motivation and Counter Intuitive Numerical Results

In the early 1950s, considering what numerical investigations could be performed on a first generation digital computer at Los Alamos National Laboratory, Enrico Fermi suggested to Stanislaw Ulam and John Pasta that the foundations of statistical mechanics could be explored. He proposed using a chain of coupled slightly nonlinear oscillators to show that the nonlinearity would lead to equipartition of energy among the degrees of freedom. The model used in the studies was a discretization of a nonlinear spring which to quartic order is given by the normalized Hamiltonian

$$H = \sum_{i}^{N} \left[\frac{p_i^2}{2} + \frac{(q_{i+1} - q_i)^2}{2} + \alpha \frac{(q_{i+1} - q_i)^3}{3} + \beta \frac{(q_{i+1} - q_i)^4}{4} \right] \qquad (2.1)$$

with N unit masses and unit harmonic coupling. The oscillator chain is known as the Fermi–Pasta–Ulam (FPU) model. The original simulations were done with only the α term present (FPU-α model) or only the β term present

(FPU-β model); most subsequent simulations and analysis were done with the FPU-β model. With periodic endpoints the chain is translationally invariant, but the original simulations, as with much subsequent work, considered fixed endpoints, related to a physical finite string. Without nonlinear terms the coordinates can be transformed to uncoupled normal modes, such that the energy is always confined to the initial modes in which it is placed. For small nonlinearities (energy in the nonlinear terms small compared to the energy in the linear terms) it is logical to place the energy in a mode (or modes) of the linear system (quasi-modes of the nonlinear system) and observe the subsequent behavior of the mode energies subject to the laws of motion described by the Hamiltonian in (2.1). The initial numerics, programmed by Mary Tsingou for the FPU-α chain with fixed ends and $N-1$ moving particles, and with all the energy placed in the first harmonic ($k = 1$) of the harmonic normal modes

$$Q_k = \sqrt{\frac{2}{N}} \sum_{i=1}^{N} q_i \sin\left(\frac{\pi i k}{N}\right) \tag{2.2}$$

(with $N = 32$ and $\alpha = 1/4$) gave, for example, the result shown in Fig. 2.1 for $Q_1 = 4$ ($E = E_1 = 0.077$). The initial energy was transferred primarily into the first four modes, with an approximate recurrence (within a few percent) occurring in a time $\omega_1 t/2\pi = 157$ fundamental periods. Similar results were obtained for other initial conditions for both the α and β models,

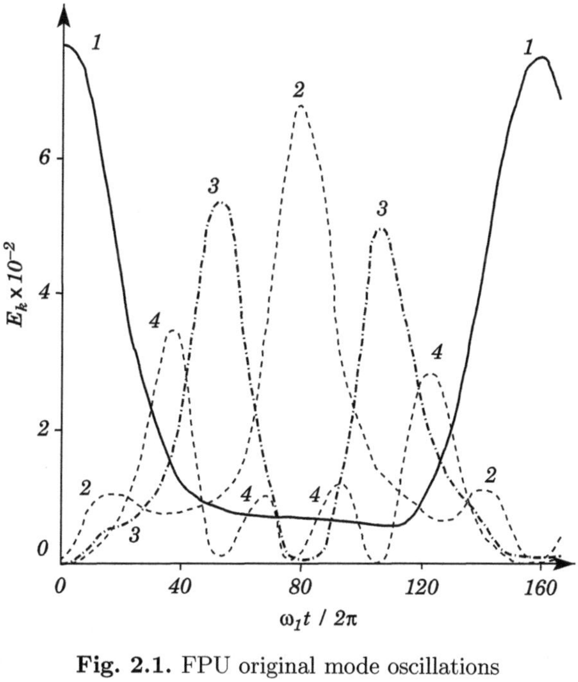

Fig. 2.1. FPU original mode oscillations

with the results presented in a Los Alamos report in 1955. Unfortunately, the untimely death of Fermi, prevented regular publication until the work appeared in Fermi's collected works [1]. The simulations did not answer the question of whether equipartition would ultimately be obtained, as was predicted from general dynamical principles of the nonexistence of global isolating integrals, see Poincaré [2], and the inferences used to support the concept of ergodicity by Fermi himself [3, 4]. Most of the near-term response to the unexpected result was to try to explain the recurrences. Using perturbative analysis Ford [5] and Jackson [6] obtained oscillations of the first few harmonics, with Jackson's approach, using nonlinearly perturbed frequencies, giving results, including the recurrence times, quite similar to the numerical observations. However, the perturbation procedures are nonconvergent, so no conclusions can be drawn from them about long-time behavior. There has been a significant body of literature concerned with these recurrences and methods of analysis. Generalizations, for example, have considered energy initially in an arbitrary mode (rather than the lowest frequency mode), the major couplings identified, and the effect of various numbers of modes used in analyzing the dynamics [7]. It was found, for example, that there is an induction period, i.e. a time during which there is little change in mode energy, if the energy is initially placed in a high-frequency mode, a condition later observed and qualitatively explained (see below).

2.1.2 Chaos Theory: KAM Isolation, Arnold Diffusion, Lyapunov Exponents, KS Entropy

Parallel to the developments, described above, for a high-dimensional Hamiltonian system, there were developments in low-dimensional Hamiltonian dynamics that informed the oscillator-chain results, and ultimately were informed by those results. In particular, the KAM theorem for coupled degrees of freedom [8, 9, 10] indicated that the generic case was a divided phase space with regular and chaotic orbits interspersed. Numerical observations, in a surface of section of a particular two degree of freedom system (the Hénon and Heiles potential), indicated mostly regular orbits at low energy, with the chaotic portion of the phase space increasing rather abruptly over a small range of increasing energy, until most of the phase space is chaotic [11]. A practical explanation of this rather abrupt increase was that local resonances between frequencies of the two freedoms, which modified the structure of the phase space in their neighborhood, would overlap with increasing energy, producing large areas of chaotic motion [12, 13]. For systems with three or more degrees of freedom KAM surfaces cannot isolate chaotic regions: leading to the possibility of "diffusion", in the sense that there are initial data which can reach points in phase space that are arbitrarily far, although such data have a small microcanonical measure when the nonintegrable perturbation is small [14]. Furthermore, a heuristic understanding of a many-dimensional system with weak coupling, backed up by simulations (see [15], Sect. 6.5), indicated the fraction of the phase volume that is stochastic continually increases with increasing number of freedoms N [16]. Another relevant theoretical result is that the upper bound on the rate of Arnold diffusion is proportional to

$$\exp\left(\frac{1}{\mu^q}\right) \tag{2.3}$$

where μ is a perturbation parameter [17] and an "optimal convergent" perturbation calculation gave a value of $q \approx 1/N$ [18, 19]. This result would indicate that the diffusion becomes large if $\mu^{1/N} \approx 1$, but the result, being of perturbative type, does not extend to so large a perturbation. Other heuristic forms will be used to estimate the diffusion rate in later Sections. Another development in low-dimensional chaos, that would inform the high-dimensional oscillator chain research, was the study of the three-particle Toda lattice [20], with Hamiltonian

$$H = \frac{p_1^2 + p_2^2 + p_3^2}{2} + \exp(-(q_1 - q_3)) + \exp(-(q_2 - q_1)) + \exp(-(q_3 - q_2)) - 3 \tag{2.4}$$

which corresponds to three particles moving on a ring with exponentially decreasing repulsive forces between them. In addition to the energy, there is a relatively obvious isolating integral, namely the total momentum, reducing the motion to two degrees of freedom. The Henon–Heiles potential [11] is a truncation of the Toda lattice. However, a surface of section of (2.4), calculated numerically, showed no chaos [21] and Hamiltonian (2.4) was subsequently proved to have a third invariant and thus was integrable, i.e. had nonchaotic phase-space trajectories [22].

In order to obtain equipartition it is sufficient for the dynamics to be "ergodic" on the energy shell, i.e. microcanonical averages over a given energy surface and time averages over motions taking place over the same energy surface must be equal. However, since it is physically relevant that the convergence to equipartition should occur on a finite time, and be possibly fast, the stronger dynamical property of "mixing" could be required. A direct numerical check of both "ergodicity" and "mixing" is impossible in systems with many degrees of freedom. However, if all trajectories are chaotic and, hence, on the average exponentially separating, positive Kolmogorov–Sinai (KS) entropy [23, 24] and exponentially fast "mixing" follow as a consequence. Therefore, an obvious quantity to be examined is the largest Lyapunov exponent, giving the average separation rate between nearby trajectories as

$$\lambda = \lim_{t\to\infty} \frac{1}{t} \ln \frac{\|\xi(t)\|}{\|\xi(0)\|} \tag{2.5}$$

where $\xi(t)$ is the tangent vector whose time evolution is described by the tangent dynamics equation described in Sect. 2.3. A positive value of λ indicates exponential separation of initially close trajectories, i.e. chaos. A difficulty of realistic Hamiltonian systems is that, in generic conditions, the energy shell is "divided" into chaotic and ordered trajectories, and hence "mixing" cannot occur everywhere in phase-space. Then, the qualitative statement might be that, if almost all of the energy surface is characterized by an invariant distribution that has a positive KS entropy, then, for all practical purposes, equipartition will be reached. The numerical calculation of Lyapunov exponents has been used extensively to test for chaotic motion, particularly after the numerical techniques were formalized [25]. It was logical that the method would be applied to the FPU chain, and became an important element in the numerical investigations in the 1980s and beyond, as discussed in more detail below.

For a more complete introduction to these topics see [15]: Sect. 3.2 (KAM Theory); Sects. 5.2 and 5.3 (concepts of stochasticity) and Sects. 6.1 and 6.2 (Arnold diffusion). See also [13] and, more rigorously, [26].

2.1.3 Geometrization of Hamiltonian Dynamics

Without attempting to be exhaustive, a few historical comments might be helpful to place the recent contributions about the geometrical approach to dynamics which are reviewed in the present Chapter, in a more general context.

The idea of looking at the collection of solutions of the Newton's equations of motion from a geometric point of view dates back to Poincaré and to the development of the qualitative theory of differential equations. Tackling the famous problem of the integrability of the three-body problem, Poincaré discovered that generic classical Hamiltonian systems, in spite of their deterministic nature, lack predictability because of their extreme sensitivity to the initial conditions. Such an instability of classical dynamics originates in homoclinic intersections, which Poincaré described in his *Méthodes Nouvelles de la Mécanique Céleste* [2] without "even attempting to draw" them. The method was later developed by Cartan among others, using what is now called symplectic geometry [27]. Although of undeniable elegance, symplectic geometry is not very helpful to advance our knowledge about the regions in phase space where the dynamics is unstable. The name of Poincaré, together with that of Fermi, is also associated with an important theorem about the nonexistence of analytic integrals of motion, besides energy, for generic nonlinear Hamiltonian systems describing at least three interacting bodies [3, 4]; this is the origin of the concept of topological accessibility of the whole constant energy hypersurface of phase space with high degree of freedom systems, with generic initial conditions.

In the 1940s, a qualitatively new attempt was made to make use of geometric concepts to relate Newtonian dynamics with statistical mechanics. Krylov [28] showed in for the first time the existence of a relationship between dynamical instability (seen as the exponential amplification of small deviations in the initial conditions of a collection of colliding objects representing idealized atoms in a gas) and phase space mixing. Phase mixing is a stronger property than ergodicity and is far more relevant to physics than ergodicity. In fact, while ergodicity assures the equality of time and phase space averages of physical quantities, phase mixing addresses the rate of approach to ensemble averages in a finite time. In modern terms, Krylov realized the necessity of chaotic dynamics to obtain fast phase mixing for the physically relevant observables and to make the connection between dynamics and statistical mechanics stronger. But Krylov also has the great historical merit of having attempted to bridge the dynamical foundations of statistical mechanics with a powerful field of mathematics, Riemannian differential geometry. Krylov knew mathematical results, concerning the properties of geodesic flows on compact negatively curved manifolds, by Hadamard [29], Hedlund [30] and Hopf [31]. He envisioned their potential interest to physics, once Newtonian dynamics is rephrased in terms of Riemannian geometric language. Such a possibility was well known since the beginning of the century, mainly due to the work of Levi-Civita; in particular that the principle of stationary action entails the close connection of a classical mechanical flow with a geodesic flow in a configuration space endowed with a suitable metric. Krylov's efforts concentrated on the analysis of the properties of physical systems which move in negatively curved regions in configuration space. For example, he discussed how the presence of an inflection point in the Lennard-Jones potential could influence the dynamics of a dilute gas (through the appearance of regions of negative scalar curvature in configuration space) and its ensuing strong instability. These attempts have been very influential on the development of the so-called abstract ergodic theory, where Anosov flows [32] (e.g., geodesic flows on compact manifolds with negative curvature) play a prominent role. Ergodicity and mixing of these flows have been thoroughly investigated. To give an example, Sinai proved ergodicity and mixing for two hard spheres by just showing that such a system is similar enough to a geodesic flow on a negatively curved compact manifold [33]. Krylov's intuitions have been worked out further by several physicists amongst whom we cite those of [34, 35, 36, 37, 38, 39, 40]. They discovered, much to their surprise, that geodesic flows associated with physical Hamiltonians do not live on negatively curved manifolds, despite their chaoticity. Only a few exceptions are known, in particular two low-dimensional models [35, 36, 41], where chaos is actually associated with hyperbolicity due to everywhere negatively curved manifolds. In fact, for certain models the regions of negative curvature of the mechanical manifolds apparently shrink by increasing the number N of degrees of freedom, thus reducing the frequency of the visits of negatively curved regions.

This somewhat biased search for negative curvature has been the main obstacle to an effective use of the geometric framework originated by Krylov to explain the source of chaos in Hamiltonian systems. On the other hand, it is true that the Jacobi equation, which describes the stability of a geodesic flow, is in practice only tractable on negatively curved manifolds. Formidable mathematical difficulties are encountered in treating the (in)stability of geodesic flows on manifolds of nonconstant and not everywhere negative curvature.

Moreover, for this kind of problem, intuition can hardly help. However, the advent of computers has been of invaluable help. As a matter of fact, during the last few years an interplay between analytic methods and numerical simulation has made it possible to overcome the difficulties, showing that the Riemannian geometric approach can be applied to dynamical systems of interest to statistical mechanics, field theory, and condensed matter physics [42]. This has extended the domain of application of geometric techniques, and has also introduced a new point of view about the origin of chaos in Hamiltonian systems, as well as new methods to describe and understand it, "new" in a sense that will be made clear in Sect. 2.4.7.

A more detailed exposition of the geometric method and its application to calculating Lyapunov exponents, which we will be summarizing in Sects. 2.4.7 and 2.8.2, can be found in [42]. See also mathematical expositions in [27, 28].

2.1.4 Development of Soliton Theory

It is somewhat ironical that the most celebrated result that came out of the investigation of the FPU chain did little to resolve Fermi's original question of whether or not the nonlinearity would lead to equipartition among the degrees of freedom. In an attempt to understand the apparent stability of the recurrences Norman Zabusky and Martin Kruskal [43, 44] found a Taylor expansion of the discreteness, valid for long wavelength modes, that recovered partial differential equations, different from the original nonlinear spring which produced the discretized chain of oscillators. The resulting equations are the Korteveg–de Vries (KdV) equation for the FPU-α chain and the modified Korteveg–de Vries (mKdV) equation for the FPU-β chain [43, 44]. The latter chain, with appropriate normalizations, gave the standard form

$$u_\tau + 12u^2 u_\xi + u_{\xi\xi\xi} = 0 \tag{2.6}$$

where $\tau = h^3 t/24$, $\xi = x - ht$, $h = L/N$, L with the length of the string and N the number of oscillators. Nonlinear equations of this and related types had been known to have stable traveling solutions, where the dispersion and nonlinearity balance to produce constant amplitude and propagation velocity. An arbitrary initial condition, such as the lowest linear mode on the FPU-β chain, breaks up initially into a set of structures each having a steady traveling solution with its own velocity. Remarkably, these structures are sufficiently stable that they pass through one another without breaking up, and the observed recurrences can be interpreted in terms of their superpositions. But these results do not improve on the best perturbation calculations, and are clearly limited to long-wavelength (low-frequency) modes by the approximations which led to (2.6). Partial differential equations, like (2.6), have an infinite number of freedoms, such that general integrability from arbitrary initial conditions requires an infinite number of invariants of the motion. The real excitement came when it was shown that such an infinite set exists for (2.6), and the new field of soliton theory and applications was born, which would take us far from the subject at hand. A final note, which is important to our overall understanding, is that a single initial nonlinear mode solution of the mKdV equation was found to become unstable as the energy is increased. A linearization around the nonlinear structure predicted the unstable wave numbers and growth rates, and showed that the values correspond to the observed mode growth for the same discretized structure on the FPU-β oscillator chain [45, 46]. The result in which one soliton decomposes into a finite number is not inconsistent with general soliton theory. The instability will give us insight into some later results.

2.1.5 Resonance Overlap Explanations

Using the concept of mode overlap to estimate the transition between regular and chaotic motion Felix Izrailev and Boris Chirikov obtained estimates for mode overlap both for low- and high-frequency modes [47]. Although there are various approximations required to obtain results, a simple numerical estimate can be made by equating the nonlinear frequency shift $\Delta\omega_k$ to the mode spacing $\delta\omega_k$, i.e. setting $\Delta\omega_k/\delta\omega_k \approx 1$. The mode overlap estimate from this approximation, in terms of energy density, is

$$\varepsilon_k = \frac{E_k}{N} = \begin{cases} 4/(3\beta k)\ , & k \ll N \\ 2k^2(N-k)/(3\beta N^2)\ , & (N-k) \ll N\ . \end{cases} \tag{2.7}$$

The result for long-wavelength modes, $k \ll N$, is not a necessary condition, as seen in many subsequent numerical experiments, but approximates another transition, discussed below, between weak and strong stochasticity (the SST). The result for short-wavelength modes, $(N-k) \ll N$, is neither necessary nor sufficient. It predicts easy overlap at short wavelengths due to mode crowding, while numerical simulations show consistently that, from a practical point of view, equipartition is more readily obtained from long-wavelength than from short-wavelength initial conditions. General theoretical arguments as to the accessibility of modes has been advanced to show that this is the case [48, 49].

We discuss a resonance overlap criterion, as presently used, in Sect. 2.4.1. The concept, initially proposed by Chirikov for two degrees of freedom and reviewed by him, including higher dimensionality, [13], can also be found in [15], Chap. 4.

2.1.6 Numerical Methods

The straightforward method of computing Lyapunov exponents, using (2.5), particularly the largest exponent, was a powerful numerical tool for statistically investigating the dynamical properties of oscillator chains. Another very useful statistical quantity is the information entropy

$$S = -\sum_k e_k \ln e_k \tag{2.8}$$

with $e_k = E_k / \sum_k E_k$, such that $S = 0$ if all the energy resides in a single mode, and has a maximum $S = \ln N$ if the energy is uniformly distributed among all modes. By using (2.5) and (2.8), detailed numerical investigations were carried out among investigators in Florence [50, 51, 52, 53], starting from long-wavelength modes of the FPU-β system, obtaining the variation of λ and S with energy density $\varepsilon = E/N$. They found a distinct break in the behavior between weak stochasticity at lower values of ε, having strong power-law dependencies of λ and S on ε, and strong stochasticity at higher ε with weak ε dependence. The transition (SST) is qualitatively related to the mode overlap criterion (2.7). Note that ε is not necessarily a small quantity.

2.1.7 Methods of Analysis and Numerical Results

It is clear from the phase space description of high-dimensional systems that mode overlap is not necessary to obtain positive Lyapunov exponents. Most generic initial conditions will lie in stochastic layers, exhibiting $\lambda > 0$. The question becomes what determines the rate of energy diffusion between the degrees of freedom? One approach to this problem is to isolate a few of the most closely coupled modes and determine if their resonant interaction results in chaos that can then couple to other resonances. This was done, as described previously, for low-dimensional chaos to understand the exponentially slow Arnold diffusion. For high-dimensional systems the situation is more complicated with rapid diffusion across overlapped resonances and slow Arnold diffusion along resonance (see [15], Chap. 6, for a detailed discussion). The method of isolating a few interacting resonances and then calculating their coupling to the larger phase space was used for another oscillator chain, a discretized sine-Gordon equation, to explore the transition from power-law (numerically observable) equipartition rates with varying ε, to exponentially slow (not numerically observable) rates [54]. It was also found, using this approach and comparison with numerics, that short-wavelength mode interactions required considerably higher energy to produce chaos. The method was then applied to the FPU-β chain, in more detail, specifically investigating the process by which stochastic interaction between a few long-wavelength modes was transferred to short-wavelength modes and calculating a transition between exponentially slow and power-law scaling of the energy transfer [55]. At about the same time, there was considerable attention given to determining the scaling of the equipartition time T_{eq} with ε, in the power-law regime, finding $T_{\mathrm{eq}} \propto \varepsilon^{-3}$ [56, 57, 58] with the latter references giving a heuristic calculation of this scaling. The numerics and method of estimation will be given in Sects. 2.5 and 2.8, respectively. Other authors have fitted the data to a "stretched exponential", $T_{\mathrm{eq}} \propto \exp(-\varepsilon^{1/4})$ [59], obtaining a better agreement over a wider ε range, but without any theoretical underpinning. Indeed, the reason for this scaling has not been explicitly explained, nor has its relation to the power-law scaling. It is very likely that, as ε is decreased, longer and longer time-scales come into play and, therefore, no definite functional form will be able to fit the increase of the time-scale over the full small ε range. The scaling $\lambda \propto \varepsilon^2$, detected at ε smaller than the threshold value of the SST transition ($\varepsilon \approx 1$), has also not been specifically related to the $T_{\mathrm{eq}} \propto \varepsilon^{-3}$ scaling in the same ε range. The scaling at higher ε, that is at ε larger than the threshold value of the SST transition, has been heuristically determined using a random matrix approximation for the tangent dynamics, intuitively suggesting that above the SST chaos is fully developed [53]. The scalings of λ with ε can now be determined, analytically, by considering the geometry of the phase space near equipartition. Making suitable assumptions about the geometry of mechanical manifolds, the scaling of λ with ε, both below and above the SST transition and the value of ε at the transition has been theoretically calculated in agreement with numerical findings [42, 60, 61]. Although the method was developed to understand the FPU-β scaling, it is applicable to oscillator chains with various force laws, as can be found in the referenced works. The mathematical procedure is outlined in Sects. 2.4.7 and 2.8.2.

2.1.8 Comparison of Different Oscillator Chains

Although the FPU-β oscillator chain has received most of the attention, there has been, from the beginning, interest in other force laws. The cubic potential in the FPU-α model is more conducive to using expansion procedures to obtain analytic estimates [62], and also, the form with periodic boundary conditions and appropriately chosen α is a third order truncation of the Toda lattice potential, which is integrable. However, the FPU-α is not energy renormalizable with varying α (does not scale with αE), and furthermore suffers from the problem of unbounded trajectories at high energy. Nevertheless, comparison with the FPU-β dynamics has added considerably to our overall understanding. If the finite time version of the Lyapunov exponent (2.5) is calculated for the N-particle Toda lattice and its FPU-α approximation, the two exponents decrease, without separation, until some "induction time" or "trapping time" $\tau_{\mathrm{T}}(\varepsilon)$, is reached, after which $\lambda_{\mathrm{FPU}-\alpha}$ attains a constant value, while λ_{Toda} continues to decay, as it must for an integrable system [61]. Plotting $\tau_{\mathrm{T}}(\varepsilon)$, with N as a parameter, in the weak stochasticity regime, it was found in [61] that $\tau_{\mathrm{T}} \propto \varepsilon^{-2}$, which is different from the $T_{\mathrm{eq}} \propto \varepsilon^{-3}$ scaling found both for the FPU-α and the FPU-β systems, i.e. the trapping time and the equipartition time scale differently with ε. A transition at some small ε to a rapid increase in τ_{T} with decreasing ε, with the transition value a function of N, was also observed and interpreted as a transition to regular motion. This phenomenon had been observed earlier in the discretized sine-Gordon system and interpreted in a similar fashion [54]. However, subsequent work with the FPU-β system elicited a different interpretation, that the transition was to the exponentially slow form of Arnold diffusion [55]. These different interpretations have not been theoretically reconciled. The Hamiltonian containing both cubic and quartic nonlinearities, as in (2.1), has also been investigated [63].

The contrast of the FPU oscillator chain with other types of chains has also led to considerable insight and some additional puzzles. The class of Klein–Gordon chains, with on-site potentials, are similar to the FPU, but more complicated, both because they lack the FPU translational invariance and because they have an additional parameter whose scaling must be determined. In addition to the sine-Gordon version, a closer comparison with the FPU-β chain employs a Klein–Gordon on-site potential having quadratic and quartic terms, with Hamiltonian

$$H = \sum_{i=1}^{N} \frac{p_i^2}{2} + \frac{(q_{i+1} - q_i)^2}{2} + \frac{mq_i^2}{2} + \frac{\beta q_i^4}{4} \,, \tag{2.9}$$

which is often called the ϕ^4 model to distinguish the quartic nonlinearity from other Klein–Gordon potentials. In the first comprehensive numerical comparison of the two systems, (2.1) and (2.9), some physical differences were observed and, qualitatively, understood. In particular, at a given ε, the ϕ^4 took significantly shorter time to obtain equipartition from long-wavelength mode initial conditions and significantly longer time from short-wavelength modes, than the FPU [64]. The fact that from short wavelengths it was generally a longer process to obtain equipartition was remarked in that early work, but little background theory had been done for these initial conditions. A later comparison for the long wavelengths provided a more complete numerical study and was able to explain quantitatively these differences [65]. We will present the numerical comparisons in Sect. 2.5 and outline the supporting theory in Sect. 2.8. The understanding of the results from short wavelengths awaited the development of new theoretical concepts, as given below. Before considering this subject we note that the emphasis on energy density, holding ε constant as N is varied, i.e. $E \propto N$, is not always the relevant way to look at a problem, as seen in calculating resonance overlap [55]. The case in which E is held fixed as N is varied has been used to analytically calculate stochasticity thresholds of the FPU and ϕ^4 models [66].

2.1.9 Dynamics at Short Wavelengths: Chaotic Breathers

Following the original numerical work, most numerical studies examined the evolution from long-wavelength (low-frequency) modes in which neighboring oscillators are nearly in phase. Zabusky and Deem [67] were the first to consider the case in which the energy is put into a high-frequency mode. In their early work, the zone–boundary mode was excited with an added spatial modulation for the FPU-α model. Our main concern here will be the FPU-β model, and spatial modulation of the mode is spontaneously created by modulational instability. Budinsky and Bountis [68] found that the zone–boundary π-mode, i.e. the mode with 180° phase shift between neighboring oscillators of the one-dimensional FPU lattice is unstable above a given energy threshold E_c which scales like $1/N$. This result was later confirmed by Flach [69] and Poggi and Ruffo [70], who also obtained the exact numerical factor relating E_c to $1/N$. These results were obtained using a direct linear stability analysis around the periodic orbit corresponding to the π-mode. Similar methods have been more recently applied to other modes and other FPU potentials by Chechin [71, 72] and Rink [73]. A technique which allows for a more general exploration of the dynamics starting from short wavelengths is to follow an envelope function of the oscillators defined by $\psi_i = (-1)^i q_i$. Since the main phase variation of the oscillator amplitudes q_i vary by nearly π from one oscillator to the next, the ψ_i vary slowly; a Taylor expansion of the envelope function in the oscillator space can produce a differential equation whose equilibrium properties, stability, and nonlinear effects can be explored (see Sect. 2.4.5). A formula for E_c, valid for all N, has been obtained in Refs. [74, 75, 76, 77] in the rotating wave approximation (RWA) given in (2.86). Besides calculating the energy threshold, the growth rates of mode amplitudes were obtained. The application to the Klein-Gordon lattices was first studied by Kivshar and Peyrard [78], following an analogy with the Benjamin–Feir instability in fluid mechanics [79]. A different approach to describe this instability had been previously introduced by Zakharov and Shabat [80], who studied the associated nonlinear Schrödinger equation in the continuum limit. Using that method for the FPU equations of motion, the instability boundary was found by Berman and Kolovskii [81] in the so-called "narrow-packet" approximation. Detailed numerics over longer times were obtained for the FPU model by Pettini [64] and for the discretized sine-Gordon equation by Goedde [54], both indicating that, for a given energy, short-wavelength (high-frequency) modes required longer times to reach equipartition than long-wavelength modes. At about the same time it was demonstrated that stable intrinsic localized modes (ILMs) could exist for anharmonic periodic structures [82]. However, from more general high-frequency initial conditions there was a tendency to form ILMs but they were not stable, breaking up and ultimately decaying toward equipartition [75, 83].

The existence of ILMs (also called breathers) on periodic chains and the complex behavior of more arbitrary high-frequency initial conditions has led to extensive study of these structures to understand their stability. A comprehensive review of these studies would lead us far from the main topic of this review (see [84] for a review and further references). The breathers can be stationary or moving, and, like low-frequency solitons, can pass through one another. Whether energy is exchanged in such an interaction depends on the system's stability properties. ILMs that are not exact solutions of the underlying system generally exchange energy, and in a particular process have been shown to transfer energy from the smaller to the larger breather [85]. This phenomenon is also observed numerically for a Klein–Gordon chain [86] and for the FPU-β chain [87, 88]. For fixed end-points, as in the original FPU studies and much subsequent work, a clearly defined instability boundary cannot be calculated for a discrete chain. Nevertheless, as described below, some approximate results are available.

The four-mode resonance overlap criterion for a stochasticity transition, described in Sect. 2.4.1 for low-frequency modes, has also been used for high-frequency modes for the discretized sine-Gordon chain [54]. It predicted the increased stability for these modes, as found numerically. Another approach to a reduced problem is to represent the main-energy containing oscillator and the immediate neighboring oscillators as a three degree of freedom system from which a mapping can be obtained [89]. In this reduced phase space, depending on the energy and the action, one observes both regular and chaotic regions. The chaotic regions are sufficient to indicate chaos in the larger system, but do not give a time-scale for equipartition to be approached. The regular regions may also be chaotic in the larger system, but are generally more weakly so. Note that the mapping presentation uses initial conditions close to those of a breather, which is narrow in oscillator space, and therefore has a broad distribution of energies in mode space. Contrarily, the mode presentation starts with a narrow distribution in mode space and therefore a broad distribution in oscillator space. The technique employing the envelope equations and the RWA has been used to describe longer time effects, as well as instability boundaries both for periodic boundary conditions at low energies [90, 91] and fixed boundary conditions at both low and high energies [92, 93, 94], and used to compare the dynamics of the FPU and ϕ^4 chains [94].

The general picture that has emerged is that if the energy is placed in a high-frequency mode or modes for which neighboring oscillators are primarily out of phase, a complicated dynamics ensues, which consists of three stages. First, there is an initial stage in which, for sufficiently high energy, the mode breaks up into a number of breather-like structures. Second, on a slower time-scale, these structures coalesce into one large unstable structure, called a "chaotic breather" (CB). Since a single large CB closely approximates a stable breather, a third and final decay stage, toward equipartition, can be very slow. One does not know whether there exists any true energy threshold to achieve equipartition, although there appears to be some numerical evidence for such a threshold in the discretized sine-Gordon system. However, as discussed extensively with respect to low-frequency mode initial conditions, the practical thresholds refer to observable timescales.

For nonlinear structures on chains having "weak spring" potentials, for which the nonlinear restoring force substracts from the linear restoring, the interaction that causes the final decay is radiation from the breathers to the propagating linear modes. For "strong spring" potentials the breather frequency is above the optical band, so a more subtle energy interchange must occur [84, 87, 95, 96]. A beat phenomenon has been postulated as the energy interchange mechanism, and used to calculate an ε-scaling that agrees with numerics [92].

2.1.10 Heat Transport in Lattice Models

A main goal of classical kinetic theory is to provide the definition of transport coefficients through phenomenological constitutive equations. The basic hypotheses of this macroscopic theory of transport phenomena are the assumption that fluxes are proportional to thermodynamic forces and that the system evolves close to equilibrium [97]. For instance, when dealing with heat transport in a solid, one defines the thermal conductivity κ through the Fourier law

$$\mathbf{J} = -\kappa \boldsymbol{\nabla} T , \tag{2.10}$$

where the heat flux $\mathbf{J}$ is the amount of heat transported through the unit surface in unit time and $T(\mathbf{x}, t)$ is the local temperature. Such a phenomenological relation was first proposed in 1808 by Fourier as an attempt to explain the phenomenon of the Earth cooling. Equation (2.10) is assumed to be valid close to equilibrium. Actually, the very definition of the local energy flux $\mathbf{J}(\mathbf{x}, t)$ and temperature field $T(\mathbf{x}, t)$ relies, in turn, on the *local equilibrium hypothesis*, i.e. on the possibility of defining a local temperature for a macroscopically small but microscopically large volume in position $\mathbf{x}$ at time t.

The first and most elementary attempt to give a microscopic foundation to Fourier's law dates back to Debye [98]. By rephrasing the results of the kinetic theory for the (dilute) phonon gas, he found that the thermal conductivity should be proportional to $Cv\ell$, where C is the heat capacity and v, ℓ are the phonon mean velocity and free path, respectively. Moreover, Debye also realized that at a microscopic level the finite thermal conductivity in crystals should be a consequence of the nonlinear forces acting among the constituent atoms [98].

Peierls further extended the conjecture of Debye and formulated a Boltzmann-like equation, which shows that anharmonicity is necessary for obtaining genuine diffusion of the energy by the so-called *Umklapp* processes, where the nonlinearity is introduced phenomenologically in the transport equation, independently of the microscopic nature of the interactions [99]. Nonetheless, the Boltzmann–Peierls approach represented an improvement in the theory of lattice thermal conductivity. It allows one to compute the dependence of κ on the temperature which agrees reasonably well with experimental data in the very low-temperature regime. However, basic questions remained, such as under which conditions is local equilibrium obtained in a physically accessible time? This kind of a problem partly inspired the numerical experiment by Fermi, Pasta and Ulam, as Fermi was aware of the conceptual difficulties concerning the possibility of constructing a satisfactory microscopic approach to transport theory. In nonlinear chains, the complex interactions among the constituent atoms or molecules of a real solid are reduced to harmonic and nonlinear springs, acting between nearest-neighbor equal-mass particles. Despite such simplifications, the basic ingredients that one reasonably conjectures to be responsible for the main physical effect (i.e. the finiteness of thermal conductivity) are contained in the model. As already described, the original study expected to verify a common belief, which had never been put to a rigorous test: an isolated mechanical system with many degrees of freedom should eventually yield equilibrium through "thermalization" of the energy. Furthermore, the measurement of the time interval needed to approach the equilibrium state, i.e. the "relaxation time" of the chain of oscillators, would have provided an indirect determination of thermal conductivity κ, since Debye's argument predicts $\kappa \propto C_v/\tau_r$, i.e. inversely proportional to the relaxation time $\tau_r \sim \ell/v$, which is assumed to represent the average time needed for a phononic excitation to relax to thermal equilibrium.

After the lack of success of the FPU numerical experiment, the first important attempt to reconsider the problem of heat transport in solids from a theoretical point of view was to consider a homogeneous harmonic chain with fixed boundary conditions in contact with stochastic Langevin heat baths [100]. The equations of motion

$$\ddot{q}_n = \omega^2(q_{n+1} - 2q_n + q_{n-1}) + \delta_{n1}(\xi_+ - \lambda\dot{q}_1) + \delta_{nN}(\xi_- - \lambda\dot{q}_N), \qquad (2.11)$$

where $\xi_\pm$ are independent stochastic processes with zero mean and variance $2\lambda_\pm k_B T_\alpha$, with $T_+ > T_-$, can be solved by a phase-space description, i.e. using the Fokker–Planck equation. However, the solutions were not successful in reproducing the Fourier law. They predicted that the heat flux was proportional to the temperature difference, rather than the temperature gradient, thus showing that homogeneous harmonic chains do not exhibit normal transport properties. Although there are many aspects of linear chains, such as the inclusion of disorder or varying masses [101] that we have not considered above, our main concern here is with nonlinear chains. Numerical studies of heat conductivity in the FPU chain were reconsidered at the end of the 1960s. In particular, nonequilibrium simulations of the FPU model (2.1) with coupling constants α and β fixed to represent the leading terms of the expansion of the Lennard-Jones potential were performed [102, 103]. These authors also considered the effect of disorder by including in the model either a disordered binary mixture of masses [102] or random nonlinear coupling constants [103]. The combination of nonlinearity and disorder did not help the researchers to obtain a clear understanding of the problem. They even found cases in which anharmonicity increases thermal conductivity. The attention was mainly focused on the form of the temperaure profile $T(x)$. They noticed that its shape depended on the existence of disorder. Although it is known that $T(x)$ is not a self-averaging observable for disordered harmonic chains, it is not known how $T(x)$ depends on disorder over long enough time-scales in anharmonic chains. Additional questions that were investigated concerned the concentration of impurities [102].

Preliminary work on homogeneous anharmonic chains considered the equal-masses FPU and Lennard-Jones chains composed of 30 particles and coupled with Langevin baths at their boundaries [104], a task that was unfeasible with the computer resources available at that time. As a consequence, several attempts of designing easy-to-simulate toy models followed these first studies. Some examples are reviewed in [105]. One was the so-called ding-a-ling model, a prototype of all models with an on-site potential, as described in (2.16) in the next Section. This model was found to exhibit normal thermal conductivity. The increase in computer power led to a revival of the heat conduction problem inbetween the mid-1980s and the mid-1990s, when nonequilibrium simulations of the FPU model [106, 107] and of the diatomic Toda chain [108, 109, 110, 111] of alternating light and heavy masses were performed. Subsenquently, there were systematic studies on the size dependence of the heat conductivity for the FPU chain with quartic [112, 113, 114] or cubic [115] nonlinear potential as well as for the diatomic Toda chain [116, 117]. They indicated a divergence of the heat conductivity with N, the number of mass points, which was interpreted as due to ballistic transport of energy through the chain. As we will comment in the following, an on-site potential determines a classical conductivity.

2.2 Formulations: Types of Oscillator Chains

2.2.1 Chains Similar to the FPU

Over the years, since the first numerical investigation by Fermi, Pasta and Ulam, many different oscillator chains have been studied. There have been various reasons for the particular choices, sometimes because they approximated physical systems, sometimes for their simplicity, and sometimes designed to bring out specific features or compare results with other chains.

In choosing an oscillation chain for the initial study, the FPU-β system was a reasonable choice, as it is a discretization of the partial differential equation for the nonlinear string with a strong nonlinear restoring force

$$\frac{\partial^2 y}{\partial t^2} - \frac{\partial^2 y}{\partial x^2}\left[1 + 3\beta\left(\frac{\partial y}{\partial x}\right)^2\right] = 0. \tag{2.12}$$

The discretization of $y(x,t)$ as $y_j(t)$,

$$\frac{\partial y}{\partial x} = \frac{y_{j+1} - y_j}{\Delta x} \quad \text{or} \quad \frac{y_j - y_{j-1}}{\Delta x}, \tag{2.13}$$

$$\frac{\partial^2 y}{\partial x^2} = \frac{y_{j+1} - 2y_j + y_{j-1}}{\Delta x^2}, \tag{2.14}$$

where $\Delta x = L/N$, with L as the length of the string and $N-1$ the number of oscillators, yields

$$\ddot{y}_j = \frac{(y_{j+1} - 2y_j + y_{j-1})}{\Delta x^2} \times \left\{(1+\beta)\left[\frac{(y_{j+1} - y_j)^2}{\Delta x^2} + \frac{(y_j - y_{j-1})^2}{\Delta x^2} + \frac{(y_{j+1} - y_j)(y_j - y_{j-1})}{\Delta x^2}\right]\right\} \tag{2.15}$$

The original work, and also the analytic investigation by Izrailev and Chirikov [47], were with fixed end points at $j = 0$ corresponding to $x = 0$, and $j = N$ corresponding to $x = L$, such that $j = 1, 2 \ldots, N-1$ for the moving oscillators. The coordinates can be rescaled at fixed N to any Δx and L to give the FPU-β part of (2.1), with Δx normalized to 1. Letting $\Delta x \rightarrow \Delta x' = L'/L$ and introducing the change of variables $y_j' \rightarrow y_j L'/L$ and $t' \rightarrow tL'/L$ leads to (2.15) again. Since $\mathrm{d}y_j'/\mathrm{d}t' = \mathrm{d}y_j/\mathrm{d}t$, the energy per mode is unchanged. Thus increasing N by adding oscillators to the end of the chain at fixed Δx is equivalent to adding oscillators by subdividing the chain at fixed L, provided the time and displacement are rescaled.

The addition of the α term to (2.1) is a logical extension to a more general restoring force. However the α and β terms have different properties, with the energy scaling differently with choices of α and β, such that the energy E is renormalizable with βE and with $\alpha^2 E$. Furthermore, the sign in the nonlinear term in (2.12) changes the behavior from a strong to a weak spring, while the α term has a directional antisymmetry. Some of the consequences of these differences will emerge in the following sections.

It also became clear in subsequent years that, while fixed endpoints were a physical condition for an actual string, periodically continued endpoints (or mass points on a circle) had some attractive features for analysis. With a periodic boundary condition (BC), waves traveling in a single direction without reflection are allowed, which is a key ingredient in the development of soliton theory, as we outline in Sect. 2.4.3. For a periodic BC, linear momentum is an exact invariant, which simplifies various analyses. If the oscillator dynamics is expressed in terms of linear modes, i.e. the modes which would be exact solutions in the absence of the nonlinearity, other differences between fixed and periodic boundaries become evident. For a periodic BC there is a highest frequency boundary mode that has exact alternation of oscillator phases, which is an exact solution of the nonlinear problem, as considered in subsequent sections.

The FPU type of oscillator chains did not realistically represent the dynamics of solid materials. A more general representation is given by the Hamiltonian

$$H = \sum_i \left[\frac{p_i^2}{2m} + U(q_i) + V(q_{i+1} - q_i) \right] \tag{2.16}$$

where U and V are on-site and inter-site potentials, respectively, which are most generally nonlinear. They can be constructed as physical models of one-dimensional crystals or by discretizations of Klein–Gordon partial differential equations. For the FPU chain $U = 0$. One form of (2.16) that is used to compare to FPU-β dynamics is the ϕ^4 chain with $V = (1/2)(q_{i+1} - q_i)^2$ and $U = (m^2/2)q_i^2 + (\beta/4)q_i^4$, as given in (2.9), and compared theoretically and numerically with the FPU in various subsequent sections. The Hamiltonian of (2.16) and the simplified form (2.9) are not rescalable as is the FPU-β, but the coefficients can be chosen to make useful comparisons.

An interesting special case of the Klein–Gordon class of partial differential equations is the sine-Gordon equation

$$y_t - y_{xx} + \sin y = 0, \tag{2.17}$$

which can discretized in space in the same manner as the nonlinear spring to obtain the system Hamiltonian

$$H = \sum_{i=1}^{N} \frac{1}{2} p_i^2 + \sum_{i=1}^{N} (1 - \cos y_i) + \sum_{i,j=1}^{N} A_{ij} y_i y_j, \tag{2.18}$$

where the coupling matrix A_{ij} is given by

$$A_{ij} = \frac{(2\delta_{ij} - \delta_{i,j-1} - \delta_{i,j+1})}{(\Delta x)^2}, \tag{2.19}$$

$p_i = \dot{y}$, $\Delta x = L/N$, and δ_{ij} is the Kronecker δ. As with the more general forms of the discretized Klein–Gordon, (2.18) is not rescalable on Δx, such that both L and N enter as essential parameters. The discretized system is of particular interest, as the partial differential equation is integrable, unlike the nonlinear spring, so the discretization, itself, becomes the only source of chaos. However, at low frequencies (long wavelengths) where the FPU approximates an integrable system, the transitions are similar [54]. One interesting feature in that work was an explicit discretization of time, so forming a $2N$-dimensional symplectic map to be analyzed. The sine-Gordon on-site potential has also been used as an interaction potential to study the Fourier heat law, as we discuss in Sect. 2.7.2. For that case the potential is known as the Frenkel–Kontorova potential.

Closely related to the FPU chain is the chain with the same interparticle potential structure but with varying masses. To explore the question of whether a fraction of the modes, in a distinguishable mode packet, could be isolated from the modes initially containing the energy, Galgani et al. [118] considered a modified FPU-β model, with fixed ends, described by the Hamiltonian

$$\mathrm{H} = \sum_{i}^{N} \left[\frac{p_i^2}{2m_i} + \frac{(q_{i+1} - q_i)^2}{2} + \beta \frac{(q_{i+1} - q_i)^4}{4} \right] \tag{2.20}$$

with $m_i = 1$ for i odd and $m_i = m < 1$ for i even. The alternation of masses separated the linear mode spectrum into branches, an acoustic branch which is only slightly modified from the usual spectrum, and an optical branch associated with the lower mass particles. The form of the spectrum is given in Sect. 2.2.2 and the implication for mode isolation in the thermodynamic limit is discussed in Sect. 2.4.2

Another oscillator chain of particular importance is the Toda lattice, which generalizes the three-particle lattice, given in (2.4), to N particles. The lattice, with exponential forces between particles, is generally thought of as constrained on a ring, which is equivalent to periodic BC. This discretized chain is exactly integrable, which makes it uninteresting in itself, but very useful for comparing to the FPU-α potential that can be considered to be a truncation of the Toda potential (see Sect. 2.4.3). The same comparison of the three particle Toda potential with its truncation, the Henon and Heiles potential [11], was very useful in understanding low-dimentional chaos, as we have already discussed briefly in Sect. 2.1.2. The N-particle Toda chain has also been used to explore the effect of alternating heavy and light masses, which is not integrable, and is discussed briefly in Sect. 2.7.1 in connection with heat conduction.

The oscillator chains described above do not exhaust the useful types that have been explored in a variety of contexts. One such chain of historical importance is the "ding-a-ling" model consisting of alternately harmonically bound and free hard-core particles, which was used to obtain the Fourier law of heat conduction [119]. It is given in (2.132), Sect. 2.7.2, and its properties are discussed there. Variants of the ding-a-ling model have also been studied in this context [120]. The less artificial potentials of the Klein–Gordon type also can produce the Fourier law, and are considered in Sect. 2.7.2.

2.2.2 Representation in Modes of the Linear System

We have already mentioned that, for linear chains, the transformation to the harmonic normal modes, as given by (2.2), gives a set of mode amplitudes Q_k that are invariant under the motion. This can be seen by applying the inverse transform, with N moving particles, to the FPU-β chain (2.1) or the ϕ^4 chain (2.9)

$$q_i = \sqrt{\frac{2}{N+1}} \sum_{k=1}^{N} Q_k \sin\left(\frac{ik\pi}{N+1}\right), \tag{2.21}$$

to obtain [121]

$$H_\beta = \sum_{k=1}^{N} \frac{1}{2}\left(P_k^2 + \omega_k^2 Q_k^2\right) + \frac{\beta}{8N+8} \sum_{i,j,k,l=1}^{N} C(i,j,k,l) Q_i Q_j Q_k Q_l, \tag{2.22}$$

where P_k are the corresponding momenta, with the mode frequencies ω_k given by

$$\omega_k = 2\sin\left(\frac{\pi k}{2N+2}\right) \tag{2.23}$$

for the FPU-β, and

$$\omega_k = \sqrt{m^2 + 4\sin^2\left(\frac{\pi k}{2N+2}\right)} \tag{2.24}$$

for the ϕ^4. It is immediately apparent from Hamilton's equations that the dynamics of the linear modes are independent of one another. The quartic terms couple the modes together, with

$$C(i,j,k,l) = \omega_i \omega_j \omega_k \omega_l \sum_P B(i+j+k+l) \tag{2.25}$$

for the FPU-β, and, for the ϕ^4,

$$C(i,j,k,l) = \sum_P B(i+j+k+l)\,. \tag{2.26}$$

The sum is over the eight permutations of the sign of i,j,k,l and the function $B(x)$ takes the value 1 if the argument is zero, -1 if the argument is $\pm 2(N+1)$, and zero otherwise. The selection rule for the couplings, which simplifies the analysis, follows from the quartic nature of the coupling (e.g., see [122]).

From (2.23) we see that frequencies spacings follow a simple sine function: they are linearly spaced (i.e. proportional to k/N) for $k \ll N$ and accumulate quadratically (i.e. as $(\pi k/2(N+1))^2$) near the highest frequency, which lies just below 2. For periodic boundary conditions, the frequencies are

$$\omega_k = 2\sin(\pi k/N), \tag{2.27}$$

which has only $N/2$ different linear frequencies, and an exact zone-boundary mode with $\omega_{N/2} = 2$. The existence of this mode with exact alternation of the phase of neighboring oscillators allows one to obtain some exact solutions, which we consider in Sect. 2.4.4. For the ϕ^4 chain, the linear part of the on-site potential results in the m^2 term in (2.24). If $m^2 \gg \pi k/N$, then there is also quadratic accumulation of frequencies above $\omega_k = m$. This bunching plays a significant role in the chaotic numerics, as described in Sect. (2.5). The FPU-α Hamiltonian can also be transformed by using the harmonic normal modes to obtain the transformed Hamiltonian

$$H_\alpha = \sum_{k=1}^{N} \frac{1}{2}\left(P_k^2 + \omega_k^2 Q_k^2\right) + \frac{\alpha}{2\sqrt{N+1}} \sum_{k,j,l=1}^{N} C(k,j,l) Q_j Q_k Q_l, \tag{2.28}$$

which is simpler, having only a product of three summations to represent the cubic term. Furthermore, it is considerably more stable than the FPU-β as it is a truncation to cubic order of an N-particle Toda lattice, as we have considered in the previous subsection.

For numerical integrations if, for example, a single mode initial condition is used, usually with all the energy in the form of potential energy, then $E = (1/2)\omega_k^2 Q_k^2$ ($\varepsilon = \omega_k^2 Q_k^2/(2N)$), and the oscillator equations are integrated with their initial values given from (2.21). Due to the nonlinearity, the energy does not remain in the initial mode but spreads through the mode spectrum, defined in terms of the instantaneous q_i by the transformation in (2.2). For energy sufficiently low that there is no resonance overlap (see Sect. 2.4.1), the energy is principally confined to the initial mode falling exponentially to other k-values, but satisfying the selection rule as given by (2.25) and following. Using perturbation theory, DeLuca et al. [55] obtained the mode energy decay in geometric progression

$$E_h \simeq \rho^2 E_{h-2\gamma}, \tag{2.29}$$

with γ as the initial mode and h the index of any high-frequency mode, and ρ is the average decay ratio between modes, 2γ apart, given by

$$\rho = \frac{3\beta E}{4\pi\gamma}. \tag{2.30}$$

The formula only holds for $\beta E \ll 1$, where resonances do not play a significant role. Numerical results for $\gamma = 3$ and 5 agreed quite well with the analytical predictions of (2.29) and (2.30).

The related problem of the FPU with alternating masses, as given in (2.20), has linear normal modes with frequencies

$$\omega_j = \frac{1 + m \pm \sqrt{1 + m^2 + 2m\cos k_j}}{m} \tag{2.31}$$

where $k_j = 2j\pi/(N+1)$ and $1 \leq j \leq N/2$ (for notational convenience we use j as the mode number). The acoustic branch has $0 < j < N/2$ and the optical branch for $N/2 < j < N$. The dispersion, calculated from (2.31), shows an optical branch that moves to higher frequencies and flattens as m is decreased; for example, the frequency separation of the minimum optical and maximum acoustic frequency is $\Delta\omega = \sqrt{2}(\sqrt{(1/m)} - 1)$, while the optical frequency spread is $\delta\omega_h = \sqrt{2/m}(\sqrt{1+m} - 1) \approx \sqrt{m/2}$, $m \ll 1$. The implication for isolated modes in the thermodynamic limit is discussed in Sect. 2.4.2.

Depending on the nonlinear forces, there are implications for the stability of nonlinear structures for the various forms of the linear modes. A weak spring ($\beta < 0$) in the FPU puts the nonlinear solution in the acoustic band which can then radiatively couple to the linear modes, destroying nonlinear stability. Similarly, for a strong spring ($\beta > 0$) but with an optical branch, a nonlinear acoustic mode can be shifted into the optical branch where it can dissipate by loosing energy to that branch. The various treatments of these phenomena are a major area for study (see, e.g., [84]), beyond the subject matter of this review. However, the phenomena will reappear in various subsequent sections.

2.3 Formulations: Methods of Numerical Analysis

Apart from the exception of integrable cases, most of the models of oscillator chains introduced in Sect. 2.2 require numerical investigation. The choice of suitable observables is then crucial to point out features of mathematical and physical interest. In this section, we introduce the description of indicators concerning both dynamical and statistical properties. In general, they are inspired by generalizations or extensions of the thermodynamic concept of entropy.

2.3.1 Measurement of Chaos Indicators

A quantitative characterization of the oscillator chains is their degree of chaoticity. This is measured by the largest Lyapunov exponent, whose positivity can be a hint to the possible equipartition of the energy among the degrees of freedom. Let us briefly recall that if

$$\dot{x}^i = X^i(x^1 \ldots x^{2N}) \tag{2.32}$$

is a generic dynamical system, the tangent dynamics to this flow is described by

$$\frac{\mathrm{d}\xi^i}{\mathrm{d}t} = J_{ik}(x(t))\xi^k \,, \tag{2.33}$$

where $J_{ik} = \partial X^i/\partial x^k$, and the largest Lyapunov exponent is given by

$$\lambda = \lim_{t\to\infty} \frac{1}{t} \ln \frac{\|\xi(t)\|}{\|\xi(0)\|} \tag{2.34}$$

for almost all choices of $\xi(0)$, under rather general assumptions. If $x = (q^1,\ldots,q^N,p^1,\ldots,p^N)$, with $X^i = (\partial H/\partial p^i)$ for $i = 1,\ldots,N$ and $X^i = -(\partial H/\partial q^i)$ for $i = N+1,\ldots,2N$ the dynamical system (2.32) represents a Hamiltonian flow. The corresponding tangent vector is $\xi = (\xi^1,\ldots,\xi^{2N}) \equiv (\xi_q^1,\ldots,\xi_q^N,\xi_p^{N+1},\ldots,\xi_p^{2N})$, and, by setting $\Lambda[x(t),\xi(t)] \equiv \{\xi^{\mathrm{T}}\, \underline{J}[x(t)]\, \xi + \xi^{\mathrm{T}}\, \underline{J}^{\mathrm{T}}[x(t)]\, \xi\}/\, 2\xi^{\mathrm{T}}\xi = [\xi^{\mathrm{T}}\dot{\xi} + \dot{\xi}^{\mathrm{T}}\xi]/2\xi^{\mathrm{T}}\xi = (\mathrm{d}/\mathrm{d}t)\ln(\xi^{\mathrm{T}}\xi)^{1/2} \equiv \frac{d}{dt}\ln\|\xi\|$, this can be formally expressed as a time average

$$\lambda = \lim_{t\to\infty} \frac{1}{t} \int_0^t \mathrm{d}\tau\, \Lambda[x(\tau),\xi(\tau)] \,. \tag{2.35}$$

Now we want to specify the more general concept of Kolmogorov–Sinai entropy [23, 24] associated with the Lyapunov exponents and discuss its relevance. Besides the largest Lyapunov exponent λ, in a dynamical system made of N degrees of freedom, each one described by a pair of canonical coordinates (position and momentum) one can define a *spectrum of Lyapunov exponents*, λ_i, where the index $i = 1, \cdots, 2N$ labels the exponents from the largest to the smallest one. An effective algorithmic procedure for evaluating the spectrum of Lyapunov exponents is discussed in [152]. Beyond rigorous mathematical definitions, an interpretation of the Lyapunov spectrum can be obtained by considering that the partial sum $h_n = \sum_{i=1}^{n} \lambda_i$ $(n \le 2N)$ measures the average exponential rates of expansion, or contraction, of a generic volume of geometric dimension n in phase space. Accordingly, $h_1 = \lambda_1 \equiv \lambda$ is equivalent to the definition given in (2.34), since a "one-dimensional volume" is a generic tangent segment in phase space; $h_2 = \lambda_1 + \lambda_2$ gives the divergence rate of a surface; $h_{2N} = \sum_{i=1}^{2N} \lambda_i$ is the average divergence rate of the whole phase space. In Hamiltonian systems, according to Liouville's theorem, any volume in phase space is conserved and $h_{2N} = 0$. Moreover, for each $\lambda_i > 0$ there exists $\lambda_{2N-i+1} = -\lambda_i$.[1] Chaotic evolution implies that a small region in phase space (for instance, the volume identifying the uncertainity region around an initial condition) is expanded and contracted with exponential rates along different directions in phase space. After a time of the order $1/\lambda$ the distance between two infinitesimally close initial conditions will have the size of the accessible phase space; accordingly, we have no means of predicting where the image of an initial point is in phase space by knowing the image of an initially close-by point, even if after a long time these points will eventually come again close to each other (for a detailed discussion see [123]). A very important conceptual achievement is that the mechanical description of a chaotic evolution can be replaced by a description in terms of a probability distribution on phase space which is invariant under time evolution and which allows one to define a *metric entropy* h. The mathematical details go beyond the scope of this manuscript; see [123, 124, 125, 126]. For our purposes, it is important to mention that Pesin later proved, under rather general assumptions, that there exist a remarkable relation between Kolmogorov's metric entropy and the positive component of the Lyapunov spectrum [127]:

$$h = \sum_{j, s.t. \lambda_j > 0} \lambda_j \ , \tag{2.36}$$

where the sum extends over all the positive Lyapunov exponents. This formula can be applied to the study of the dynamics of Hamiltonian systems, like the FPU chain. In this respect, it is particularly interesting to check this formula in the thermodynamic limit, in which the number of oscillators tends to infinity. In general, this limit does not commute with the limit $t \to \infty$ in (2.5), i.e. the measurement of λ and h may depend on the order in which these limits are performed. Numerical evidence of the existence of a limit curve for the spectrum of Lyapunov exponents in the thermodynamic limit for the FPU chain was later obtained ([128]; see also Fig. 2.2). Further numerical evidence of the existence of such a limit for a variety of physical systems have been subsequently obtained. However, a rigorous mathematical proof is still lacking, although some attempts in this direction exist [129, 130]. The value of h is expected to depend on some typical parameters, like the energy density ε for a Hamiltonian chain of oscillators. For instance, the Lyapunov spectrum of the FPU-β model shown in Fig. 2.2 is obtained for $\beta\varepsilon = 10$, which is sufficiently large to yield a strongly chaotic dynamics. By decreasing ε sufficiently to enter the almost-recurrent dynamical regime observed by

[1] For each conserved quantity like the energy, momentum etc., there is a pair of conjugated exponents that are zero. Stated differently, each conservation law amounts to a geometrical constraint that limits the access of the trajectory to a submanifold of phase space. Integrability has the consequence that all λ_i are zero, i.e. there can be as many conservation laws as the number of degrees of freedom; the converse is in general not true.

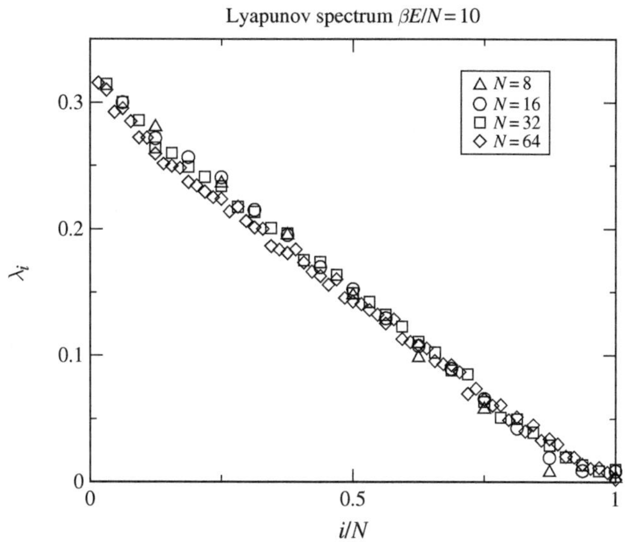

Fig. 2.2. The spectrum of positive Lyapunov exponents of the FPU-β model for different chain lengths, from 8 up to 64 oscillators

Fermi, Pasta, and Ulam in their original numerical experiment, the shape of the spectrum also changes significantly. In this weakly chaotic regime, the maximum Lyapunov exponent is found to decrease and the positive component of the Lyapunov spectrum approaches the horizontal axis. Still the only null exponents are those corresponding to the conserved quantities, although the others take significantly smaller values and the value of h is drastically reduced. According to this description, one is led to conclude that, in the thermodynamic limit, all possible chaotic degrees of freedom should remain chaotic for arbitrarily small values of ε, despite that beyond a certain value it will become practically impossible to distinguish them from zero. In this respect, h cannot provide a characterization of the weakly chaotic regime in terms of an effective number of active degrees of freedom, as discussed in the following Section. Nonetheless, the Lyapunov analysis can provide a clear quantitative characterization of the strong and weak chaotic regimes observed in the FPU-chain. Actually, the maximum Lyapunov exponent of the FPU-β model has been analytically estimated [131] on the basis of the geometrical approach, sketched in Sect. 2.4.7. It has been found that there is a transition value of the energy density, ε_c, at which the scaling of λ with ε changes from a strong ε-scaling, $\lambda(\varepsilon) \propto \varepsilon^2$, to a weaker one $\lambda(\varepsilon) \propto \varepsilon^{1/4}$. The numerics is given in Sect. 2.5 and the calculation of the scaling is outlined in Sect. 2.8. This steep scaling of $\lambda(\varepsilon)$ below ε_c implies that the typical *relaxation time*, i.e. the inverse of λ, may become exceedingly large for very small values of ε. It is worth stressing that this result seems independent on the size N of the system, thus indicating that the different relaxation regimes represent a statistically relevant effect.

2.3.2 Equipartition Indicators: Information Entropy, Effective Number of Modes

In the numerical experiment by Fermi, Pasta and Ulam, the initial energy was placed in a single low-k mode and the authors aimed at studying how this energy would eventually flow to the other modes. The description of the dynamics in terms of Fourier modes was a natural approach at least for small specific energy, despite the fact that they are not the proper modes of the chain. They expected that the nonlinearity would yield a fast decay towards equipartition of the energy among the Fourier modes as a natural condition to be fullfilled at thermodynamic equilibrium. The existence of the two dynamical regimes in the FPU problem for low and high values of the energy density, ε, has been characterized in this context by introducing a suitable equipartition indicator among the Fourier modes [51, 52]. This indicator is inspired by information entropy, but, at variance with Kolmogorv's metric entropy, it relies upon a heuristic definition. In a chain made of N oscillators with periodic boundary conditions there are $N/2$ independent Fourier modes. A *spectral entropy* $S(t)$ can be defined as

$$S(t) = -\sum_{n=1}^{N/2} p_n(t) \ln p_n(t) \ , \tag{2.37}$$

where $p_n(t) = E_n(t)/\sum_n E_n(t)$, $E_n(t)$ being the harmonic energy of the Fourier mode with wave vector $k_n = 2\pi n/N$ at time t. When only a single Fourier mode is excited $S(t)$ vanishes, and it takes its maximum value $S_{\max} = \ln(N/2)$ when equipartition of the energy among the Fourier modes is obtained. Numerical studies showed that this quantity exhibits good statistical properties, while it can describe the approach to energy equipartition starting from either single-mode or multimode initial excitations. To compare chains of different lengths, a normalized quantity was defined:

$$\eta(t) = \frac{S_{\max} - S(t)}{S_{\max} - S(0)} . \tag{2.38}$$

Notice $\eta(t)$ tends towards zero when the system approaches equipartition and that it keeps a value close to 1 when the initial spectral entropy is maintained during time evolution. In the long time limit $\eta(t)$ was found numerically to approach an asymptotic average value $\bar{\eta}$, which was used for identifying the equipartition thresholds of the FPU-α and -β models [51, 52]. Moreover, it has been also observed that the very dynamics of $\eta(t)$ provides a qualitative characterization of the different dynamical regimes observed in these chain models [132]. The regular, quasi-recurrent dynamics of η or of $n_{\text{eff}} = N_{\text{eff}}/N$

(see below) observed for small values of ε turns to a fast decay towards small η ($n_{\text{eff}} \sim 1$) for large values of ε.

A more physically transparent measure is what we call the effective number of modes containing energy, which can be defined as

$$N_{\text{eff}} \equiv \exp S, \tag{2.39}$$

which is conveniently normalized as

$$n_{\text{eff}} = N_{\text{eff}}/N\ . \tag{2.40}$$

For oscillators, the same definitions (2.37)–(2.40) can be used, with the energy of each oscillator taken directly from the Hamiltonian, by assigning half of the difference potential to each neighbor, to obtain the normalized effective number of oscillators containing energy, n_{osc}, which we will use in the numerics from short-wavelength mode initial conditions. The instantaneous values of n_{eff} do not asymptote to one, at equipartition, due to fluctuations. A simplified calculation of the effect of fluctuations introduces a deviation δe_i from equipartition $e_i = \bar{e}_i + \delta e_i$. Expanding the logarithmic function in S in (2.37) as $\ln(1+\delta e_i/\bar{e}_i) = \delta e_i/\bar{e}_i - (1/2)(\delta e_i/\bar{e}_i)^2$ and performing the summation over i yields

$$n_{\text{eff}} = n_{\text{osc}} = \frac{1}{N}\exp\{-N\bar{e}\ln\bar{e} - N\overline{\delta e^2}/(2\bar{e})\} = \exp\{-N\overline{\delta e^2}/(2\bar{e})\}. \tag{2.41}$$

Taking $\bar{e} = 1/N$ and making the assumption of normal statistics, that for each normal mode $\overline{\delta e^2} = \bar{e}^2$ (this is confirmed by calculations), we see that N cancels giving an asymptotic value $n_{\text{eff}} = n_{\text{osc}} = \exp(-0.5) = 0.61$, at equipartition, for both modes and oscillators. More accurate calculations have been made separately for modes and oscillators, including the nonlinear terms in the oscillator calculation, yielding at equipartition, ([92] Appendix D),

$$n_{\text{eff}} = 0.65 \qquad n_{\text{osc}} = 0.74. \tag{2.42}$$

To obtain some smoothing of the numerical values of $n_{\text{eff}}(t)$ and $n_{\text{osc}}(t)$, various short-time averages of these quantities have been used, yielding somewhat different values from those predicted in (2.42).

2.4 Formulations: Analytic, Low-Energy and Short-Time Results

2.4.1 Transformations and Low-Dimensional Calculations

We have seen in Sect. 2.2 that a transformation to the coordinates of harmonic normal modes decouples the modes if only linear forces are present. For small values of βE, the smaller nonlinear terms couple all of the modes together.

Taking the FPU-β system with energy initially placed in a long-wavelength mode, which we consider here, the selection rule for the couplings results in a geometric progression of the energy fall-off to shorter-wavelength modes [55]. The strongest interactions are therefore among neighboring modes, with the initial energy in a long-wavelength mode interchanging energy most strongly with its nearest neighbors. The resulting beat oscillations, as observed numerically in the original and much subsequent work (see Fig. 2.1), involved primarily a few modes. The predominant localization among a few modes allows a useful investigation of a reduced problem, involving some minimum number of modes. To look at "resonance overlap," a four-mode subsystem is examined, which contains two three-mode resonances. This was done for the sine-Gordon chain by Goedde et al. [54] and then in more detail for the FPU-β chain by DeLuca et al. [55]. Summarizing the analytic method, a transformation of the four-mode Hamiltonian to action-angle variables exhibits two slow angles of the major resonances $\theta_s = \theta_1 + \theta_3 - 2\theta_2$ and $\theta_{\mathrm{sp}} = \theta_2 + \theta_4 - 2\theta_3$. A second transformation is performed to the new variables θ_s and θ_{sp} followed by employing the method of averaging over the two remaining fast angles. The resultant averaged Hamiltonian has two additional approximate constants of the motion, which are the actions related to the averaged-over angles and thus is reduced to two freedoms. The resulting Hamiltonian has the approximate form

$$
\begin{aligned}
H_4 = \left(\frac{\pi}{N}\right)^2 E_\gamma \Big[& -(J_s + J_{\mathrm{sp}})/4J_{\mathrm{c}} \\
& + \frac{R}{8J_{\mathrm{c}}^2}(3J_s^2 + 3J_{\mathrm{sp}}^2 - 4J_sJ_{\mathrm{sp}} + J_{\mathrm{c}}J_s - 2J_{\mathrm{c}}J_{\mathrm{sp}}) \\
& + \frac{R}{16J_{\mathrm{c}}^2}\sqrt{J_1J_2J_3J_4}\cos(\theta_{\mathrm{sp}} + \theta_s) \\
& + \frac{R}{16J_{\mathrm{c}}^2}\left(\sqrt{J_1J_2^2J_3}\cos\theta_s + \sqrt{J_2J_3^2J_4}\cos\theta_{\mathrm{sp}}\right)\Big] ,
\end{aligned} \tag{2.43}
$$

where J_{c}, J_d, J_s, and J_{sp} are the transformed actions and J_1, J_2, J_3, and J_4 are the original actions, related to the transformed actions by the canonical transformation

$$J_1 = J_s \tag{2.44}$$

$$J_2 = J_{\mathrm{sp}} \tag{2.45}$$

$$J_3 = J_s - 2J_{\mathrm{sp}} + J_{\mathrm{c}} \tag{2.46}$$

$$J_4 = J_{\mathrm{sp}} + J_d \, . \tag{2.47}$$

J_{c} and J_d are new constants of the motion, resulting from the averaging, and J_{c} was chosen such that $J_{\mathrm{c}} = E_\gamma/\Omega_\gamma$, i.e. the action corresponding to the initial energy, primarily in mode γ. The concept of "resonance overlap" is taken from low-dimensional chaos theory, which considers separately the phase space motion $H(J_s, \theta_s)$ with $J_{\mathrm{sp}} = \text{const.}$ and $H(J_{\mathrm{sp}}, \theta_{\mathrm{sp}})$ with $J_s = \text{const.}$, with overlap being the condition that for some values of θ_s and $\theta_{\rm sp}$ we obtain $J_s = J_{\rm sp}$. The variable actions J_s and $J_{\rm sp}$ are numerically studied by looking at the phase space of one degree of freedom in a surface of section of the other freedom, with area filling trajectories indicating resonance overlap. The overlap is governed by

$$R = (N+1)\frac{6\beta}{\pi^2}E_\gamma \geq 1 \,, \tag{2.48}$$

where R measures the ratio of nonlinear to linear energy in the resonant degrees of freedom, analogous to the energy ratios used to calculate the mode overlap condition in (2.7). As in that calculation, from our understanding of low-dimensional chaos, we expect significant stochasticity to appear for $R > 1$. $R = 1$ has recently been shown to be the transition to instability for periodic solutions of the full chain [133].

The results for four R-values are given in Fig. 2.3, showing the transition to stochasticity in the reduced system. From the same four-mode calculation, the frequency of a typical resonant trajectory is given by

$$\Omega_{\rm B} \approx \mu\gamma\beta E_\gamma \left(\frac{\pi}{N}\right)^2 \tag{2.49}$$

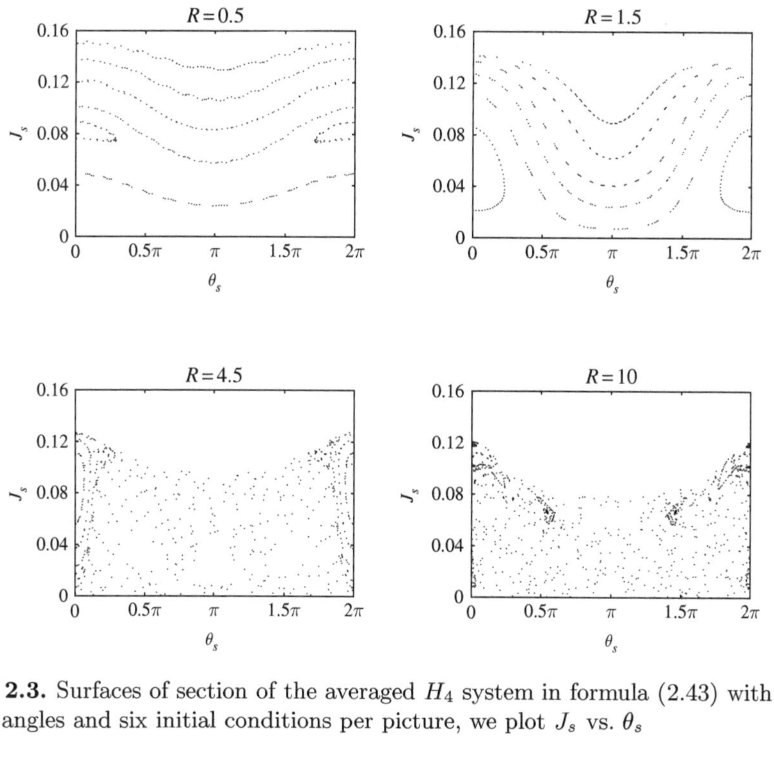

Fig. 2.3. Surfaces of section of the averaged H_4 system in formula (2.43) with two slow angles and six initial conditions per picture, we plot J_s vs. θ_s

with μ a constant of order unity, dependent on the particular initial conditions. (Here and afterward, we use the approximation $N + 1 \approx N$.)

The transition to stochasticity in a reduced system is neither necessary to ultimately reach equipartition, nor sufficient to produce equipartition on numerically observable time-scales. We note from (2.48), as the number of freedoms increases overlap occurs at decreasing energy. However, from (2.49), we see that the stochasticity also exhibits itself on increasingly slow time-scales. Furthermore, there is a *competition* between local resonance overlap, which spreads energy among neighboring modes, and the process of Arnold diffusion, which transports energy along guiding resonances to modes in other parts of the phase space. This latter process is exponentially slow at low energy. Although a rigorous upper bound on Arnold diffusion has the form given in (2.3), this does not determine the diffusion rate from the long wavelengths to the short wavelengths. The appropriate calculation is made from a three-resonance model ([13]; see also [15], Sects. 6.1 and 6.2). We have already considered the two resonances, which produce the local stochasticity. The third resonance, called the guiding resonance, links two short-wavelength modes to the low frequencies via the selection rule (2.25). Again, following [55], the calculation yields a rate of energy increase in the short-wavelength mode proportional to $\exp(-\pi/\varepsilon)$, with $\varepsilon \approx \Omega_{\mathrm{B}}/\delta\Omega_h$ where $\delta\Omega_h$ is the short-wavelength resonance frequency. Thus we expect the diffusion to be numerically observable if $\Omega_{\mathrm{B}} > \delta\Omega_h$, i.e. the low frequency beat becomes comparable to a high-frequency resonance that it can couple to, that is, one for which $B \neq 0$ in (2.25). The smallest $\delta\Omega_h$ (largest ε) is $\delta\Omega_h = \gamma(\pi/N)^2$. Substituting for Ω_{B} from (2.49), together with this $\delta\Omega_h$ yields the inequality

$$\mu\beta E_\gamma > 1 \tag{2.50}$$

for diffusion along resonances to compete with diffusion across resonances. Here, as in all other equations βE appears as a product, which measures the nonlinearity. The implications of (2.50) can be seen in numerical calculations in Fig. 2.4 at small values of R for some relatively small oscillator chains, for which the lower edge gives a long-time asymptotic value of N_{eff}. Considering that for $R > 1$ there is strong local coupling among modes, then as R increases and the energy interchange spreads to more modes, there is an increase of $N_{\mathrm{eff}} \propto R$, given by this lower edge. However, at some value of $E_\gamma = E_{\mathrm{c}}$, satisfying (2.50) the values of N_{eff} leave this asymptote, and, in fact, approach equipartition over longer times. This scaling, first found numerically in [56], is physically explained by the direct transfer of energy through the guiding resonances to high-frequency modes (see [55] for a more detailed calculation). We illustrate the spreading to higher modes in Fig. 2.5 at $R = 2.9$ for $N = 32$, below the E_{c} transition as found in Fig. 2.4. The increase in energy in some high-frequency modes, specified from the selection rules, is above the background, but does not increase with time. We will contrast this result with the spectrum for $E > E_{\mathrm{c}}$ in Fig. 2.9, which approaches equipartition as time increases.

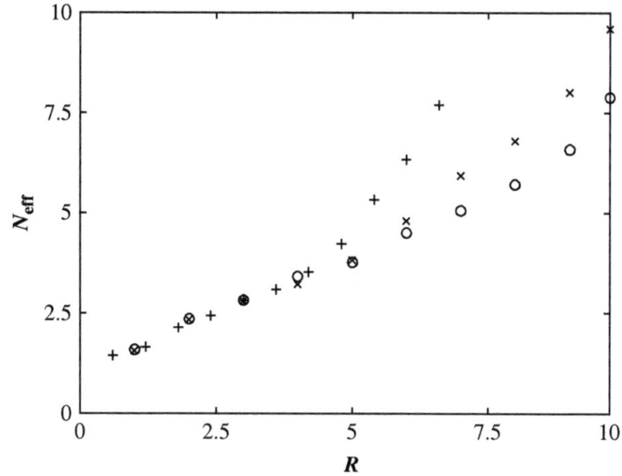

Fig. 2.4. Effective number of modes N_{eff} vs. R after $t = 2000(3/(\gamma(\pi/N)^2)$; *pluses* $N = 16$; *crosses* $N = 32$; *circles* $N = 64$

As with the FPU-β oscillator chain, the FPU-α can also be analyzed in terms of overlapping resonances to determine the onset of large stochastic layers among the long-wavelength modes. Because the nonlinear term is cubic, rather than quartic, the resonances are simpler, involving only three terms, and the scaling, with ε, for resonance overlap, is different. Shepelyansky [62] has used the same averaging procedure as described in the four-mode approximation of the FPU-β, to analyze the FPU-α chain, obtaining the Hamiltonian

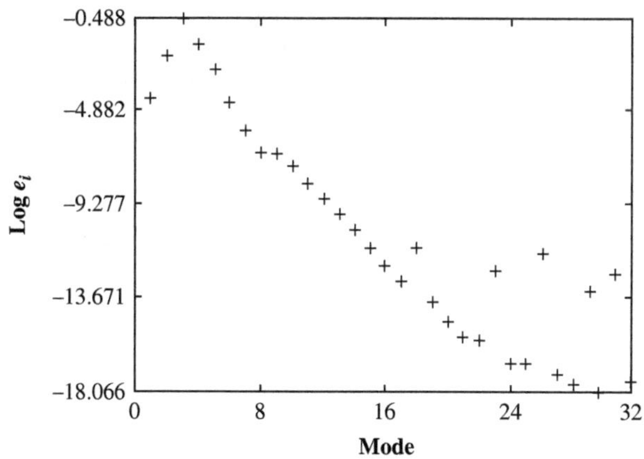

Fig. 2.5. Log of average energies at $R = 2.9$ for $N = 32$ ($E = 1.4$) after $t = 2000(N/\pi)^2$

$$H = \sum_{k=1}^{N} \omega_k I_k + \frac{\alpha}{2\sqrt{N+1}} \cdot \sum_{k_1,k_2,k_3=1}^{N} (\omega_{k_1}\omega_{k_2}\omega_{k_3} I_{k_1} I_{k_2} I_{k_3})^{1/2}$$
$$\times \cos(\theta_{k_3} - \theta_{k_2} - \theta_{k_1}) \delta_{k_3,k_1+k_2} \tag{2.51}$$

where all the angles have been averaged over, except the resonant ones for which $k_3 = k_1 + k_2$ in the long-wavelength spectrum. For these wavenumbers from (2.23), $\omega_k \approx \pi k/(N+1)$ such that $\omega_3 \approx \omega_1 + \omega_2$. The Is and θs are the action-angle variables, as in (2.43), before the final transformation to the resonant coordinates. Because of the lower cubic products, Shepelyansky was able to examine the full Hamiltonian, and after making two simplifying further transformations he derived the approximate chaos border at long wavelengths, $\gamma \ll N$, where γ is the k-value at the center of the resonance

$$\alpha N^{3/2} E^{1/2} / \gamma^2 > 1. \tag{2.52}$$

Comparing (2.52) with (2.48), which has $R > 1$ for resonance overlap, we see that the scaling with the perturbation strengths α or β are the same, as is the scaling of $\varepsilon \propto N^2$ if we substitute for the energy density $\varepsilon = E/N$ in both cases. Thus for fixed energy density (fixed temperature), both formulas predict a resonant transition to local chaos in the thermodynamic limit, $N \to \infty$. The energy-dependence with quartic or cubic nonlinearities is, of course, different. Shepelyansky investigated the transition of (2.52) numerically, using the largest Lyapunov exponent, finding reasonable agreement. He also fits the distribution of linear mode energies to the distribution

$$E_k \propto \frac{1}{k_c \exp(k/k_c - c) + 1} \tag{2.53}$$

(with the best fit for $c = 2.65$) such that k_c is a measure of the number of modes containing energy, similar to N_{eff}, but for early times for which the energy distribution still decreases exponentially with mode number, i.e. the energy has not significantly diffused to the high frequencies through the Arnold Web. The numerical estimate for the scaling is

$$k_c \sim (N^3 \alpha^2 E)^{1/4} \tag{2.54}$$

which the author was able to predict analytically. This is contrasted with the result from Fig. 2.4, which indicates that

$$N_{\text{eff}} \sim (N\beta E)^m , \tag{2.55}$$

i.e. is governed by the number of modes that can satisfy the local overlap condition $R > 1$ with R given in (2.48). Shepelyansky [62] has analytically estimated $m = 1/2$.

The FPU-α model can be obtained as a third order truncation of the power series expansion of the Toda lattice potential, defined by the Hamiltonian:

$$H(p,q) = \sum_{k=1}^{N} \frac{p_k^2}{2} + \frac{a}{b} \sum_{k=1}^{N} \left[\exp(-b(q_{k+1}-q_k)) + b(q_{k+1}-q_k) - 1\right] . \quad (2.56)$$

Since the Toda lattice is integrable, i.e. does not exhibit stochastic behavior, the FPU-α is more stable than the FPU-β. However, because the nonlinear potential is cubic the trajectories become unbounded at high energy. Therefore, it is restricted to examining low-energy phenomena, as was described above. Using (2.35) for calculating λ for neighboring trajectories, and choosing the constants a and b in (2.56) to correspond to the FPU-α given in (2.1), Pettini and co-workers [61] compared the variation with time of the integrable and nonintegrable systems, with the result as shown in Fig. 2.6. The initial conditions, starting on separate orbits, separate linearly (see [15]) from which, calculating from (2.35) over short times, a large Lyapunov exponent is obtained. However this effect continually diminishes in the averaging process and, after a long time, only an average exponential divergence of the trajectories remains. In Fig. 2.6 we show the value of λ stabilizing at the average exponent for the FPU-α system, while it vanishes for the Toda system. As an aside remark, we point out that the stabilized value of λ, shown in this figure, is not necessarily the asymptotic value, but may correspond to a value in a more localized region of the phase space. Without exploring this possibility in detail we note that the numerical values of λ presented in Sect. 2.5, and compared to calculations in Sect. 2.8, have been obtained in a way that should be close to the infinite-time average.

Considerable effort has been directed toward the comparison of the FPU-β chain with oscillator chains constructed from discretization of the Klein–Gordon equation, particularly the ϕ^4 chain, with the nonlinear term being

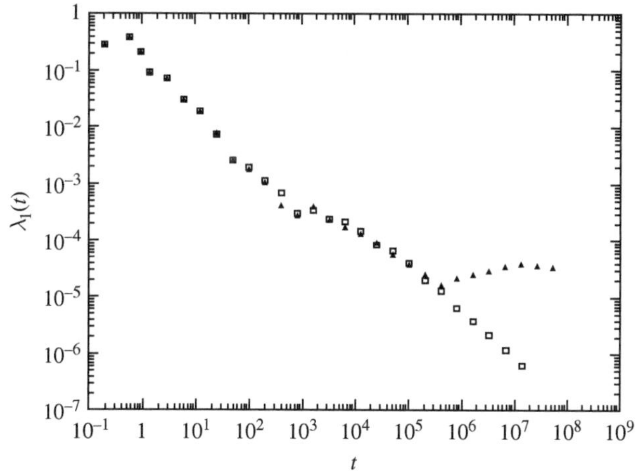

Fig. 2.6. Maximal Lyapunov exponent vs. time for the Toda lattice (open squares) and for the FPU-α model (solid triangles) for $N = 32$ and $\varepsilon = 0.0217$

an on-site potential. However, little attention was given to the comparison on shorter time-scales, from long-wavelength initial conditions. Comparing the coefficients in (2.25) and (2.26), for small m, we see that the nonlinearity is much weaker for long wavelengths (small ω for the FPU potential than for the ϕ^4 potential). The opposite holds for short wavelengths, where the ωs are about 2 (for small m). Physically this is easily understood, as the forces between neighboring oscillators are quite small for the FPU at long wavelengths: neighboring oscillators are in phase, with nearly the same amplitudes, while at short wavelengths the nearly out of phase amplitudes amplifies the forces between them, as compared to the nonlinear self-force of the ϕ^4. The consequences for the times to achieve equipartition, starting from either low- or high-frequency initial conditions, will be presented in Sect. 2.8. The strong nonlinearity, coupled with the weak dispersion at short wavelengths, which is evident from either (2.23) or (2.24), leads to narrow structures in the oscillator space, which exhibit the short-time characteristics of breathers. These structures called chaotic breathers (CB's) are introduced in Sect. 2.4.4, and investigated in some detail in Sects. 2.6 and 2.7.2.

2.4.2 The Thermodynamic Limit

The analysis described in Sect. 2.4.1 of considering a few modes which contain most of the energy, to understand the subsequent behavior, is appropriate for finite, relatively small, values of N. We have already seen that for fixed ε and increasing N, (2.48), with $R > 1$, for the FPU-β and (2.52) for the FPU-α indicate local stochasticity and therefore diffusion throughout the phase space in the Arnold web, but without specifying the time-scale. For fixed $\varepsilon = E/N$, approaching the thermodynamic limit ($N \to \infty$), other questions arise.

In Sect. 2.4.1 we saw, from various perspectives that at fixed E the number of modes forming an energy-containing packet, in a reasonably short time, would increase with N, while the fraction of modes $n_{\text{eff}} = N_{\text{eff}}/N$ would remain constant. This implied that for fixed E equipartition would not be reached, at least for computationally observable times, for large values of N. However, at fixed E, ε decreases with increase in N, so the question of what happens for fixed ε in the thermodynamic limit was not addressed.

In Sect. 2.3.1, we numerically indicate that the value of the largest Lyapunov exponent decreases with the power law $\lambda \propto \varepsilon^2$, but for low values of N there are faster drop-offs, which may be exponentially varying, as suggested in the previous section. However, the drop-off value of ε occurs at increasingly small ε as N is increased. Similarly, in Sect. 2.5.2, the time to equipartition T_{eq} increases as a power law in ε at smaller N, with any faster increases appearing at smaller ε as N is increased. The implication is that $T_{\text{eq}} \propto \varepsilon^{-3}$ at the thermodynamic limit.

From a different perspective, Galgani and coworkers [48, 49] used a convergent perturbation theory to rigorously show that two groups of oscillators, well separated in frequency space, would transfer energy exponentially slowly from the low to the high frequencies, i.e. energy would be "frozen out" of the high-frequency oscillators over times $\tau \propto \exp(a\omega_h/\omega_l)$ the ratio of the high to low frequencies. Using these ideas, they estimated transfer times from low to high frequencies for the FPU chain, but came to the conclusion that the energy could not be bounded away from the high frequencies for exponentially long times in the thermodynamic limit [118]. In the same paper they returned to the concept of two well-separated groups of oscillators by employing the alternating mass chain given in (2.20). As described in Sect. 2.2.2, with the lighter mass m much smaller than the heavier unit mass, an optical branch becomes thin with $\delta\omega_h \approx (m/2)^{1/2}$ and can be treated in perturbation theory as occurring at a single high frequency $\omega_h \approx (2/m)^{1/2}$. The upper edge of the acoustic band is $\omega_{l(\max)} \sim 1$. The ratio appears in the rate of change of the high-frequency action proportional to $\exp(-B(\omega_h/\omega_{l(\max)}))$ at fixed N and ε, in qualitative agreement with their previous work. However, B vanished inversely with a power of N, and thus the high frequency modes were not isolated in the thermodynamic limit. Their numerical studies with $0 < N < 200$ were not conclusive, but indicated a weak N-dependence.

A similar oscillator chain to that given in (2.20), but with the heavy and light masses distributed randomly, also resulted for finite N in only a partial filling of the modes [134]. A calculation of n_{eff}, starting from long-wavelength initial conditions, indicated that only the acoustic modes came to equipartion in the time-scales investigated, with very little energy in the optical branch. This is also consistent with exponentially slow transfer to the optical branch, but definitive answers in the thermodynamic limit cannot be obtained from numerically observable times. We present results in Sect. 2.5.

2.4.3 Long-Wavelength Approximations: KdV, mKdV; Stability; Exact Periodic Solutions

In the introduction, we described, briefly, that Taylor series expansions for long wavelengths of the FPU-α and FPU-β chains result respectively in the KdV and mKdV partial differential equations. We illustrate the method for the most extensively studied case of the mKdV approximation to the FPU-β chain. Starting from the differential form of the oscillator equations

$$\frac{\partial^2 y_j}{\partial t^2} = (y_{j+1} - y_j) - (y_j - y_{j-1}) + \frac{1}{3}\left[(y_{j+1} - y_j)^3 - (y_j - y_{j-1})^3\right] , \quad (2.57)$$

where the displacements of the lattice sites have been rescaled to obtain the nonlinear coefficient 1/3, Zabusky and Kruskal [44] used a Taylor expansion to make y into a continuous variable $y(x,t)$, using a integro-differential form

$$u \equiv -\frac{y_t}{2h} + \frac{1}{2}\int_0^{y_x} \sqrt{1 + h^2\eta^2}\,\mathrm{d}\eta \quad (2.58)$$

to transform (2.57) in lowest order in $\Delta x \equiv h = L/N$ into the equation

$$u_\tau + 12u^2 u_\xi + u_{\xi\xi\xi} = 0, \tag{2.59}$$

where L is the chain length, N the number of oscillators; the time and length variables have been rescaled by $\tau = h^3 t/24$ and $\xi = x - ht$, and the subscripts t,x,τ,ξ denote differentiation with respect to that variable.

Defining u by a differential in t and an integral in x reduced the time derivatives by one and increased the space derivative by one to obtain the well-known mKdV equation, which is integrable as we show below. We note, however, that the choice of a unique direction in the variable ξ implies a traveling solution and thus a solution only exists as a single nonlinear wave on an infinite or periodic chain.

Periodic solutions, stationary in the frame $\xi - C\tau$, can be obtained by integrating (2.59) twice, giving

$$\frac{1}{2}u_\xi^2 + u^4 - \frac{1}{2}Cu^2 - Bu + A = \frac{1}{2}u_\xi^2 + P(u) = 0, \tag{2.60}$$

where A and B are constants of integration. For a periodic lattice the mean of u must be zero which implies $B = 0$. Equation (2.60) is in the form of a one-degree-of-freedom Hamiltonian, which is therefore integrable. Equilibrium solutions to (2.60) have been obtained in terms of the Jacobi elliptic functions (or cnoidial waves) $cn(\xi, q)$, with q^2 (the modulus) taken as a parameter with $0 \le q \le 1$. Driscoll and O'Neal [45, 46] examined the solutions for stability, both analytically for long-wavelength perturbations, and numerically. They determined the unstable modes of any solution u_0 by numerically solving the linearized eigenvalue problem

$$\mathrm{i}\nu v - Cv_\xi + 12(u_0 v)_\xi + v_{\xi\xi\xi} = 0, \tag{2.61}$$

where $u(\xi, \tau) = u_0(\xi) + v(\xi)\exp(\mathrm{i}\nu\tau)$. The waves are found to be stable if all four roots of the associated polynomial $P(u)$ are real, and unstable if two roots are real and two are complex. Although the normalized time τ is scaled by $h \propto 1/N$, the complete rescaling of the normalized equations back to unnormalized variables gives growth rates independent of N. Driscoll and O'Neil then compared the growth rates with those found from numerical integration of the equations of motion for the FPU chain with various values of N. For wavenumber $\pi k/L$ they obtain the growth rate as a function of the modulus q^2 of the elliptic function, as shown in Fig. 2.7. In this situation the growth of the unstable modes in the continuous limit is found to be an upper bound on the mode growth of the finite chain. A similar relation was also found to exist between the existence of instability in the sine-Gordon equation and equipartition in an oscillator chain corresponding to the discretized sine-Gordon equation [95]. For the FPU system we would not expect the mKdV instability to be directly related to equipartition among the high frequency modes, as the mKdV differential equation does not describe these modes. However, mixing of low-frequency modes in the continuous system corresponds to stochasticity among low-frequency modes in the discrete system. To explore this further,

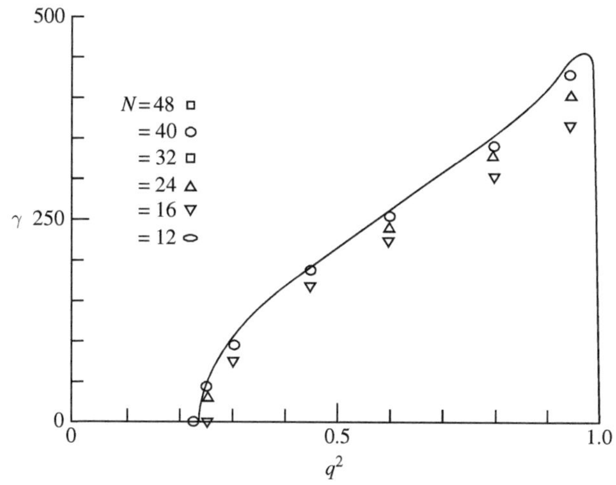

Fig. 2.7. Scaled lattice instability rate γ (*data points*) and mKdV prediction (*solid curves*) for cnoidal waves vs. modulus q^2

we compare the onset of instability in the rescaled parameters of the mKdV equation, as found by Driscoll and O'Neal, to the parameters governing the interaction among the low-frequency modes in the four-mode analysis. The rescaling of R from (2.48) gives the relationship $R = (8/\pi^2)(\gamma/2)^2 q^2 K^2(q^2)$, where γ is the number of nodes of the cnoidal function and $K(q^2)$ is the complete elliptic integral. The instability appears for $q^2 \approx 0.25$ ($K = 1.7$) for $\gamma = 2$, which corresponds to $R_c \approx 0.58$. This is close to the value which produced a separatrix layer in the four-mode resonance interaction found in Sect. 2.4.1, so we conclude that the mechanisms are related. The single-mode initial conditions give rise to beat phenomenon, corresponding to stable solitons. It is only when the solitons become unstable that this manifests itself as chaos in the discrete system.

For the FPU-α oscillator chain a similar Taylor expansion and transformation to new variables, gives the KdV equation

$$u_\tau + u u_\xi + \frac{1}{24} u_{\xi\xi\xi} = 0 \tag{2.62}$$

which is also integrable. However, for the KdV equation, linearization of the solution does not exhibit unstable eigenvalues, and thus chaos appears when the Taylor expansion breaks down at larger values of energy when the discreteness leads to diffusion among the low-frequency modes. As with the mKdV equation, a single soliton, which travels in a given direction, cannot satisfy fixed boundary conditions so that more than one soliton is required to describe any prescribed initial condition.

For either the KdV or mKdV equation an initial condition consisting of a long-wavelength linear mode, say

$$u(\tau = 0, \xi = x) = u_0 \cos\left(\frac{2\pi x}{L}\right) , \tag{2.63}$$

where $u \sim \varepsilon^{1/2}$ (ε is the energy per degree of freedom in the FPU) will break up into a set of solitons. The shortest wavelength of the resulting solitons can be estimated from dimensional analysis [44, 135]. For long wavelengths, the dispersion is weak, therefore the dynamics is dominated by the nonlinearity. One can therefore neglect the dispersive term, which for the KdV equation (2.62) satisfies $(1/24)u_{\xi\xi\xi}/(uu_\xi) < 1$ or dimensionally

$$\frac{l^{-3}}{(\sqrt{\varepsilon}/l)} \leq 1 \,\text{or}\, l_{\text{min}} \sim \varepsilon^{-1/4} . \tag{2.64}$$

By reintroducing α, using the scaling $\varepsilon \to \alpha^2\varepsilon$, and by considering that the fraction of degrees of freedom $n_{\text{eff}} \simeq 1/l_{\text{min}}$, we obtain

$$n_{\text{eff}} \sim \alpha^{1/2}\varepsilon^{1/4} . \tag{2.65}$$

This result agrees with the scaling estimated and found numerically by Shepelyansky [62], if $n_{\text{eff}} = N_{\text{eff}}/N$ is substituted in (2.54) and also coincides with later studies of Biello et al. [136] and Berchialla et al. [137]. In a manner similar to that described above, the scaling of n_{eff} with $\beta\varepsilon$, for the FPU-β, can be determined from the ratio of dispersive to nonlinear terms in (2.59), giving

$$\frac{l^{-3}}{\varepsilon/l} \geq 1 \,\text{or}\, l_{\text{min}} \sim \varepsilon^{-1/2} . \tag{2.66}$$

Changing ε to $\varepsilon \to \beta\varepsilon$ and $n_{\text{eff}} \simeq 1/l_{min}$, we obtain

$$n_{\text{eff}} \sim (\beta\varepsilon)^{1/2} . \tag{2.67}$$

This result is in agreement with the estimate by Shepelyansky [62], but different from the scaling found numerically in Fig. 2.4, which, however, has not been examined for large N, where (2.67) applies.

It has also been recently shown [133] that it is possible to construct exact periodic solutions of the FPU chains, for fixed $\beta\varepsilon$ at finite N, by a Newton method. The authors called the solutions q-breathers (QBs) in analogy to short-wavelength solutions of a few oscillators, since they designated the linear mode number by q. These solutions are complementary to those obtained from continuous approximations. For the FPU-α and FPU-β, asymptotic expansions in the small parameters $\rho = \alpha/\sqrt{2(N+1)}$ and $\sigma = \beta/(2(N+1))$, respectively, produce exponentially decaying linear-mode spectra, similar to those described in Sect. 2.4.1, and similar to expansions obtained from soliton solutions with periodic boundary conditions. Unlike the solitons, the periodic solutions cannot be summed to produce the initial conditions of a single linear mode. Nevertheless, for small perturbation parameters, they are sufficiently close to single-mode initial conditions that interesting results can be obtained from them to compare to the usual numerics. For the FPU-β, the periodic solutions become unstable, similar to the instability we have seen in the mKdV equation. A very interesting result is that the bifurcation to unstable solutions of the FPU-β chain occurs at a value of $R = 1 + \mathrm{O}(1/N^2)$; from Sect. 2.4.1 we recall that $R = 6\beta E(N+1)/\pi^2$ relates nonlinear to linear terms in the Hamiltonian. The condition $R \simeq 1$ was a semi-quantitative transition for resonance overlap, leading to local chaos. Here we find that the condition becomes precise for the onset of an instability that also leads to local chaos. We note again, as in Sect. 2.4.1, that with increasing N the value of βE at which the instability occurs continually shrinks to reach zero in the thermodynamic limit.

2.4.4 Short-Wavelength (high-frequency) Initial Conditions

Let us consider the equations of motion in Fourier space for the FPU-β model with periodic boundary conditions $q_i = q_{i+N}$

$$\ddot{Q}_r = F_r(Q_1, \ldots, Q_{N-1}) \,, \qquad r = 1, \ldots, N-1 \,, \tag{2.68}$$

where

$$F_r(Q_1, \ldots, Q_{N-1}) = -\omega_r^2 Q_r - \frac{\beta\omega_r}{2N} \sum_{j,k,l=1}^{N-1} \omega_j \omega_k \omega_l \, C_{rjkl} \, Q_j Q_k Q_l \,, \tag{2.69}$$

and the frequencies are in this case

$$\omega_k = 2\sin\left(\frac{\pi k}{N}\right) \qquad k = 0, \ldots, N-1. \tag{2.70}$$

The coupling coefficients, analogous to those in (2.25) for fixed boundaries, are given by

$$C_{ijkl} = -\Delta_{i+j+k+l} + \Delta_{i+j-k-l} + \Delta_{i-j+k-l} + \Delta_{i-j-k+l} \,, \tag{2.71}$$

where

$$\Delta_r = \begin{cases} (-1)^m & \text{for } r = mN \text{ with} m \in Z \\ 0 & \text{otherwise} \,. \end{cases} \tag{2.72}$$

The center of mass motion is decoupled; this is why the sum in (2.68) extends up to $N-1$. A natural question that arises is whether a set of modes exists which is decoupled from the others. If we put the energy *only* in this set, this is not shared by the others. Such a set is an invariant manifold in Fourier space. The question of existence has been positively solved [70, 71]. For instance, modes

$$k = \frac{N}{4}; \; \frac{N}{3}; \; \frac{N}{2}; \; \frac{2N}{3}; \; \frac{3N}{4} \tag{2.73}$$

are decoupled and are the only one-mode solutions with this property. The time-dependence of such periodic solutions is given in [70]. Moreover periodic and quasi-periodic solutions evolving on two-mode manifolds have been derived, and a full classification of higher dimensional invariant manifolds has been obtained. The existence of these invariant manifolds is related to spatial symmetries [71, 72].

The question of linear stability is more difficult and it has been solved analytically only for the zone boundary mode $k = N/2$. In this case, one finds the critical energy E_c ($\beta = 1$)

$$E_{\mathrm{c}} = \frac{2N}{9} \sin^2\left(\frac{\pi}{N}\right) \frac{7\cos^2(\pi/N) - 1}{[3\cos^2(\pi/N) - 1]^2} \, . \tag{2.74}$$

Above this energy, the zone-boundary mode solution looses stability by developing a spatial modulation. The initial zig-zag spatial pattern deforms in such a way to create a smooth long-wavelength envelope, with many bumps. Since the critical energy vanishes as N increases, in the thermodynamic limit the zone-boundary mode is always unstable. However, the rate at which the instability develops diverges with system size N [70]. The development of the instability leads to the creation of a "chaotic breather," as discussed in Sects. 2.4.5 and 2.4.6

Such instabilities exist also for other invariant modes and set of invariant modes, but have not yet been studied carefully neither analytically nor numerically. For instance, it is well known that if the energy is initially put in even (odd) modes in a FPU chain with an even number of oscillators and periodic boundary condition, energy remains in the set of even (odd) modes forever, until a critical energy is reached above which energy is exchanged among the two sets. In fact the set of even (odd) modes is an invariant set, according to our definition.

Let us sketch the derivation of formula (2.74) following Dauxois et al. [138]. Due to periodic boundary conditions, the normal modes are plane waves of the form

$$q_n(t) = \frac{a}{2}\left(\mathrm{e}^{\mathrm{i}\theta_n(t)} + \mathrm{e}^{-\mathrm{i}\theta_n(t)}\right) \tag{2.75}$$

where $\theta_n(t) = qn - \omega t$ and $q = 2\pi k/N$ ($k = -N/2, \ldots, N/2$). The dispersion relation of nonlinear phonons in the RWA given by (2.86) is $\omega^2(q) = 4(1 + \Delta)\sin^2(q/2)$, where $\Delta = 3a^2\sin^2(q/2)$ takes into account the nonlinearity. Modulational instability is investigated by studying the linearized equation associated with the envelope of the carrier wave (2.75). Therefore, one introduces infinitesimal perturbations in the amplitude and phase and looks for solutions of the form

$$\begin{aligned} q_n(t) &= \frac{a}{2}[1 + b_n(t)]\ \exp(\mathrm{i}[\theta_n(t) + \psi_n(t)]) + \frac{a}{2}[1 + b_n(t)]\ \exp(\mathrm{i}[\theta_n(t) + \psi_n(t)]) \\ &= a[1 + b_n(t)]\ \cos[qn - \omega t + \psi_n(t)]\, , \end{aligned} \tag{2.76}$$

where b_n and ψ_n are reals and assumed to be small in comparison with the parameters of the carrier wave. Substituting (2.76) into the equations of motion, one obtains for the real and imaginary part of the secular term $\exp(\mathrm{i}(qn - \omega t))$

$$-\omega^2 b_n + 2\omega\dot{\psi}_n + \ddot{b}_n = (1+2\Delta)\left[\cos q\,(b_{n+1}+b_{n-1}) - 2b_n\right]$$
$$-\Delta\,(b_{n+1}+b_{n-1}-2b_n\cos q) - (1+2\Delta)\sin q\,(\psi_{n+1}-\psi_{n-1}) \quad (2.77)$$
$$-\omega^2\psi_n - 2\omega\dot{b}_n + \ddot{\psi}_n = (1+2\Delta)\left[\cos q\,(\psi_{n+1}+\psi_{n-1}) - 2\psi_n\right]$$
$$+(1+2\Delta)\sin q\,(b_{n+1}-b_{n-1}) + \Delta\,(\psi_{n+1}+\psi_{n-1}-2\psi_n\cos q)\,. \quad (2.78)$$

Further assuming $b_n = b_0\ \mathrm{e}^{\mathrm{i}(Qn-\Omega t)} + \mathrm{c.c.}$ and $\psi_n = \psi_0\ \mathrm{e}^{\mathrm{i}(Qn-\Omega t)} + \mathrm{c.c.}$ one gets the two following equations for the secular term $\mathrm{e}^{\mathrm{i}(Qn-\Omega t)}$

$$b_0[\Omega^2+\omega^2+2(1+2\Delta)(\cos q\cos Q-1) - 2\Delta(\cos Q-\cos q)]$$
$$-2\mathrm{i}\psi_0\left[\omega\Omega + (1+2\Delta)\sin q\sin Q\right] = 0 \quad (2.79)$$
$$\psi_0[\Omega^2+\omega^2+2(1+2\Delta)(\cos q\cos Q-1) + 2\Delta(\cos Q-\cos q)]$$
$$+2ib_0\left[\omega\Omega + (1+2\Delta)\sin q\sin Q\right] = 0. \quad (2.80)$$

Nontrivial solutions for the linear system of (2.79)–(2.80) can be found only if the equations determinant vanishes, i.e. if the following equation is fulfilled:

$$\left\{(\Omega+\omega)^2 - 4(1+2\Delta)\sin^2\left(\frac{q+Q}{2}\right)\right\}\times$$
$$\times\left\{(\Omega-\omega)^2 - 4(1+2\Delta)\sin^2\left(\frac{q-Q}{2}\right)\right\} = 4\Delta^2\left(\cos Q-\cos q\right)^2. \quad (2.81)$$

This equation admits four different solutions when the wavevectors q of the unperturbed wave and Q of the perturbation are fixed. If one of the solutions is complex, an instability of one of the modes $(q \pm Q)$ is present, with a growth rate equal to the imaginary part of the solution. Using this method, one can derive the instability threshold amplitude for any wavenumber. A first interesting case is $q = \pi$, the zone-boundary mode. One can easily see that (2.81) admits two real and two complex conjugate imaginary solutions if and only if

$$\cos^2\frac{Q}{2} > \frac{1+\Delta}{1+3\Delta}\,. \quad (2.82)$$

The first mode to become unstable when increasing the amplitude a corresponds to the wavenumber $Q = 2\pi/N$. Therefore, the critical amplitude a_{c} above which the $q = \pi$-mode looses stability is

$$a_{\mathrm{c}} = \left(\frac{\sin^2(\pi/N)}{3\left[3\cos^2(\pi/N)-1\right]}\right)^{1/2}. \quad (2.83)$$

Since for the π-mode the energy is given by $E = N(2a^2+4a^4)$, one obtains the critical energy given in (2.74). The asymptotic behavior for large N of this formula gives the same threshold as (2.99) in Sect. 2.4.6, see also [81, 139]. This critical energy is also very close to the Chirikov threshold for short wavelength (2.7).

2.4.5 Expansions for Generic Discrete Systems-Envelopes

If we excite a high-frequency mode, γ, with $n \equiv N+1-\gamma \ll N+1$, then the instantaneous oscillator amplitude alternates from one oscillator to the next. As in previous studies [90, 91, 92], to remove this fast variations an envelope function $\psi_i(t) = (-1)^i q_i(t)$ is introduced, giving a smoothed spatial profile. The smoothed profile allows the oscillator to be described by a continuous variable from a Taylor expansion, giving

$$\psi_{tt} + 4\psi + 16\beta\psi^3 + \{\psi_{xx} + 12\beta(\psi\psi_x^2 + \psi^2\psi_x x)\} + \cdots = 0 \,, \tag{2.84}$$

where subscripts t and x stand for temporal and spatial derivatives of $\psi(x,t)$. Linear terms with spatial derivatives describe the dispersion, the dependence of the frequency ω on effective wave number $\pi k/(N+1)$ in (2.84), while nonlinear terms produce a frequency shift, that steepens the envelope function tending to form localized states (CBs). This process qualitatively explains why relaxation is accompanied by the formation of sharply localized states if energy is initially deposited in the high-frequency part of the spectrum, where the effect of dispersion is small, while only broad nonlinear structures are formed if the energy is initially in the low frequency modes where the dispersion is large [92]. Keeping the leading terms proportional to powers of degree zero and two and assuming a monochromatic dependence $\psi(x,t) = \psi(x)\cos(\omega t)$, leads to an equation for $\psi(x)$

$$(-\omega^2 + 4)\psi + \psi_{xx} + \beta(12\psi^3 + 9\psi\psi_x^2 + 9\psi^2\psi_{xx}) = 0 \,, \tag{2.85}$$

where we have used the RWA, i.e. the expansion

$$\cos^3(\omega t) = (3/4)\cos(\omega t) + (1/4)\cos(3\omega t), \tag{2.86}$$

and dropped terms proportional to $\cos(3\omega t)$ [90, 91]. Neglecting terms proportional to β yields a linear equation for the eigenmodes:

$$(-\omega^2 + 4)\psi + \psi_{xx} = 0 \,. \tag{2.87}$$

Solving (2.87) with zero boundary conditions at $x = 0$ and at $x = N+1$ gives eigenmodes for $n = N+1-k \ll N+1$ which correspond to the high-frequency linear normal modes of the discrete FPU chain

$$\psi_n^{(0)}(x) = \psi_{\max,n}\sin(q_n x) \,, \qquad \omega^2 = 4 - q_n^2 \,, \qquad q_n = \frac{\pi n}{N+1} \,. \tag{2.88}$$

The nonlinear equation (2.85) has exact analytical solutions, $\psi(x)$, which are periodic functions of x. There are three types of solutions:

(i) the infinite chain having a single localized breather with $\psi(x) \to 0$ as $x \to \pm\infty$, with frequency $\omega = \omega_{\mathrm{B}}$

$$\omega_{\mathrm{B}}^2 = 4 + 6\beta\psi_m^2 \,, \tag{2.89}$$

where ψ_m is the breather maximum amplitude given in (2.91);

(ii) the chain with periodic boundary conditions including the π-mode for which each oscillator has opposite phase and equal amplitude as its neighbors, and, correspondingly, the envelope function $\psi = \psi_{\text{max}} = \psi_{\text{min}} \equiv \psi_m$, a constant, for which the nonlinear frequency shift reaches a maximum value

$$\omega_{\text{B}}^2 = 4 + 12\beta\psi_m^2, \tag{2.90}$$

(iii) and the case of fixed zero displacement at $x = 0$ and $x = N + 1$, with intermediate values of ω_{B}.

For a single breather, $n = 1$, the breather structure is similar to a breather on an infinite line which has analytic approximations for small and large amplitude. For $9\beta\psi_m^2 \ll 1$

$$\psi_{\text{B}}(x) = \frac{\psi_m}{\cosh(\sqrt{6\beta}\psi_m x)}, \tag{2.91}$$

while in the large amplitude case $9\beta\psi_m^2 \gg 1$, the breather has a finite width of 4–5 oscillators

$$\psi_{\text{B}}(x) = \psi_m \cos\sqrt{\frac{2}{3}}x \qquad |x| < \pi\sqrt{\frac{3}{8}}. \tag{2.92}$$

For most numerical studies of oscillator chains the initial state imposed on the system is that of a single linear mode. This state is generally not close to an equilibrium. The initial state rapidly relaxes, governed by the nonlinear equations. The evolution may be influenced by the underlying stability of nearby equilibria, but cannot be analyzed directly as perturbations around those equilibria. It is also possible to prepare the initial condition to be close to an equilibrium and consequently to directly analyze linear stability. The envelope solutions are fast oscillating functions of time which are subject to parametric (modulation) instability, i.e. an instability which is driven by the periodic variation of the frequency that appears in the linear equation for a perturbation. The frequency shift is caused by the nonlinearity in the unperturbed envelope solution. For the usually applied modal initial conditions, unstable breakup of modes is observed [87, 88]. However, numerical calculations show that the nonlinear stage of this instability leads to the formation of long-living self-organized localized structures, the chaotic breathers, which appear to be marginally stable with respect to a fast modulational instability.

Another question is how many breathers appear after the relatively short time of evolution from an initial state. In this context, fixed zero boundary conditions are significantly different from the π-mode initial values for periodic boundary conditions. In the periodic case, the π-mode is simultaneously a normal mode of the linear problem and an exact solution to the nonlinear envelope equation. Evolution from this equilibrium state is initiated by a modulation instability, and the wavelength of the fastest growing mode of the linearized equations gives an estimate of the number of breathers generated during the nonlinear phase of instability. In the case of zero boundary conditions, the high-frequency normal modes do not satisfy the nonlinear envelope equation. When used as initial conditions at low energy, they relax toward or around a few nearest stable equilibrium solutions. We expect that the linear analysis could, at best, only qualitatively describe their evolution.

2.4.6 Instability from Short-Wavelength Initial Conditions

For analysis of nonstationary envelopes, which describe relaxation, instability, or breather translational motion, it is convenient to rewrite the basic equation (2.84) in the form of two coupled equations for amplitude $q(x,t)$ and phase $\phi(x,t)$

$$\psi(x,t) = q(x,t)\cos(\omega t + \phi(x,t)) \,. \tag{2.93}$$

Substituting (2.93) in (2.84) and collecting terms proportional to $\sin(\omega t + \phi(x,t))$ and $\cos(\omega t + \phi(x,t))$ leads to coupled equations

$$q\phi_{tt} + 2q_t(\omega + \phi_t) + 2q_x\phi_x + q\phi_{xx} + 12\beta q^2 q_x \phi_x + 3\beta q^3 \phi_{xx} = 0 \,, \tag{2.94}$$

$$q_{tt} - (\omega + \phi_t)^2 q + 4q + q_{xx} - q\phi_x^2 + 12\beta q^3 + 9\beta q(q_x)_x - 6\beta q^3 \phi_x^2 = 0 \,. \tag{2.95}$$

The frequency ω is a constant given approximately by (2.89) and determined by the amplitude of the unperturbed solution. When the amplitude is slightly varied, $q(x,t) = \psi(x) + \delta q(x,t)$, the frequency of the fast nonlinear oscillation is also varied. As ω is taken to be contant this effect is represented by the time-varying phase, $\phi(x,t) = \delta\phi(x,t)$. Since (2.94) depends on derivatives of $\delta\phi(x,t)$, but not the phase itself, it can be linearized by considering the derivatives of $\delta\phi(x,t)$ as first order corrections. This yields two coupled linear equations

$$2\omega\delta q_t + \psi\delta\phi_{tt} + 2\psi_x(1 + 6\beta\psi^2)\delta\phi_x + \psi(1 + 3\beta\psi^2)\delta\phi_{xx} = 0 \,, \tag{2.96}$$

$$\delta q_{tt} + ((1+9\beta\psi^2)\delta q_x)_x + (4-\omega^2+36\beta\psi^2+18\beta\psi\psi_{xx}+9\beta\psi_x^2)\delta q - 2\omega\psi\delta\phi_t = 0 \,. \tag{2.97}$$

These equations have been solved numerically under various assumptions, with the result being that breather equilibria in chains with fixed ends are probably marginally stable to parametric instabilities [93]. This probably accounts for the long-time stability of the breathers that are formed from the parametric instabilities of mode initial conditions. To explore the latter situation a useful approximation is to consider the case of constant spatial profile of the envelope $\psi(x,t) = \psi_m \cos\omega t$. This corresponds to the π-mode with periodic boundary conditions, which has the highest nonlinear frequency shift (2.90) [91]. This mode is a solution to (2.85) but does not belong to the envelope solutions with zero boundary conditions. Setting the spatial derivative of $\psi(x)$ equal to zero, (2.96) and (2.97) reduce to coupled equations for $\delta\phi(x,t)$ and $\delta q(x,t)$, with constant coefficients. They can be solved by letting

$\delta q(x,t) \propto \delta\phi(x,t) \propto \exp(st + ikx)$ which gives a biquadratic equation for s. Substituting for ω from (2.90), the result is

$$s^4 + 2[36y + 8 - k^2(1+6y)]s^2 = k^2(1+3y)[24y - k^2(1+9y)] ,\quad y = \beta\psi_m^2 . \tag{2.98}$$

This gives a threshold for the modulation instability of the π-mode

$$6\beta\psi_m^2 \frac{(N+1)^2}{\pi^2} > 1 . \tag{2.99}$$

There is a most unstable wavenumber k_m which corresponds to the maximum of the growth rate, s_m. In the limit of small ψ_m, $9y \ll 1$, the value of k_m and s_m is found by dropping s^4, then setting $\mathrm{d}(s^2)/\mathrm{d}(k^2) = 0$, to obtain [91]

$$k_m = \sqrt{12\beta}\psi_m , \qquad s_m = 3\beta\psi_m^2 . \tag{2.100}$$

For intermediate amplitude envelopes all terms are included in (2.98). In the limit of large amplitudes $9y \gg 1$, the fastest growing mode has wavenumber and maximum growth rate [92]

$$k_m = 1.23 , \qquad s_m = 0.93\sqrt{\beta}\psi_m . \tag{2.101}$$

Comparing (2.100) and (2.101), the transition from small to large amplitude takes place at $\beta\psi_m^2 \simeq 1/9$ which corresponds, for $\beta = 0.1$, to $\psi_m \simeq 1$.

These results have been checked by numerical calculations, starting from various mode initial conditions, obtaining reasonable agreement [93]. For example, in one way of forming initial conditions one can take a set of Fourier modes to approximate a square wave, using $a_n = 4/\pi n$, n odd, with $n = N + 1 - \gamma$ (γ is an initial k-value). Considering the nine highest frequency modes, for $E = 16$, one obtains the evolution as shown, at three times, in Fig. 2.8. The initial nine ripples, with large end-values, in Fig. 2.8a, are characteristic of the Fourier sum. The evolution through various transitions, e.g., Fig. 2.8b at $t = 100$, leads to a large amplitude fastest growing mode at $t = 220$ with a wavelength of $\lambda = 16$, Fig. 2.8c, which is predicted from (2.100). The growth time from the first emergence of the fastest growing mode (not shown) is also consistent with the observations. The fastest growing k-value is established from either a smaller or a larger number of initial modes than the corresponding value $n_m = 128(2\pi/k_m)$. In either case, the subsequent time evolution, on a slower time-scale, is to form nonlinear chaotic breathers which coalesce and then decay to equipartition, on slower time-scales.

For the ϕ^4 oscillator chain, starting from (2.9), one can make the same Taylor expansion used for the FPU chain, followed by the rotating wave approximation, to obtain [94]

$$(\omega^2 + m^2 + 4)\psi + \psi_{xx} + \frac{3}{4}\beta\psi^3 = 0, \tag{2.102}$$

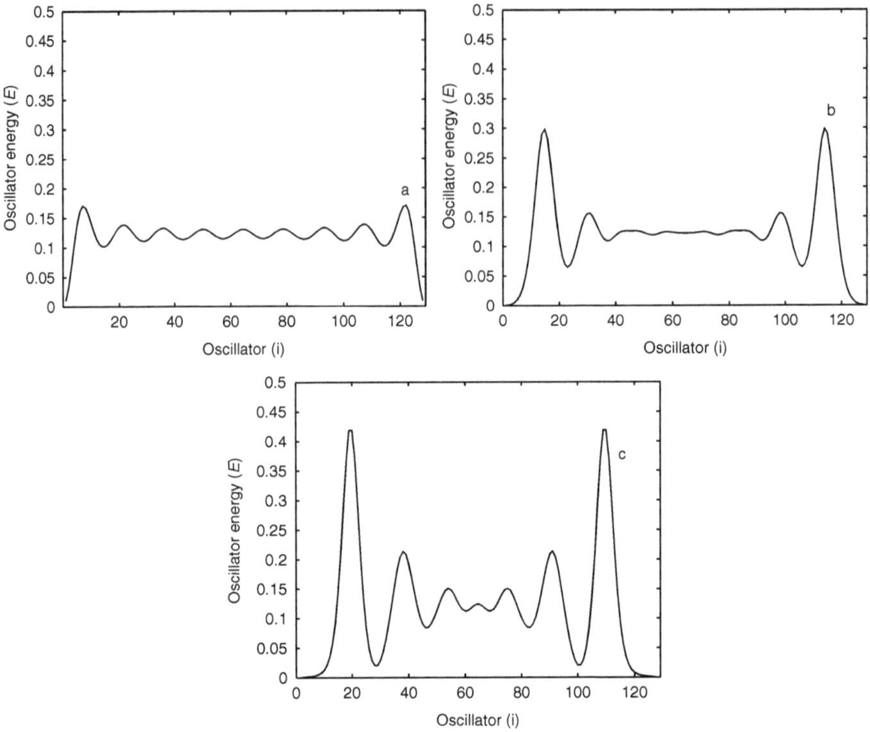

Fig. 2.8. Illustrating a set of initial conditions fixed at the ends but with a 9-mode Fourier spectrum approximating a π-mode distribution in the central region, at times: (a) $t = 10$, (b) $t = 100$, (c) $t = 220$. The predicted fastest growing mode has $\lambda_m = 16$, in agreement with the results seen in (c).

which is to be compared with (2.85). One observes three differences: the extra linear coefficient m^2, the smaller coefficient of the ψ^3 term (3/4 rather than 12), and the absence of the mixed cubic terms. For most comparisons, we choose m small ($m = 0.1$), which is negligible compared to 4, and the mixed cubic term in the FPU plays only a minor role in the dynamics. The remaining clear difference is the factor of 16 that the ψ^3 term is smaller in the ϕ^4 envelope than the FPU envelope. This factor is clearly understood by a comparison of the original FPU and ϕ^4 Hamiltonians (2.1) and (2.9). Since the nonlinear ϕ^4 potential results from a self force, the potential arises only from the extension of the mass point q_i and is therefore quartic in that extension. In contrast, the FPU quartic potential arises from the difference of neighboring oscillators $(q_{i+1} - q_i)$. Since these oscillators are approximately π out of phase, there is an increase of 4 factors of 2 in the quartic potential, i.e. 16. The effect of the factor of 16 is reflected in all subsequent calculations, appearing linearly in the instability threshold (2.99), as a square-root in the most unstable wavenumber k_m (2.100) and linearly in the maximum growth rate s_m (2.100), which were found numerically in comparisons between the two chains [94]. Care must be taken in the comparisons if the energy E is used as the parameter, because there is a range of scalings of ψ_m with E from low to high energy (see [93, 94], for a detailed treatment). We will return to these comparisons of scalings in later sections describing the stochastic behavior.

In Sect. 2.4.5 we saw that the envelope approximation with periodic boundary conditions has an equilibrium solution that is exactly constant, the boundary or π-mode. This equilibrium becomes unstable at an envelope amplitude given by (2.99), and, using the low amplitude approximation for the energy,

$$E \approx 2N\psi_m^2 \,, \tag{2.103}$$

then (2.99) yields the energy border of instability

$$\beta E > \frac{\pi^2}{3N} \,. \tag{2.104}$$

This result can also be found directly from the discrete π-mode [70], so does not depend on the envelope expansion. For the FPU-β chain with fixed ends, the instability still exists, but is only approximately given by (2.104), as the envelope equilibrium no longer has an expansion that yields an exact border of instability. Although the envelope equilibrium of the π-mode results from breather-like equations, it is an extended mode, rather than the usually studied intrinsically localized modes (ILMs), which are much more stable. This general class of short-wavelength modes, near the π-boundary mode, had previously been studied by Berman and Kolovskii [81] using a different technique. Starting from the mode representation for the FPU-β chain, and assuming that only a few modes neighboring the π-mode were present, they expanded about that wave number k_π as $n = k - k_\pi$ ($n \ll k_\pi$), and removing the fast oscillations by factoring out $\exp\left[(i\omega_\pi + \lambda n)t\right]$, $\lambda = 2(\pi/N)\cos(\pi k_\pi/N)$, from the modes, they arrive at the nonlinear Schrödinger (NLS) equation

$$i\frac{\partial \Phi}{\partial t} = \Omega \frac{\partial^2 \Phi}{\partial \theta^2} + V_0|\Phi|^2\Phi \,, \tag{2.105}$$

where $V_0 = (3\beta/N)\sin^2(\pi k_\pi/N)$, $\Omega = (\pi/N)^2\sin(\pi k_\pi/N)$, and $\Phi(\theta,t) = \Phi(\theta+2\pi,t)$. The NLS equation, like the KdV equation, is completely integrable with soliton-like solutions. However, like the mKdV equation, the solitons are subject to an instability which can be calculated by linearization of the equilibrium, as outlined for the envelope approximation in Sect. 2.4.5. The result, for instability, is the same as in (2.104), but now involves a narrow mode packet, rather than the single π-mode. It is also close to the Chirikov mode overlap criterion for local chaos, which from (2.7) can be written as $\beta E \approx 2\pi^2/3N$. As already discussed, all of these criteria signal the onset of local chaos in the discrete chain, but do not inform us about equipartition on nonexponential time-scales. However, the Berman–Kolovskii paper considers a second transition at which the narrow packet approximation breaks down and therefore the NLS approximation is no longer valid. For this higher energy, the packet size is not contained, which may lead to power-law time-scales to equipartition. The value they estimate is

$$\beta E > \frac{2\pi^2}{3} , \qquad (2.106)$$

which is independent of N. This high-frequency estimate is quite close to the low-frequency estimate (2.50), both predicting nonexponentially slow diffusion, which survives in the thermodynamic limit. Both (2.106) and (2.50) give results for extended modes, which is the relevant result for low frequencies; but for high frequencies (2.106) does not take into account the later developments, predicting the formation of long-lived but not fully stable ILMs (the CBs). The original work included qualitative numerical support for both (2.104) and (2.106). A more detailed numerical investigation, including the formation and slow decay of the CBs, is given in Sect. 2.6. Some recent comments on, and reference to extensions of, the narrow packet approximation can be found in [140].

2.4.7 Geometric Formalism and the Method of Estimating the Largest Lyapunov Exponent

As already mentioned in the Introduction, classical perturbation theory (CPT) is inadequate to describe the properties of Hamiltonian dynamics when the phase space is formed by chaotic orbits, even if a perturbative description is justified when the time-scales are short with respect to the instability time-scales. For the transition from weak stochasticity to strong stochasticity in high-dimensional systems, the perturbative treatment is completely inadequate, due to the energies involved which are much larger than the values for which it is meaningful to consider the systems as quasi-integrable. Moreover, the canonical transformation from natural coordinates to angle-action variables, which is a prerequisite to tackle chaos from the point of view of homoclinic intersections, is very complicated and necessarily approximated, not to speak of the lack of the generalization of the Poincaré–Birkhoff theorem at arbitrary N concerning the fate of resonant tori, another necessary prerequisite for the standard description of chaos through homoclinic intersections.

A problem that naturally arises is how to explain the origin of a Strong Stochasticity Threshold (SST) and how to compute, the crossover energy. According to the above arguments, one has to look for some *nonperturbative* method. The only rigorous theoretical framework dealing with the opposite situation of CPT, i.e. with completely chaotic trajectories, is ergodic theory. We have already mentioned in Sect. 2.1.3 that it was Krylov who first realized the relevance of mixing for statistical mechanics, and the relevance of the stability properties of geodesics on Riemannian manifolds of negative curvature for mixing. More recently, the geometric approach has been reconsidered with the aid of numerical simulations, finding out that the dominant mechanism for dynamical instability in physically relevant geodesics flows is *parametric instability* due to curvature variations along the geodesics, instead of the negative curvature [141, 142, 143, 144].

For a dynamical system described by the Lagrangian function

$$L(q,\dot{q}) = \frac{1}{2} a_{ik}(q)\dot{q}^i\dot{q}^k - V(q) \,, \tag{2.107}$$

according to Maupertuis' principle of stationary action, among all the possible isoenergetic paths $\gamma(t)$ with fixed end points, the paths which make the first variation of the action functional vanish, which is such that

$$\delta\mathcal{A} = \delta \int_{\gamma(t)} p_i \,\mathrm{d}q_i = \delta \int_{\gamma(t)} \frac{\partial L}{\partial \dot{q}_i} \dot{q}_i \,\mathrm{d}t = 0 \tag{2.108}$$

are natural motions.

Since $2W = \dot{q}_i \partial L/\partial \dot{q}_i$, is the kinetic energy, Maupertuis' principle reads

$$\delta \int_{\gamma(t)} 2W \,\mathrm{d}t = \delta \int_{\gamma(t)} \left(g_{ik}\dot{q}^i\dot{q}^k\right)^{1/2} \,\mathrm{d}t = \delta \int_{\gamma(s)} \mathrm{d}s = 0 \,. \tag{2.109}$$

The last integral indicates that if the configuration space M of a system with N degrees of freedom is given a proper Riemannian structure by introducing the metric [42, 145]

$$g_{ik} = 2[E - V(q)]a_{ik} \tag{2.110}$$

so that $\mathrm{d}s^2 = 4[E-V(q)]a_{ij}\mathrm{d}q^i\mathrm{d}q^j$ is its arclength, then the trajectories of the Newtonian motions coincide with the geodesics of the manifold M endowed with the metric tensor (2.110). This metric is known as the Jacobi metric and is defined in the region of the configuration space where $E > V(q)$. In local coordinates, the geodesic equations on a Riemannian manifold are given by

$$\frac{\mathrm{d}^2 q^i}{\mathrm{d}s^2} + \Gamma^i_{jk} \frac{\mathrm{d}q^j}{\mathrm{d}s}\frac{\mathrm{d}q^k}{\mathrm{d}s} = 0 \tag{2.111}$$

where s is the proper time and Γ^i_{jk} are the Christoffel coefficients of the Levi-Civita connection associated with g_{ik} [146]. By direct computation, it can be easily verified that the geodesic equations, together with the relation between s and t, i.e. $\mathrm{d}s^2 = 4[E - V(q)]\mathrm{d}t^2$, yield

$$\frac{\mathrm{d}^2 q^i}{\mathrm{d}t^2} = -\frac{\partial V}{\partial q^i}, \tag{2.112}$$

i.e. Newton's equations associated with the Lagrangian (2.107).

The stability of an orbit is related to the curvature of the Riemannian manifold. If we define the curvature K at a point x relative to a tangent plane π_a which is spanned by the vectors u, v at x, then

$$K(u,v) = K(x,\pi) = \frac{\langle R(v,u)u, v\rangle}{|u \wedge v|^2} \,, \tag{2.113}$$

where R is the Riemann–Christoffel curvature tensor [146]. K turns out to be independent of the choice of the two vectors u, v in π. The knowledge of K for the $N(N-1)$ planes π spanned by a maximal set of linearly independent vectors completely determines R at x. If $\dim(M) = 2$ then K coincides with the Gaussian curvature. A manifold is *isotropic* if $K(x, \pi)$ does not depend on the choice of the plane π. The remarkable result (Schur's theorem [146]) is that in this case K is also constant, i.e. it also does not depend on the point x. For a constant-curvature, i.e. isotropic, manifold (2.113) reduces to a constant

$$K = \frac{1}{N(N-1)} \mathcal{R} \,, \tag{2.114}$$

where N is the number of degrees of freedom and $\mathcal{R}$ is the scalar curvature.

For a congruence of geodesics $\{\gamma_\tau(s) = \gamma(s, \tau) \,|\, \tau \in R\}$ issuing from a neighborhood $\mathcal{I}$ of a point of a manifold [for more details see [142]], dependent on the parameter τ, fixing a reference geodesic $\bar{\gamma}(s, \tau_0)$, if $\dot{\gamma}(s)$ is the vector field tangent to $\bar{\gamma}$ in s, and $J(s)$ the vector field tangent in τ_0 to the curves $\gamma_s(\tau)$ for a fixed s, then the evolution of J contains the information on the stability (or instability) of the reference geodesic $\bar{\gamma}$; if $|J|$ grows exponentially, then the geodesic will be unstable in the Lyapunov sense, otherwise it will be stable. It is remarkable that such an evolution is completely determined by the curvature tensor R, which is a consequence of the fact that J is a Jacobi field, i.e. it obeys the equation

$$\nabla^2_{\dot{\gamma}} J(s) + R(J(s), \dot{\gamma}(s))\dot{\gamma}(s) = 0 \,. \tag{2.115}$$

Among several Riemannian geometrizations of Newtonian dynamics, a very interesting one is defined in an enlarged configuration spacetime $M \times R^2$, with local coordinates $(q^0, q^1, \ldots, q^i, \ldots, q^N, q^{N+1})$, endowed with a nondegenerate pseudo-Riemannian metric whose arc-length [147]

$$\mathrm{d}s^2 = g_{\mu\nu}\, \mathrm{d}q^\mu \mathrm{d}q^\nu = a_{ij}\, \mathrm{d}q^i \mathrm{d}q^j - 2V(q)(\mathrm{d}q^0)^2 + 2\, \mathrm{d}q^0 \mathrm{d}q^{N+1} \tag{2.116}$$

is called the *Eisenhart metric.* The natural motions are obtained as the canonical projection on the configuration space-time of those geodesics for which the arclength is positive-definite and given by $\mathrm{d}s^2 = (\text{const.})^2 \mathrm{d}t^2$. The geometric formulations of Newtonian dynamics, based on Jacobi and Eisenhart metrics, respectively, are equivalent. The interest in the Eisenhart metric is that the instability equation for the geodesic spread (2.115) written in this metric yields the standard tangent dynamics equation which is commonly used in numerical computations of Lyapunov exponents.

We note, parenthetically, that the two basic topological conditions for the onset of chaos in any deterministic dynamics are *stretching* and *folding* of volumes in phase space [148]. In the case of Hamiltonian chaos, these two conditions are fulfilled by the existence of homoclinic intersections [15, 148]. In the Riemannian description of Hamiltonian chaos, *stretching* of nearby trajectories is provided by *instability*, and *folding* by not allowing the distance to grow indefinitely, i.e. by *compactness*. In this way, the phase trajectories forget the initial conditions; their evolution becomes unpredictable in the long run. In the majority of systems of physical interest, the configuration space is a bounded domain so that the instability of nearby trajectories, studied by means of (2.115), implies chaos.

In the particular case of isotropic (or constant curvature) manifolds, (2.115) becomes very simple: choosing a geodesic frame, i.e. a reference frame transported along a reference geodesic, the Jacobi equation is written as

$$\frac{\mathrm{d}^2 J}{\mathrm{d}s^2} + K\,J = 0\,, \tag{2.117}$$

and has either bounded oscillating solutions $\|J\| \propto \cos(\sqrt{K}\,s)$ or exponentially unstable solutions $\|J\| \propto \exp(\sqrt{-K}\,s)$ according to the sign of the constant sectional curvature K. If the curvatures are negative, the geodesic flow is unstable even if the manifold is no longer isotropic. Equation (2.117) is valid only if K is constant. Nevertheless, for $\dim M = 2$ (surfaces), the Jacobi equation, again written in a geodesic reference frame for the sake of simplicity, takes a form very close to that of isotropic manifolds,

$$\frac{\mathrm{d}^2 J}{\mathrm{d}s^2} + \frac{1}{2}\mathcal{R}(s)\,J = 0\,, \tag{2.118}$$

where $\mathcal{R}(s)$ denotes the scalar curvature of the manifold at the point $P = \gamma(s)$. This equation helps in understanding the origin of geodesic instability besides hyperbolicity. In fact, the solutions of (2.118) may exhibit an exponentially growing envelope even if the curvature $\mathcal{R}(s)$ is everywhere positive but non constant. This is the case, for example, of two harmonic oscillators coupled through cubic or quartic terms [143, 149].

In many physically relevant systems (typically a set of coupled anharmonic oscillators on a lattice in d space-dimensions) the curvatures are neither constant nor everywhere negative, and the straightforward approach based on (2.117) does not apply. This is the main difficulty in extending the methods of abstract ergodic theory to physically relevant models. The key point is to realize that negative curvatures are not strictly necessary for chaos, with the *bumpiness* of the manifold, being responsible for curvature fluctuations along the geodesics, that can trigger *parametric instability* and hence exponentially growing solutions of the stability equation (2.118).

In the large N case, under a set of suitable hypotheses [for details see [60]], it is possible to derive a scalar effective stability equation resembling (2.118), where the role of $R(s)$ is played by a random process, from which an analytic estimate of the largest Lyapunov exponent can be obtained. The theory leads to the stochastic equation

$$\frac{\mathrm{d}^2 \psi}{\mathrm{d}s^2} + \langle k_R \rangle_{\Sigma_E}\,\psi + \langle \delta^2 k_R \rangle_{\Sigma_E}^{1/2}\,\eta(s)\,\psi = 0\,, \tag{2.119}$$

where ψ denotes any of the components of J in (2.115) because all of them are assumed to obey the same effective equation of motion. Here $\langle k_R \rangle_{\Sigma_E}$ is the microcanonical average of the $1/N$ fraction of the Ricci curvature K_{R} of the mechanical manifold[2], and $\langle \delta^2 k_R \rangle_{\Sigma_E} \equiv \frac{1}{N-1} \langle \delta^2 K_{\mathrm{R}} \rangle_{\Sigma_E}$ is the variance of the Ricci curvature of the mechanical manifold averaged on the constant energy manifold $\Sigma_E = H^{-1}(E)$; finally, $\eta(s)$ is a Gaussian δ-correlated random process of zero mean and unit variance.

Equation (2.119) is a scalar equation which, *independently of the knowledge of dynamics*, provides a measure of the average degree of instability of the dynamics through the growth-rate of $\psi(s)$. The peculiar properties of a given Hamiltonian system enter (2.119) through the global geometric properties $\langle k_R \rangle_{\Sigma_E}$ and $\langle \delta^2 k_R \rangle_{\Sigma_E}$ of the ambient Riemannian manifold. Moreover these averages are functions of the energy E of the system (and of the energy density $\varepsilon = E/N$, which is the relevant quantity for $N \to \infty$), so that from (2.119) one can obtain the energy dependence of the geometric instability exponent.

Equation (2.119) is of the form

$$\frac{\mathrm{d}^2 \psi}{\mathrm{d}s^2} + \Omega(s)\, \psi = 0 \tag{2.120}$$

representing a stochastic oscillator where the squared frequency $\Omega(s)$ is the above described stochastic Gaussian process. The process $\Omega(t)$, with proper time s replaced by physical time t, is assumed to be stationary and δ-correlated, that is its time correlation function $\Gamma_\Omega(t_1, t_2)$ is such that $\Gamma_\Omega(t_1, t_2) = \Gamma_\Omega(|t_2 - t_1|)$ and $\Gamma_\Omega(t) = \tau\, \sigma_\Omega^2\, \delta(t)$, where τ is a characteristic time-scale of the process. The evaluation of this time-scale is still a rather delicate point, where some arbitrariness enters the theory. In studying various models an estimate has been successfully introduced, which combines the evaluation of the time needed to join two successive conjugate points along a geodesic (conjugate points are those points where the Jacobi field vanishes) with another time-scale which can be inferred by means of dimensional arguments. In [60] arguments are given which lead to the following two time-scales

$$\tau_1 = \left\langle \frac{\mathrm{d}t}{\mathrm{d}s} \right\rangle \frac{\pi}{2\sqrt{\Omega_0 + \sigma_\Omega}} \tag{2.121}$$

and

$$\tau_2 = \left\langle \frac{\mathrm{d}t}{\mathrm{d}s} \right\rangle \frac{l^2 \Omega_0}{6} \frac{2\pi}{\sqrt{\Omega_0}} \simeq \frac{\Omega_0^{1/2}}{\sigma_\Omega}\ , \tag{2.122}$$

respectively, where l is defined as $l = 1/\sqrt{\sigma_\Omega}$, and $\Omega_0 = \langle \delta^2 k_R \rangle_{\Sigma_E}$, $\sigma_\Omega = \langle k_R \rangle_{\Sigma_E}$, whence τ in $\Gamma_\Omega(t) = \tau\, \sigma_\Omega^2\, \delta(t)$ is obtained by combining τ_1 with τ_2 as follows

[2] The Ricci curvature is the sum of the $N-1$ curvatures K, given in (2.113), relative to the $N-1$ planes spanned by a given vector v and $N-1$ other (unit) vectors orthogonal to v. In components, $K_{\mathrm{R}} = R^j_{ijk} v^i v^k$, where R^j_{ijk} are the components of the Riemann curvature tensor.

$$\tau^{-1} = 2\left(\tau_1^{-1} + \tau_2^{-1}\right) . \tag{2.123}$$

Whenever $\Omega(s)$ in (2.120) has a nonvanishing stochastic component, the solution $\psi(s)$ is exponentially growing on the average [150]. By considering the proper time parametrization as a function of the physical time t, the estimate of the largest Lyapunov exponent from the physical time growth-rate of $[\psi^2(t) + \dot{\psi}^2(t)]$ is then given by

$$\lambda = \lim_{t\to\infty} \frac{1}{2t} \log \frac{\psi^2(t) + \dot{\psi}^2(t)}{\psi^2(0) + \dot{\psi}^2(0)} . \tag{2.124}$$

The ratio $(\psi^2(t)+\dot{\psi}^2(t))/(\psi^2(0)+\dot{\psi}^2(0))$ is computed by means of a technique developed by Van Kampen [150] and summarized in [60]. The main point is to compute the time evolution of the second moments of ψ and $\dot{\psi}$, averaged over the realizations of the stochastic process. In general, the solutions of a stochastic oscillator equation as the one we are dealing with, are unstable. The envelope of an unstable solution exponentially grows in time with a growth rate λ which—with the aid of Van Kampen's method—is found to be

$$\lambda(\Omega_0, \sigma_\Omega, \tau) = \frac{1}{2}\left(\Lambda - \frac{4\Omega_0}{3\Lambda}\right) ,$$
$$\Lambda = \left(2\sigma_\Omega^2\tau + \sqrt{\left(\frac{4\Omega_0}{3}\right)^3 + (2\sigma_\Omega^2\tau)^2}\right)^{1/3} . \tag{2.125}$$

The quantities $\Omega_0 = \langle k_R\rangle_{\Sigma_E}$, $\sigma_\Omega = \langle \delta^2 k_R\rangle_{\Sigma_E}$ and τ can be computed as static, i.e. *microcanonical* averages. Therefore (2.125) gives an analytic, though approximate, formula for the largest Lyapunov exponent *independent* of the numerical integration of the dynamics and of the tangent dynamics.

An important remark is that this analytic formula for λ is derived under the geometric assumption of *quasi-isotropy* of the mechanical manifolds (for details see [60]), but this assumption is inadequate to tackle those systems whose mechanical manifolds are topologically nontrivial, in this case the theory has to take into account the role of an additional instability mechanism originating in the neighborhoods of the critical points of the potential function.

2.5 Numerical Results: Relaxation to Equilibrium from Low-Frequency Modes

2.5.1 Observations of Diffusion: Numerical Determination of λ

We have already discussed the early observations of periodicities among a few low-frequency modes, at low energies, and their explanation in terms of beat oscillations. We have also considered that solitons could be constructed from low-frequency mode initial conditions, which also produce the observed beats, i.e. another way of looking at the same phenomenon. Instability of single solitons in the mKdV equations was found to occur, for a given FPU-β chain, at about the same energy at which the interaction of the beat oscillations generates stochastic layers (the concept of resonance overlap), and at the instability of an exact periodic solution. For large systems, at not too low energy, generic initial conditions would be expected to lie in the stochastic portion of the phase space and be able to diffuse to all portions of it by the Arnold diffusion mechanism. A stochasticity threshold (ST), to observable positive Lyapunov exponent, λ, and (possibly) an observable time to equipartition, $T_{\rm eq}$, would be expected. In Sect. 2.3, we have discussed the numerical determination of the statistical quantities λ and $n_{\rm eff}$ from which $T_{\rm eq}$ (or some fraction of it) is found. In the higher energy regime, these and other statistical measures of the diffusion have been explored in the late 1970s and 1980s, particularly from the group in Firenze (see [50, 51, 52, 123, 128, 151]) and in Milano (see [25, 48, 49, 152, 153]). Here, and in Sect. 2.5.2, we present some of the numerical results, which are then used to guide further analytic studies. We restrict our numerical observations, in these sections, mainly to initial conditions of energy in low-frequency modes. The numerics starting from high-frequency modes have a somewhat different character, and will be treated separately in Sect. 2.6. Before presenting the results using the statistical measure, we first present in Fig. 2.9, a series of mode spectra, for $R = 8$, $N = 32$, above the $E_{\rm c}$

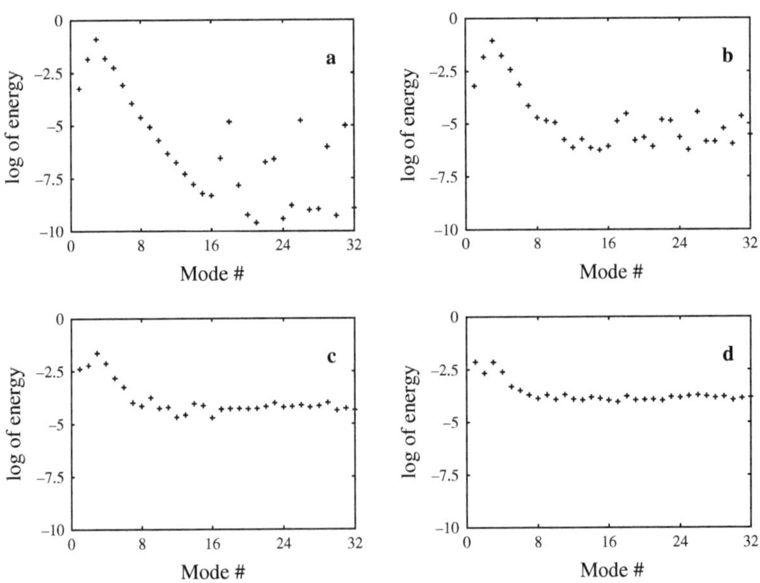

Fig. 2.9. Log of average energies at $R = 8.0$, $N = 32$ ($E = 4$) and four consecutive times $\tau = \gamma(\pi/N)^2 t/3, \gamma = 3$; **a** $\tau = 2000$, $n_{\rm eff} = 6.24$; **b** $\tau = 10,000$, $n_{\rm eff} = 8.67$; **c** $\tau = 40,000$, $n_{\rm eff} = 18.99$ and **d** $\tau = 78,000$, $n_{\rm eff} = 25.48$

transition as found in Fig. 2.4. The increase in energy in some high-frequency modes, specified from the selection rules, is well above the background, and continues to increase with time, finally resulting in equipartition. Note that the first frame in Fig. 2.9, with $\tau = 2000$, corresponds to $N = 32$ at $R = 8$ in Fig. 2.4. The time has been normalized to scale away the number of oscillators in the chain. The values of $n_{\rm eff}$, at each time, are listed in the figure caption.

The early statistical results, for the FPU-β chain with β chosen to be 0.1, can be summed up with a graph from Pettini and Landolfi [50]. Choosing a range of N from 64 to 512, and the initial cluster of low-frequency modes Δk, with $\Delta k/N$ fixed at 1/16, they numerically calculated $\lambda(t)$ and $\eta(t)$ (see Sect. 2.3.2). To calculate for times that were not as long as $T_{\rm eq}$ they fitted $\eta(t)$ to the stretched exponential

$$\eta(t) = \begin{cases} \exp(-(t/\tau)^{\nu}), & t < \tau_{\rm R}, \\ \exp(-(\tau_{\rm R}/\tau)^{\nu}), & t > \tau_{\rm R} \end{cases} \tag{2.126}$$

with $\tau_{\rm R}$ taken as a measure of $T_{\rm eq}$. Their results for an asymptotic λ and for $\tau_{\rm R}$ are shown together in Fig. 2.10, for a particular N and Δk, but confirmed for other values of N and Δk with $\Delta k/N =$ const.. As shown in the figure, the slopes are power laws in $\varepsilon = E/N$, with a transition in the slope of λ from $\lambda \propto \varepsilon^2$ at low ε to $\lambda \propto \varepsilon^{2/3}$ at higher ε. They interpreted this transition as that from weak to strong stochasticity (the SST). The weaker scaling at higher ε was predicted from an assumption of a fully random process, which might be expected to follow from strong mode overlap (see Sect. 2.8.1). We note from (2.7) that the transition roughly coincides with the prediction of mode overlap. Although not discussed in the original work, we also see the scaling $\tau_{\rm R} \propto \varepsilon^{-3}$, a scaling later confirmed for $T_{\rm eq}$ from extensive calculations

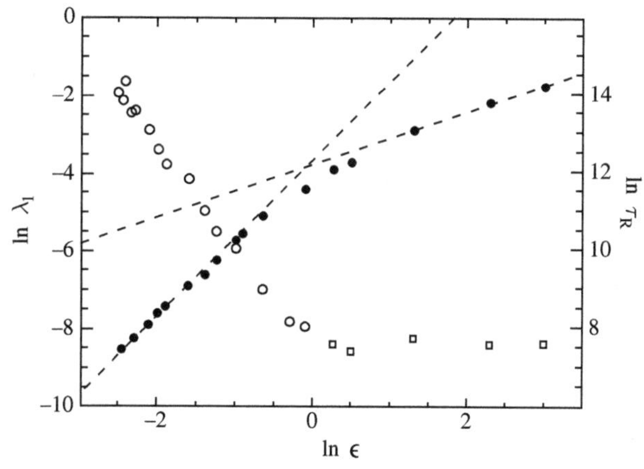

Fig. 2.10. The relaxation time $\tau_{\rm R}$ and the maximum Lyapunov exponent λ vs. the energy per mode $\varepsilon = E/N$, for $N = 128$ and $\Delta k = 4$. *Open circles* and *squares* are relaxation times, and *closed circles* are the Lyapunov exponents

as presented in the following subsection. The λ-scaling has been placed on a firm theoretical basis, as described in Sect. 2.4.7, with the theory compared to numerics in Sect. 2.8.2. Pettini and coworkers [61] have made extensive calculations of $\lambda(\varepsilon)$, using various initial conditions, for FPU-α, FPU-β, and for the combination of the two, as given in (2.1). The FPU-α chain was particularly useful to explore a transition from power-law dependence of λ on ε to a condition in which λ was not obtainable on computer accessible time-scales, which they called the stochasticity transition (ST). Their interpretation was that the ST gave a transition from essentially regular motion to chaotic motion. However, numerical determinations of $T_{\rm eq}$ for the FPU-β chain, as reviewed in the next section, interprets the more rapid increase of $T_{\rm eq}$ with decreasing ε as a transition from power-law $T_{\rm eq} \propto \varepsilon^{-q}$ to exponential $T_{\rm eq} \propto \exp(-a\varepsilon^{b})$ variation of $T_{\rm eq}$ with ε, as predicted by the Arnold diffusion mechanism (see [55, 57, 58, 65, 121, 134]). We return to this unresolved question of whether or not there is a transition at small ε to regular motion (no diffusion) after presenting the numerics here and in the next subsection.

The type of initial conditions typically used for the FPU-α system are one or a few modes, for which both short-time and long-time dynamics are observed. In Fig. 2.11 from Casetti et al. [61], we present values of $\lambda(\varepsilon)$ on log–log scales for three values of $N = 32$ (squares), 64 (triangles) and 128 (circles), starting from energy initially in the longest wavelength mode. The dominant scaling of $\lambda \propto \varepsilon^{5/3}$ is significantly weaker than the FPU-β of $\lambda \propto \varepsilon^2$. At low values of ε there is a clear drop-off to smaller values of λ with a

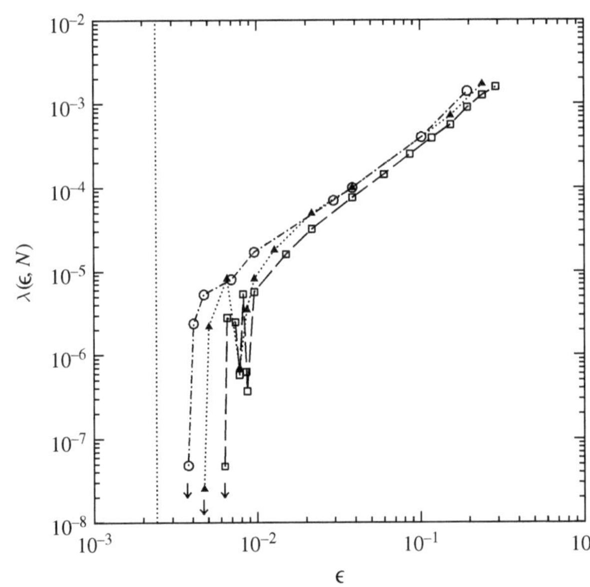

Fig. 2.11. Lyapunov vs. ε. Long-wavelength initial condition

weak N-dependence, such that the drop-off occurs at larger ε for smaller N. The results are slightly confused by an apparent resonant dip in λ occurring near $\varepsilon = 10^{-2}$. Another way of choosing initial conditions for measuring λ is random over oscillators, which is also random over modes. This places the initial coordinates near equipartition to begin with, and therefore, should approximate the long-time average more rapidly. The results, shown in Fig. 2.12, include two small N-values, $N = 8$ and 16, which cannot be compared directly with the previous figure, and may also include some "small-N" effects. For the cases of $N = 32$ (stars), and 64 (square stars), in the power-law regime, the same scaling $\lambda \propto \varepsilon^{5/3}$ is found, agreeing with the previous figure. Again, just considering the highest two N-values, the drop-off to significantly smaller λs have a scaling of $\varepsilon_c N^2 \approx$ const. (or $E_c N \approx$ const.), where ε_c is the approximate break from the main scaling. This separation by N-value is much larger than in the previous figure. We note parenthetically that the scaling of this transition is the same as found for overlap of resonances in the FPU-β system described in Sect. 2.4.1. We discuss these results further after obtaining numerical values of T_{eq} in the following Subsection.

2.5.2 Numerical Determination of T_{eq} Scaling with ε

Considerable numerical effort has gone into the determination of the scaling of T_{eq} for the FPU-β system and comparisons with systems with other force laws. For the FPU-β, Deluca et al. [55, 57, 58], showed for a fixed $\varepsilon = E/N = 0.5$,

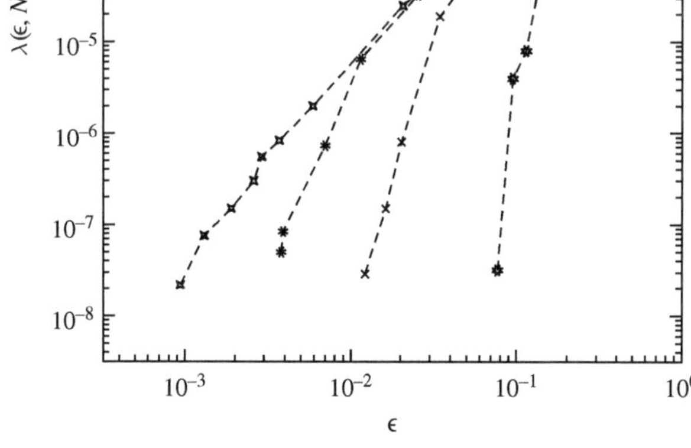

Fig. 2.12. Lyapunov vs. ε. Random initial condition

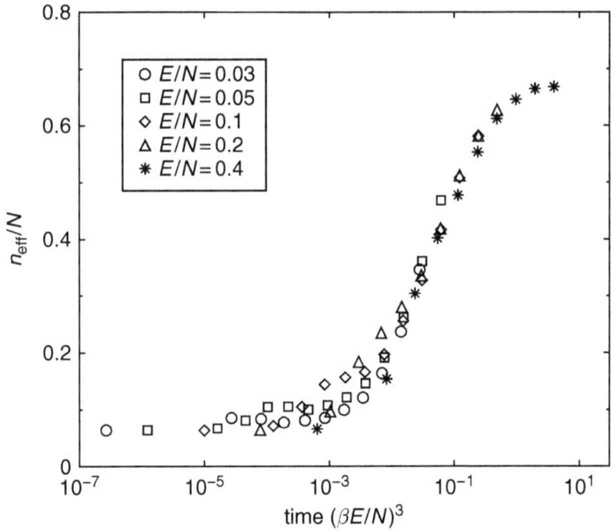

Fig. 2.13. Universal dependence of n_{eff}/N on ε^{-3}

over a range of relatively large N-values, $N = 256 - 4096$, with a fixed percentage of initial modes, $\delta k/N = 1/16$, that $n_{\text{eff}}(t)$ fell on a universal curve, i.e. was independent of N, and therefore of E. For these relatively large N-values a previous weak dependence on N with δk held constant, was partly resolved by showing that $n_{\text{eff}}(t)$ has an initial transient, which mostly disappears for larger values of n_{eff}. This is a reasonable consequence of phase space ergodic mixing. With smaller values of N investigated in other studies, there are significant weak N-dependences in the time to obtain equipartition. The universal dependence of n_{eff} on ε^{-3} is illustrated in Fig. 2.13, where the time is normalized to $\tau = (\beta\varepsilon)^3 t$. Since the times to equipartition become very long as ε becomes small, only the largest values of ε were integrated to T_{eq} at $n_{\text{eff}} = 0.65$. Because of the universal nature of the result shown in Fig. 2.13, the scaling of T_{eq} with ε should also be obtainable from a smaller value of n_{eff}, say $n_{\text{eff}} = 0.4$. This was done, finding $t(n_{\text{eff}} = 0.4) \propto \varepsilon^{-3}$, but with some suggestion of steeper scaling at the smallest ε, interpreted in subsequent longer integrations $n_{\text{eff}} = 0.65$ as a breakdown of power-law scaling if the initial driving frequency becomes too small [65]. A heuristic calculation of the observed scaling was given in [58], which we will summarize in Sect. 2.8.3. The results of DeLuca et al. [58] were reexamined by Berchialla et al. [59] for $N = 511$, again for $t(n_{\text{eff}} = 0.4)$, obtaining ε^{-3} scaling with ε in the same ε range as previously, but extended to smaller ε-values where steeper scaling was observed. In addition to the log–log plot, giving the $t \propto \varepsilon^{-3}$ scaling over part of the ε range, a best-fit stretched-exponential dependence, $t \propto \exp(\varepsilon^{-1/4})$, was also plotted showing a good fit to the smaller values of ε, as shown in Fig. 2.14, with the vertical dashed lines indicating the range of the data from

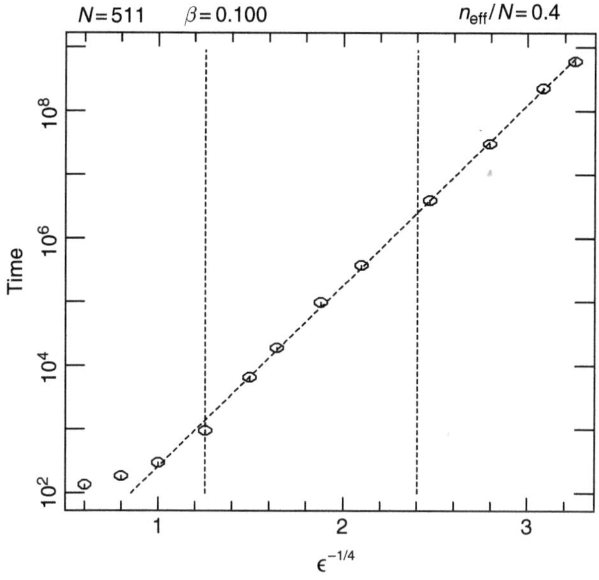

Fig. 2.14. Stretched exponential fit to the relaxation time

the DeLuca et al. [58] paper. (The values for $\varepsilon > 1$ are considered to be above the SST and not of interest in the comparison). These results will be discussed further, after presenting additional data from other oscillator chains. An early comparison of the FPU-β chain with the ϕ^4 chain was made by Pettini and Cerruti-Sola [64]. Both oscillator chains have quartic nonlinear potentials, but the ϕ^4 quartic term is on-site, while the FPU-β quartic term is between oscillators. As already discussed in Sect. 2.2, the FPU-β chain is simpler and easier to analyze. The early numerical comparison of the chains covered long-wavelength (low-frequency), short-wavelength (high-frequency) and intermediate-wavelength initial conditions. The recent theory, elucidating the more complex behavior, starting from high frequencies, had not been developed, but some general observations could be made. For long-wavelength initial conditions, at a given ε, the time to equipartition was shorter for the ϕ^4 system than for the FPU-β system. The opposite held at short wavelengths with T_{eq} shorter for the FPU-β than the ϕ^4 system. It generally took longer to reach equipartition from short-wavelengths at a given ε. We have already discussed reasons for the different behavior and some short-time results in Sect. 2.4.1. Numerical results for short wavelengths will be given in Sect. 2.5. Here we present the detailed long-wavelength comparison as given in [65]. In Fig. 2.15 we compare the scaling of $T_{\text{eq}}(\varepsilon)$ for the FPU-β and ϕ^4 chains, each for two values of N. The nonlinear coefficient is $\beta = 0.1$ for both chains, and $m = 0.1$ for the ϕ^4 chain. Except for the FPU-β chain with $N = 500$ (or 512), the other cases used $\Delta k/N = 1/16$, as with the numerical results presented

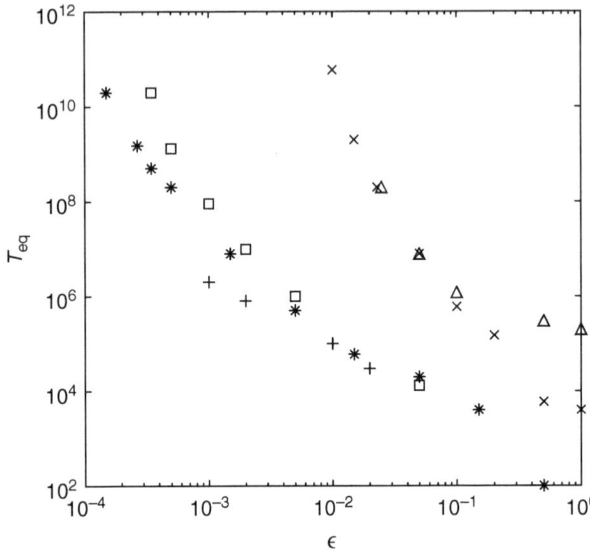

Fig. 2.15. Comparison of equipartition times for the FPU-β [$N = 128$ (*crosses*), $N = 512$ (*triangles*)) and the ϕ^4 model ($m = 0.1$, $N = 64$ (*squares*), $N = 128$ (*stars*)]. The ϕ^4 data with plusses are taken from [64]

in previous figures. For the $N = 500, 512$ data, some $\Delta k/N = 1/16$ initial conditions and some $\Delta k = 8$ ($\Delta k/N = 1/64$) initial conditions were used. In all cases, the initial conditions used the lowest modes, equally excited. [For the two pairs of FPU-β results, with large N, at $\varepsilon = 0.5$ and $\varepsilon = 1$, the lower points used $\Delta k/N = 1/16$ ($N = 512$) while the upper points used $\Delta k = 8$ ($N = 500$).] However, although there is some initial $\Delta k/N$ dependence, the main reason for the separation was due to a statistical (and perhaps physical) anomaly that sometimes occurs at higher N-values. The criterion for determining T_{eq} was the first time $n_{\text{eff}}(t) = 0.65$. As discussed in Sect. 2.3.2 there are fluctuations on short time-scales which lead to $n_{\text{eff}} = 0.65$ at equipartition, but these fluctuations also appear on the $n_{\text{eff}}(t)$ curves. Usually, $n_{\text{eff}}(t)$ crosses 0.65 and then fluctuates around that value. In the two upper values of T_{eq}, discussed above, the first fluctuation occurred just below $n_{\text{eff}} = 0.65$, such that the first actual crossing was significantly later in time. Returning to discussion of the main results, the central portions of all curves display power-law scaling, with FPU-β scaling giving $T_{\text{eq}}(\text{FPU} - \beta) \propto \varepsilon^{-3}$. This is also seen when $t(n_{\text{eff}} = 0.4)$ is used to determine the scaling. The somewhat flatter scaling at large ε is again interpreted as crossing the strong stochasticity threshold (SST). For the ϕ^4 chain the central portion of the ε-values gives $T_{\text{eq}}(\phi^4) \propto \varepsilon^{-2.5}$, slightly flatter than the FPU-β scaling. We will discuss this difference, together with a heuristic calculation of the scaling in Sect. 2.8.3. The $T_{\text{eq}}(\varepsilon)$ slopes become steeper at the lowest values of ε. For the ϕ^4 chain, comparing the values of T_{eq} for $N = 64$ (squares) with the values of $N = 128$ (stars), we see that for intermediate ε-values the two curves lie close together, agreeing with the expectations that T_{eq} is a function of ε, only, [58, 121], as found in Sect. 2.8.3. However, at lower values of ε we see that the $N = 64$ points break away from the $N = 128$ values, producing larger values of T_{eq} with decreasing ε. This behavior was also found for the FPU system, and understood in terms of a critical value of energy, E_c for which the transition occurs. If the value of E_c is the same for all N, as calculated in Sect. 2.4.1 for the FPU system, then the value of ε_c at which the transition takes place would vary inversely with N and therefore occur at a factor of 2 higher ε for $N = 64$ than for $N = 128$, as observed in the figure. The values of T_{eq} for the FPU chain for $N = 128$ (crosses) and $N = 512$ and 500 (triangles), would have the equivalent break at higher values of T_{eq}, which were not numerically reached. In Fig. 2.16, we plot T_{eq}, on a log scale, vs. $1/\varepsilon$, for $N = 128$ (stars) and for $N = 64$ (squares). In the range in which the diffusion is exponentially slow (Arnold diffusion), then from the exponential scaling with $\delta\omega_h/\Omega_B \propto 1/E$, if (2.54) is not satisfied, we expect to obtain a straight line for $\log(T_{eq})$ vs. $1/\varepsilon$ if N is held constant. This is, indeed, found for 4 of the values, with a transition between $\varepsilon = 10^{-3}$ and 5×10^{-4}, for $N = 128$, and between 2×10^{-3} and 10^{-3} for $N = 64$, indicating that the change in dependence occurs at a fixed value of $E = E_c$.

Because of the additional on-site parameter m, the numerical investigation of the ϕ^4 chain involves a larger parameter space than the FPU chain. We can see from the linear mode frequencies, given in Sect. 2.2.2, that for $m \gg 1/N$,

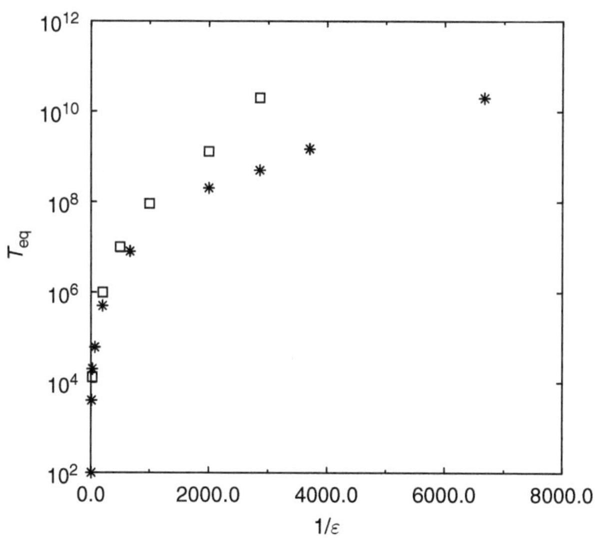

Fig. 2.16. Equipartition time for the FPU-β model in the exponentially slow diffusion regime

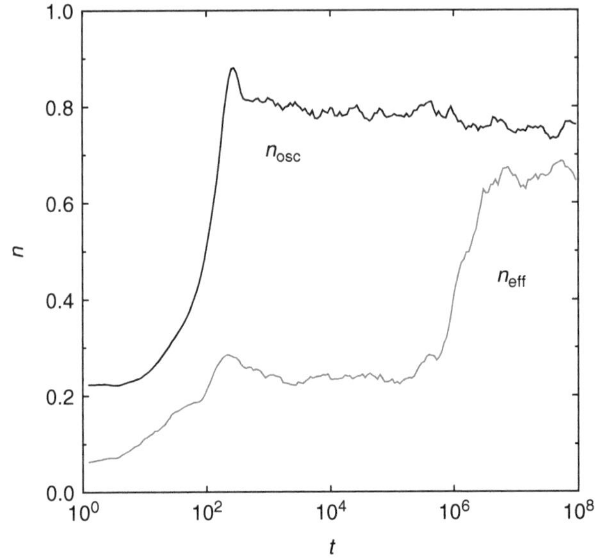

Fig. 2.17. Trapping in an intermediate state before the relaxation to equipartition begins, as illustrated by the time evolution of n_{eff} and n_{osc}

the mode frequencies bunch together for long-wavelength modes, similar to high-frequency bunching. The result, starting from long-wavelength initial conditions, similar to what we shall see for both FPU-β and ϕ^4 at short wavelengths, is that there is trapping in a group of closely coupled modes that leads to a plateau in n_{eff}, before the continuation toward equipartion. We illustrate the plateau in Fig. 2.17, calculating both n_{eff} and n_{osc}, for typical parameters with $m = 1$ and $\Delta k/N = 1/16$. The energy is distributed among many oscillators, and a plateau at $n_{\text{eff}} \approx 0.2$ exists for a long time before the rise to the equilibrium value of $n_{\text{eff}} = 0.65$. The n_{eff} plateau is similar to an effect seen for initial conditions in short wavelengths as in Fig. 2.23. However, the physics is quite different at short wavelengths, where the energy concentrates into a few oscillators, which can be seen by comparing the $n_{\text{osc}}(t)$ plots in Figs. 2.17 and 2.23. In Fig. 2.18 we scale T_{eq} to the best fit of $\varepsilon/m^{2.75}$ which we will discuss in Sect. 2.8.3.

2.6 Numerical Results: Relaxation to Equilibrium from High-Frequency Modes

2.6.1 Dynamical Studies of Self-Organization into Chaotic Breathers and Their Interaction

In a systematic study of chaotic breathers (CBs) in a FPU oscillator chain with periodic boundaries, starting from the boundary mode, that has 180

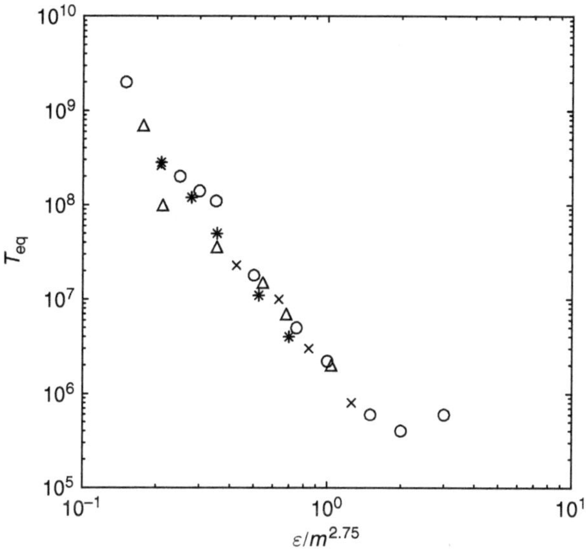

Fig. 2.18. Rescaling of the equipartition time for the ϕ^4 model; each symbol corresponds to a different m-value

phase shift between neighboring oscillators, Cretegny et al. [87] illustrated the dynamics with a three-dimensional presentation first used by Burlakov and Kiselev [83]. The result shown in Fig. 2.19a, with accompanying explanatory figures, shows on a gray intensity scale, the CBs emerging from the fastest growing mode, moving spatially and interacting with one another, with the larger CBs absorbing the smaller ones, until only a single moving CB remains. Figure 2.19b, c illustrate this with time snapshots. Finally there is a longer time period in which the CB can continue to grow by taking energy from background modes with similar symmetry, but ultimately decaying to an equipartition state, as seen in Fig. 2.19d. For this study an intermediate value of $\varepsilon \simeq 0.35$ was used. This result is quantified with statistical measures, which we save for the following section.

As already described in Sect. 2.4.6, it is not necessary to start from the π-mode to obtain results of this nature. In fact, for fixed boundaries there is no exact π-mode. More generic initial conditions with fixed boundaries were used by Ullmann et al. (2000), obtaining results very similar to Fig. 2.19. For example for $N = 128$, choosing $\gamma = 120$, $n = N + 1 - \gamma = 9$ initial peaks, a result close to Fig. 2.19a was obtained, except that there is a competition between the initially fixed number of peaks and the wavenumber of the fastest growing mode. This roughly led to the result that, even for energies for which $n_m \equiv N k_m / 2\pi < 9$ the number of initial CBs $n_{\mathrm{B}} \geq 9$. They also explored a wide range of initial conditions, finding with $N = 128$, that proto-breathers began to form for $\gamma \gtrsim 80$, and became fully formed CBs for $\gamma \gtrsim 100$. They

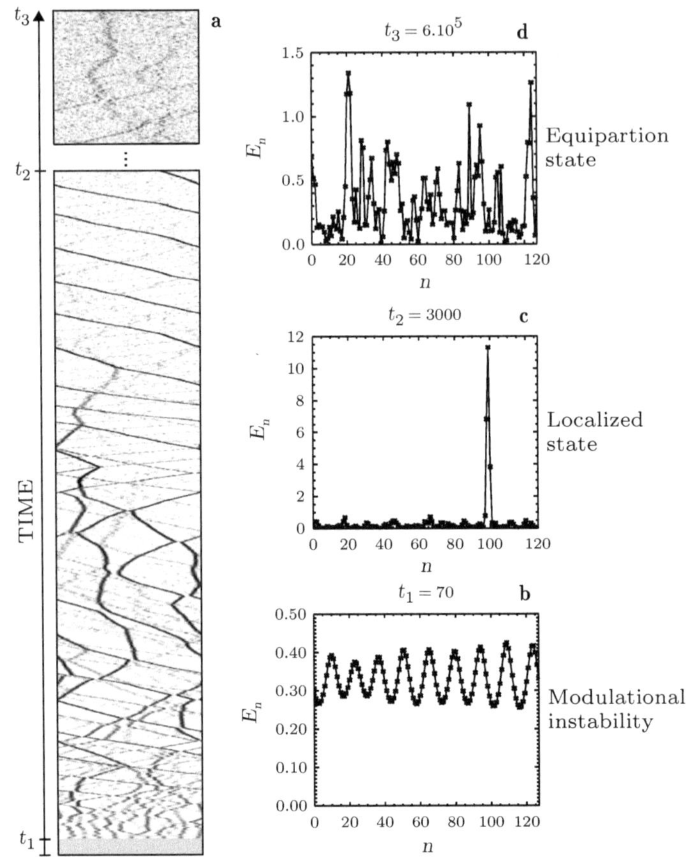

Fig. 2.19. Evolution of the local energy E_n along the chain. In panel **a**, the horizontal axis indicates the position along the chain and the vertical axis corresponds to time (time is going upward). The grey scale goes from $E_n = 0$ (white) to the maximum E_n-value (black). The lower rectangle corresponds to $0 < t < 3000$ and the upper one to $5.994 \times 10^5 < t < 6 \times 10^5$. Panels **b, c**, and **d** show the instantaneous E_n along the $N = 128$ chain at three different times. Note the difference in vertical amplitude

used a particular value of $E = 50$ ($\varepsilon = 0.39$). This latter result was obtained by direct observation of the spectrum, and also by the use of macroscopic quantities, as described in the next subsection.

Various numerics of CB formation and decay were repeated in papers employing the analytic methods using envelope functions, in order to compare analytic formulae with numerical results. For example, for the FPU chain the logarithmic rate of decay of the number of CBs $1/\tau \equiv n_{\mathrm{B}}^{-1}(\mathrm{d}n_{\mathrm{B}}/\mathrm{d}t)$ was constant at fixed energy, indicating that the time-scale for coalescence

$$\tau_{\mathrm{B}} \sim (n_{\mathrm{B}}\sigma_{\mathrm{B}}v_{\mathrm{B}})^{-1} \tag{2.127}$$

with $\tau_{\rm B} \propto \tau$, preserved the constancy of the product CB density×CB interaction cross-section×CB velocity during the coalescence [92]. At lower initial energies, for the FPU chain, $\tau_{\rm B} \propto E^{-2}$ [91], while at higher energies $\tau_{\rm B} \propto E^{-1}$ [88]. These scalings are understood, and the underlying theory will be outlined in Sect. 2.9.

The dynamics of the ϕ^4 chain has some significant differences from the FPU dynamics. Nevertheless, the overall features have much in common. For example, for the lowest mode of the envelope, $n = 1$ ($\gamma = 128$), the development of the parametric instability is quite similar for the ϕ^4 and FPU, provided the ϕ^4 energy is a factor of 16 higher to account for the factor of 16 in the nonlinear term, with $E \propto \psi_m$ in the energy range considered. The initial nonuniform growth of the instability is shown in the time snapshots in Fig. 2.20, and the coalescence shown in four time snapshots in Fig. 2.21.

By scaling either the energy or the time, the early development of the instability can be obtained reasonably close to that in Fig. 2.20. Similarly, the nearly complete coalescence into a single CB in Fig. 2.21d at $t = 5000$, can be roughly compared with the localized state at $t = 2000$ from the observation of a residual second breather seen in Fig. 2.19a, and at $t = 3000$ in Fig. 2.19c. Note that the faster evolution times in the FPU example reflect reasonably well the effective higher energy.

2.6.2 Transitions and Time-Scales to Equipartition

To systematically study the transitions and time-scales for the creation of breathers, their coalescence, and their destruction leading to equipartition,

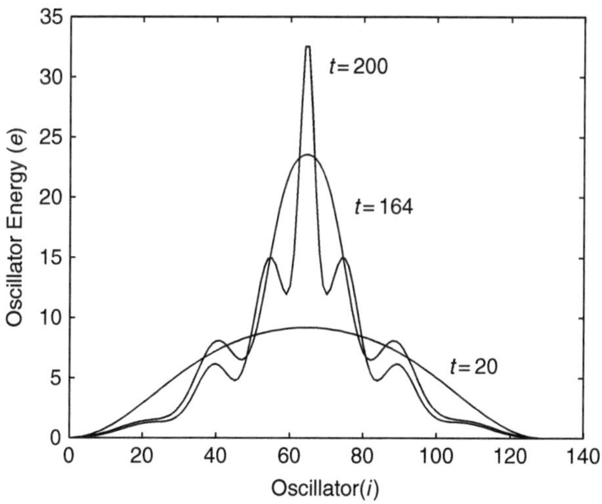

Fig. 2.20. Development of the fastest growing mode from the initial mode $\gamma = 128\,(n = 1)$ for the ϕ^4 model

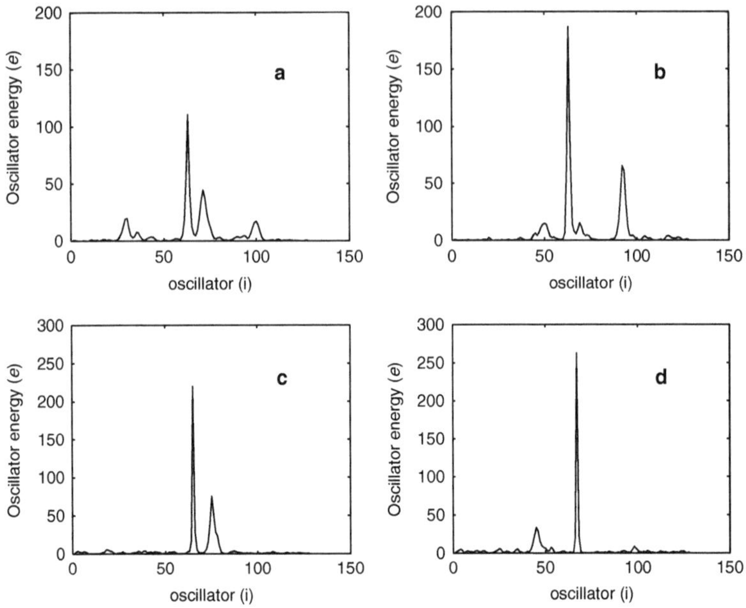

Fig. 2.21. Coalescence of a few chaotic breathers into primarily a single breather fot the ϕ^4 chain: **a** $t = 300$, **b** $t = 1200$, **c** $t = 4000$, and **d** $t = 5000$

macroscopic quantities are most useful. We have introduced these macroscopic quantities in Sect. 2.3.2 and used them to study the evolution from long-wavelength modes to equipartition in Sect. 2.5.2. For long-wavelength initial conditions the quantity n_{eff} is most useful, while n_{osc} gives little additional information. In contrast, for short-wavelength initial conditions, in which the intermediate CB state occurs, n_{osc} is the most useful macroscopic indicator but n_{eff} also serves a useful function; other macroscopic indicators can also be used. In Fig. 2.22a we plot the evolution of $n_{\text{osc}}(t)$ during the evolution, using a linear time-scale, beginning after the multiple CBs have formed through their coalescence time of about $t \simeq 2 \times 10^4$, and then through a longer time interval to approximately $t \simeq 2.5 \times 10^5$. Finally, an approximate equipartition is reached at $T_{\text{eq}} \sim 5 \times 10^5$, as the breather disintegrates, between $t = 2.5 \times 10^5$ and $t = 5 \times 10^5$. It is quite difficult to extract a definite T_{eq} from n_{osc}, which is most useful for describing the evolution of the CB states. We also present the results for n_{eff} on a linear time-scale, but over twice the time period, in Fig. 2.22b, observing that $T_{\text{eq}} \simeq 10^6$. The approximate numerical asymptotes are slightly lower than the theoretical values, given in Sect. 2.3.2, which are $n_{\text{osc}} = 0.74$ and $n_{\text{eff}} = 0.655$.

The overall time dependence is better seen on a scale which is logarithmic in time. We illustrate the full time dependence for the ϕ^4 oscillator chain, in Fig. 2.23, starting from the same initial conditions as in Figs. 2.20 and 2.21. We note here that the asymptotes correspond more closely to the theoretical

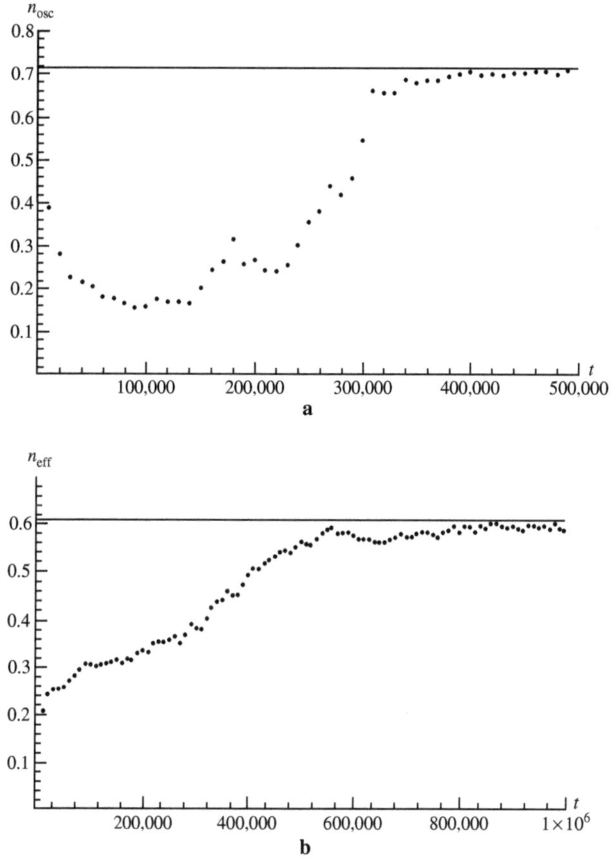

Fig. 2.22. Time dependence of **a** $n_{\rm osc}$ and **b** $n_{\rm eff}$. The horizontal lines are asymptotes

values. Again, most of the relevant information is contained in $n_{\rm osc}(t)$. The logarithmic dependence of the time-dynamics is seen by estimating that the instability growth is in the range $30 \lesssim t \lesssim 300$, the breather formation and coalescence between $300 \lesssim t \lesssim 3 \times 10^3$, and the single breather exists between $3 \times 10^3 \lesssim t \lesssim 3 \times 10^4$. In the final decade the breather is destroyed, leading to equipartition.

The scalings for short-wavelength initial conditions in the regions of lower and higher energy density ε can be obtained similarly to that in Sect. 2.5.2 for long-wavelength initial conditions. As for long wavelengths there is a transition between two distinct regions of power-law scaling with energy density ε , the lower energy region which has steeper ε-scaling, and the higher energy region with flatter ε-scaling. Returning to the numerics of the FPU, in Fig. 2.24 we plot $n_{\rm eff}$ vs. normalized time for a series of values of ε, showing that in the lower energy regime $0.33 \leq \varepsilon \leq 1.44$ in Fig. 2.24a, the data

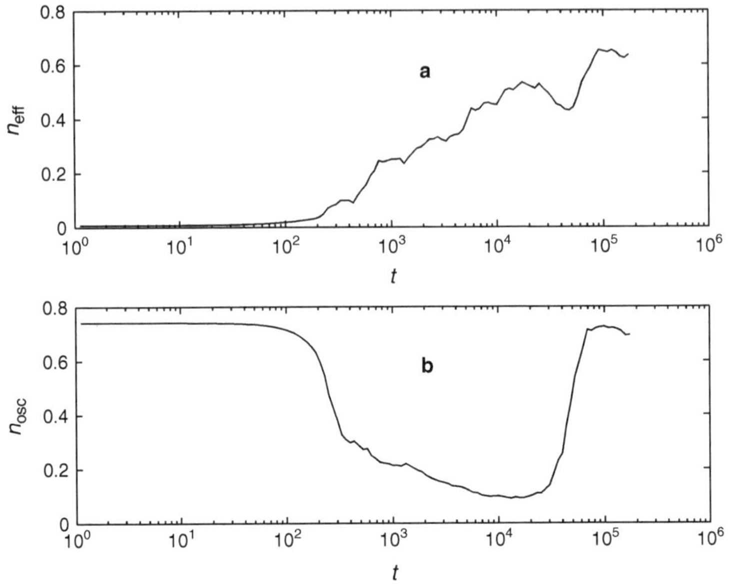

Fig. 2.23. Time dependence of (a) n_{eff} and (b) n_{osc} for the ϕ^4 chain. The initial energy density is $\varepsilon = 2$ and $\gamma = 128$ $(n = 1)$

collapses onto a single curve using ε^{-2} as a normalizing factor. At very high values of ε, $10^3 \leq \varepsilon \leq 10^6$ the dynamics has become purely random over the mode space, and the weak normalizing factor $\varepsilon^{-1/4}$ is observed (Fig. 2.24b).

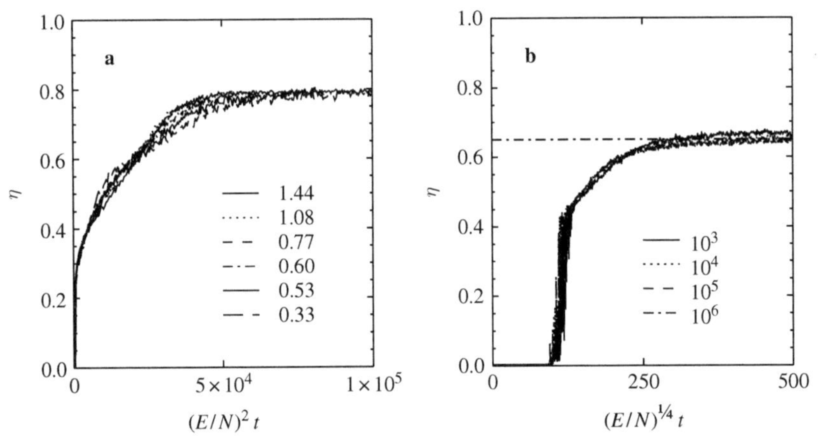

Fig. 2.24. In panel **a** [resp. **b**] the evolution of $n_{\text{eff}}(t)$ vs. the rescaled time $t\,(E/N)^2$ (resp. $t\,(E/N)^{1/4}$) is reported for different energy densities. Each *curve* in (a) [resp. **b**] corresponds to the average over 20 (resp. 50) different initial conditions for a chain with $N = 512$ sites. The *dot-dashed line* reported in (b) indicates the asymptotic value $n_{\text{eff}} \simeq 0.655$

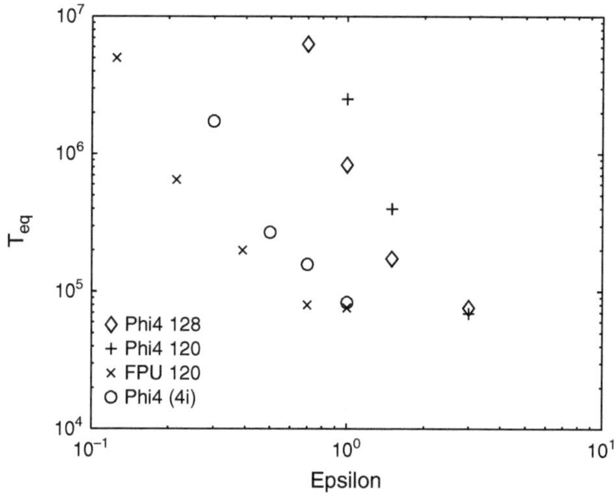

Fig. 2.25. Comparison of the equipartition time for the FPU and the ϕ^4 lattice

The lower ε scaling, starting from short wave-lengths, is different from the ε^{-3} scaling obtained in Sect. 2.5.2 from long wave-lengths. Estimates for the scalings, as well as values of T_{eq} in the various regimes will be presented in Sect. 2.9.2.

Finally, we compare in Fig. 2.25 the times to equipartition, for a few distinct initial conditions of the FPU and ϕ^4 chains at relatively low values of ε, with $N = 128$. All values of T_{eq} were taken at the first crossing (or touching) of $n_{\mathrm{eff}} = 0.65$. The slopes become steeper, for both chains and all initial conditions, at the lowest values of ε. This may be associated with the transition to exponential scaling, as previously discussed for low-frequency initial conditions. We also see the basic separation of a factor of 4 in ε at a given T_{eq}, between the FPU and ϕ^4 results, due to the difference in the nonlinear forces. There are also more subtle differences, discussed in the original paper [94].

2.7 Numerical Results: Stationary Nonequilibrium Properties

2.7.1 Numerical Studies of the Divergence of Heat Conductivity

As a result of a number of studies of the heat conductivity in FPU and diatomic Toda models the conductivity of long but finite chains diverges as

$$\kappa(N) \propto N^{\alpha} . \tag{2.128}$$

In Table 2.1, we compare the available estimates of the exponent α determined by different authors in various models and conditions. The acronym NEMD refers to nonequilibrium molecular dynamics simulations, where the chain is in contact at its boundaries with thermal baths at different temperatures. The heat conductivity is estimated by (2.10). The acronym GK indicates equilibrium simulations, where the chain is at equilibrium and the heat conductivity is computed by the Green–Kubo formula: for an isotropic homogeneous solid made of atoms placed on a regular cubic lattice of volume V in d dimensions, the thermal conductivity κ is expressed as a scalar quantity (see [154])

$$\kappa = \frac{1}{k_{\mathrm{B}} T^2 d} \lim_{t \to \infty} \int_0^t \mathrm{d}\tau \lim_{V \to \infty} V^{-1} \langle J(\tau) \cdot J(0) \rangle \,, \tag{2.129}$$

where k_{B} is the Boltzmann constant; T, the temperature of the solid; $\langle\ \rangle$, the equilibrium average; and J is the heat current vector.

This formula applies when the integral on the r.h.s. is finite. In a chain of atoms, J is a scalar quantity and a divergent heat conductivity can be signalled by the slow decay in time of the heat current autocorrelation function $\langle J(t)J(0)\rangle \sim t^{-\beta}$ with $\beta < 1$. In fact, the integral in (2.129) diverges. In finite chains one expects that an exponential decay eventually sets in, so that simulations should be performed for different chain lengths in order to be sure to pick up the truly asymptotic scaling behavior.

A quantitative comparison with the nonequilibrium measurements, based on the Fourier law (2.10), can be performed by noticing that energy propagates with the constant sound velocity v_{s}. This can be understood by, for example looking at the spatio-temporal correlation function $C(i,t) = \langle j_i(t) j_0(0)\rangle$ of the local heat flux plotted in Fig. 2.26 (see [114]). Accordingly, one can turn the time divergence of κ as determined from the Green–Kubo formula into a divergence with N by restricting the integral in formula (2.129) to times smaller than the "transit time" Na/v_s. This amounts to ignoring all the contributions from sites at a distance larger than N. With the above estimate of C, one obtains that $\kappa \propto N^{1-\beta}$, i.e. $\alpha = 1 - \beta$. The latter exponent is the

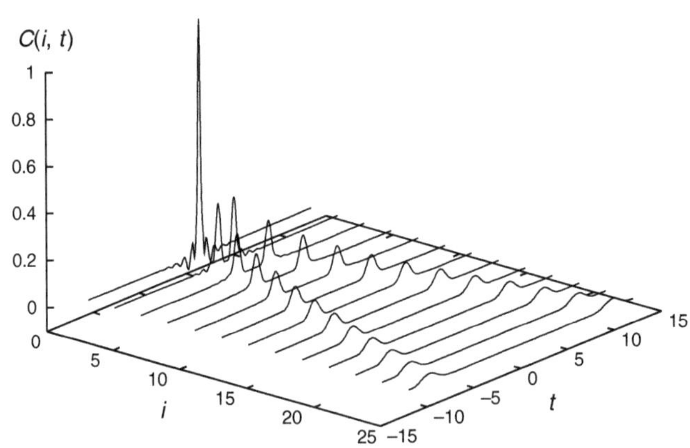

Fig. 2.26. The spatio-temporal correlation function $C(i,t) = \langle j_i(t) j_0(0)\rangle$ of the local flux for the FPU-β model. Microcanonical simulations, energy density 8.8

Table 2.1. The estimated exponent α of divergence of the conductivity with size N, as obtained from both nonequilibrium molecular dynamics (NEMD) simulations and through Green–Kubo (GK) equilibrium studies. Only the significative digits are reported as given in the quoted References

Model	Reference	α (NEMD)	α (GK)
FPU-β	[113, 114]	0.37	0.37
FPU-α	[115]	~ 0.4	–
Diatomic FPU r=2	[117]	0.43	compatible
Diatomic Toda r=2	[116]	0.35–0.37	0.35
	[117]	0.39	compatible
Diatomic Toda r=8	[117]	0.44	compatible
Diatomic hard points	[116]	0.35	–

one reported in the last column of Table 2.1. All numerical values there range between 0.35 and 0.44, suggesting a nontrivial universal behavior. It is also remarkable to notice the overall consistency among the results obtained with different thermostat schemes (ranging from deterministic to stochastic ones).

In order to better appreciate the quality of the divergence rate that can be numerically obtained, in Fig. 2.27 we have plotted the finite-length conductivity $\kappa(N) = JN/(T_+ - T_-)$ vs. the number of particles in the FPU-β

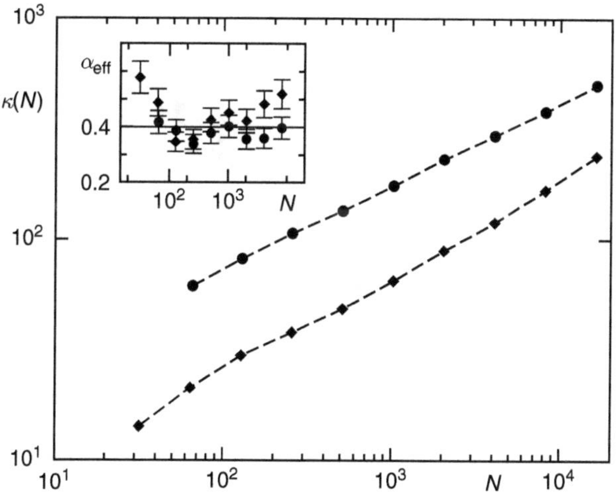

Fig. 2.27. Thermal conductivity of the FPU-β model vs. lattice length N for $T_+ = 0.11$, $T_- = 0.09$. The inset shows the effective growth rate α_{eff} versus N. *Circles* and *diamonds* correspond to free and fixed b.c., respectively

model for fixed and free boundary conditions. In the inset, one can see that the effective growth rate α_{eff}, which corresponds to the logarithmic derivative of $\kappa(N)$, is basically the same in both cases, despite the clear differences in the actual values of the flux itself. The value of α_{eff} seems to be quite close to 0.4, a value which has been predicted by different theoretical approaches (e.g., see [113]).

Once the divergence is clearly established, the next question concerns the universality of the divergence rate. The discussion of this point involves considering a possible dependence on the temperature as well as on the leading nonlinearities [115]. Both questions are addressed in Fig. 2.28, where $\kappa(N)$ is computed in the FPU-α model at a relatively low temperature.

One can see that changes in the temperature gradient, without modifying the average $T = (T_+ + T_-)/2$, modify the effective conductivity only at relatively small sizes. Moreover, the two sets of measures corresponding to $\Delta T = 0.1$ and 0.02 (triangles and circles, respectively) approach each other for N larger than 10^3. In both cases $\kappa(N)$ increases linearly with N for $N < 10^3$ and no sizeable temperature gradient forms along the chain. Both facts hint at a weakness of anharmonic effects up to this time/length scales. This is confirmed by the comparison with the results for a pure harmonic chain (with the same setup and same parameters) that exhibit a clean linear growth of κ with N (see the solid line in Fig. 2.28) and a few-percent differences in the initial size range. The fact that κ is smaller for larger ΔT can be thus attributed to a stronger boundary scattering that reduces the conductivity. From the inset of

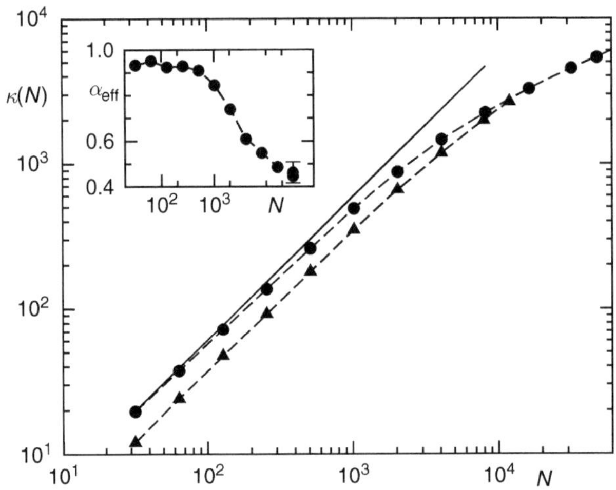

Fig. 2.28. Thermal conductivity of the FPU-α model vs. lattice length N. *Triangles* and *circles* refer to $\Delta T = 0.1$ and $\Delta T = 0.02$, respectively. The *solid line* corresponds to the linear divergence observed in a harmonic chain with the same temperatures. The inset shows the effective divergence rate α_{eff} vs. N for the data corresponding to *full circles*

Fig. 2.28, one can only conclude that for chain sizes up to $O(10^4)$ the exponent α seems to approach the value 0.4, although nothing prevents the possibility that it converges to a larger value. In particular, there is evidence that the scaling properties of $\kappa(N)$ in the FPU-α model exhibit much different features with respect to the FPU-β model, at least in the explored range of temperatures and sizes. Recent theoretical estimates based on the mode-coupling approach, indicate that, as a matter of symmetry, heat conductivity in the FPU-α and -β models should diverge with different values of the exponent in (2.128). Moreover, these values are supposed to represent the two universality classes characterizing the power-law divergence of the heat conductivity for any FPU-like model (including more phenomenological nearest-neighbor potentials, like the Lennard-Jones ones). It must be pointed out that such a theoretical prediction is very difficult to be checked numerically. Typically, finite size/time effects, as those observed in the FPU-α model, are found to last over extremely long integration times and for very large system sizes which do not allow one to confirm or disprove the theoretical expectations. Anyway, a power-law divergence of the heat transport coefficient is always present in such a class of models, although in finite systems the value of the exponent α varies significantly with temperature and size. It is worth mentioning that this could be a relevant point for what concerns possible comparison with experiments, performed on almost-one-dimensional systems, like carbon nanotubes or polymers.

Finally, we want to discuss the role of the boundary resistance in connection with the temperature dependence of conductivity. In fact, an interesting application of (2.128) has been proposed with reference to the FPU-β model [155]. There, it has been empirically found that the bulk conductivity scales with N and T as

$$\kappa \simeq \begin{cases} 1.2\, N^{\alpha}T^{-1} & (T \sim 0.1) \\ 2\, N^{\alpha}T^{1/4} & (T > 50) \end{cases} . \tag{2.130}$$

According to kinetic theory, the conductivity can also be expressed as $\kappa = \ell v_s C_v$, where ℓ is the mean free path of phonons. Since C_v and v_s are almost constant and of order 1 in a wide temperature range, $\ell \sim \kappa$. Hence, at low temperatures the boundary jumps dominate the thermal profile up to the size N_* that can be estimated according to (2.130). At low temperatures this effect is very strong with $N_* \sim (2\varepsilon/T)^{(1/1-\alpha)}$, while smaller boundary resistances are found at large temperatures, with $N_* \sim (2\varepsilon T^{1/4})^{(1/1-\alpha)}$, where ε is the energy density.

2.7.2 Force Laws That Predict Classical Heat Conductivity:Coupled-Rotors, Ding–a–Ling and Klein–Gordon Chains

A simple example of a classical-spin one-dimensional model with nearest neighbor interactions has a potential

$$V(x) = 1 - \cos x \, , \tag{2.131}$$

i.e. is a chain of N coupled pendula. It has been extensively studied [151, 153, 156] as an example of a chaotic dynamical system that becomes integrable both in the low- and high-energy limits, when it reduces to a harmonic chain and to free rotors, respectively. In the two integrable limits, the relaxation to equilibrium slows down very rapidly for most of the observables of thermodynamic interest (e.g., the specific heat: see [151, 156]). As a consequence, the equivalence between ensemble and time averages is established over accessible time-scales only inside a limited interval of the energy density ε. Here, we focus our attention mainly on heat conduction in the strongly chaotic regime.

It has been shown that, contrary to the expectations, this model exhibits a finite conductivity in spite of the existence of an acoustic branch in its spectrum in the harmonic limit [157, 159]. Simulations have been performed for $T_+ = 0.55$, $T_- = 0.35$, and chain lengths ranging from $N = 32$ to 1024 with fixed boundary conditions and Nosé–Hoover thermostats [157]. The equations of motion have been integrated with a fourth order Runge–Kutta algorithm and a time step $\Delta t = 0.01$. The results, reported in Fig. 2.29 clearly reveal a convergence to a value of κ approximately equal to 7 (see the circles). The dotted line is the best fit with the function $a + b/N$ and indicates a constant κ for $N \to \infty$. This is the first system where normal heat conduction was convincingly ascertained in the absence of a local potential and was confirmed by calculating an exponential decay of heat flux correlation appearing in the Green–Kubo formula (2.129).

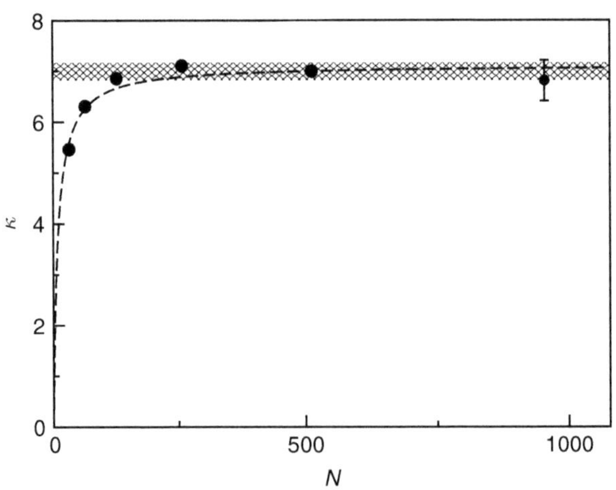

Fig. 2.29. Conductivity κ vs. chain length N as obtained from nonequilibrium molecular dynamics. *Circles* correspond to the rotator model with temperatures $T_+ = 0.55$ and $T_- = 0.35$. The *dashed line* represents the best fit with the function $a + b/N$. The *shaded region* represents the uncertainty about the conductivity on the basis of the Green–Kubo formula

One explanation of the striking difference in the transport behavior exhibited by the model compared to, for instance, the FPU models is to notice that the pair potential $V(q_{i+1} - q_i)$ possesses infinitely many equivalent valleys. As long as $(q_{i+1} - q_i)$ remains confined to the same valley, there is no reason to expect any qualitative difference with, for example, the FPU-β model. Phase slips (jumps of the energy barrier), however, may very well act as localized random kicks, that contribute to scattering of the low-frequency modes, thus leading to a finite conductivity. In order to test the validity of this conjecture, one can study the temperature dependence of κ for low temperatures when jumps across barriers become increasingly rare. The data plotted in Fig. 2.30 indicate that the thermal conductivity behaves as $\kappa \approx \exp(\eta/T)$ with $\eta \approx 1.2$. The same scaling behavior is exhibited by the average escape time τ (see triangles in Fig. 2.30) though with a different $\eta \approx 2$. The latter behavior can be explained by assuming that the phase slips are the results of activation processes. Accordingly, the probability of their occurrence is proportional to $\exp(-\Delta V/T)$, where ΔV is the barrier height to be overcome. The behavior of τ is thus understood, once we notice that $\Delta V = 2$. In the absence of phase slips, the dependence of the conductivity on the length should be the same as in FPU-β model, i.e. $\kappa \approx N^\alpha$, with $\alpha \approx 0.4$. In the presence of phase slips, it is natural to expect that the conductivity is limited by the average distance $\overline{N}$ between consecutive phase slips. Under the further assumption of a uniform distribution of the slips, their spatial and temporal separation has to be of the same order, thus implying that $\kappa(T)$ exhibits the same divergence as τ for $T \to 0$, though with a different rate $\kappa \approx \exp[\alpha\Delta V/T]$. Therefore, at least

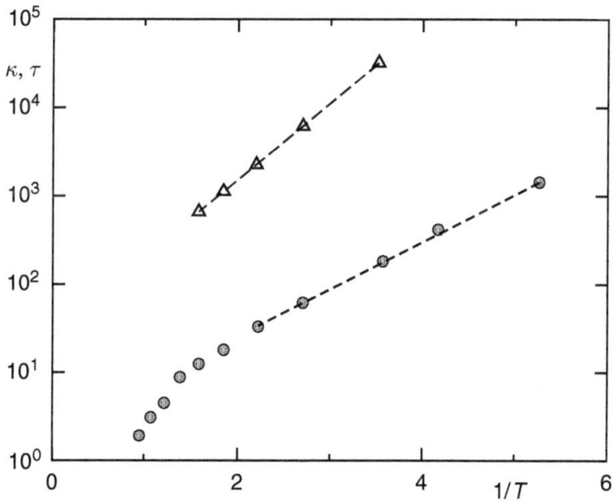

Fig. 2.30. Thermal conductivity κ vs. the inverse temperature $1/T$ in the rotor model (*open circles*). *Triangles* correspond to the average time separation between consecutive phase slips in the same system

on a qualitative level, one can indirectly confirm that phase slips are responsible for the normal heat transport. However, there is a discrepancy between the observed and the expected value of the exponent η (1.2 vs. 0.8). We now consider the class of models whose Hamiltonian contains a local "substrate" potential $U(q_i)$. This means that translational invariance breaks down and the total momentum is no longer a constant of the motion. Accordingly, the dispersion relation is such that $\omega(k) \neq 0$ for $k = 0$.

The so-called "ding-a-ling" model was first introduced as a toy model for a one-dimensional plasma [158]. It can refer in different contexts to : (i) a set of identical charge-sheets embedded in a fixed neutralizing background; (ii) a system of harmonic oscillators with the same frequency and equilibrium positions sitting on a periodic lattice and undergoing elastic collisions that exchange their velocities. Notice that in the low-energy limit, it reduces to a set of independent harmonic oscillators with equal frequency (no dispersion). A modified version of this model, where the harmonic oscillators (say the even-numbered particles) alternate with *free* particles of the same (unit) mass, was later introduced for studying heat transport [119]. The free particles are only constrained to lie between the two adjacent oscillators. The Hamiltonian can be symbolically written as

$$\mathcal{H} = \frac{1}{2}\sum_{l}^{N} \left[p_l^2 + \omega_l^2 q_l^2\right] + \text{“hard point core”} \ , \tag{2.132}$$

where $\omega_l = \omega$ for even l and zero otherwise. A common feature of this class of models is that within collisions the motion of the particles can be determined analytically so that the basic requirement is the computation of the occurrence times of the collision events. Therefore, the dynamics naturally reduces to a discrete mapping. For an isolated system (e.g., a chain with periodic boundary conditions) the dynamics depends only on the dimensionless parameter $\varepsilon/(\omega\, a)^2$ where ε is the energy per particle and a the lattice spacing. The dynamical behavior of the model was studied by fixing $\varepsilon = 1$ and changing ω [119]. When ω and N were large enough the dynamics was found to be strongly chaotic and soliton-like pulses are sufficiently attenuated [160]. This renders the model a good candidate to check the validity of Fourier's law.

The Fourier law was first confirmed with the performance of a series of non-equilibrium simulations, where the freely moving end-particles were put in contact with two Maxwellian reservoirs. The average flux J was then computed by summing the amounts of energy δE exchanged with one of the reservoirs in all collisions during the simulation time. The average temperature gradient was estimated with a linear fit (to get rid of boundary effects). By evaluating the thermal conductivity as a function of the lattice length up to $N = 18$ for $T_+ = 2.5$, $T_- = 1.5$ and $\omega = 1$, it was concluded that $\kappa(N)$ attains a constant limiting value for $N > 10$ (see Fig. 2.31). After having established the existence of a finite value of the transport coefficient, the value of κ was compared with the result of linear response theory, using a Green–

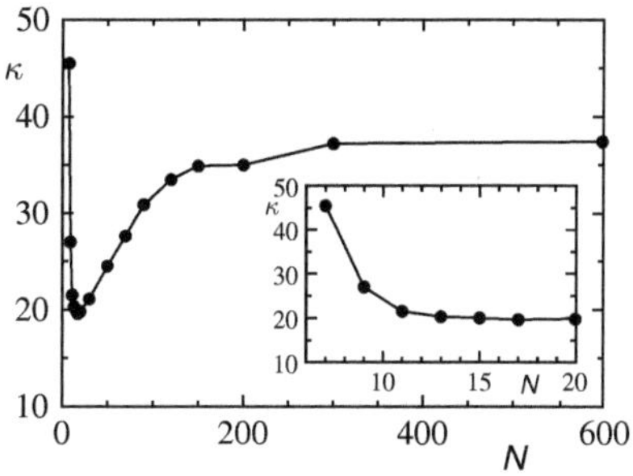

Fig. 2.31. Thermal conductivity of the ding-a-ling model. Size dependence of κ for $\omega = 1$ and $\varepsilon = 1.5$ (from [162]). In the inset an expanded view is presented in the range of sizes considered in [119]

Kubo formula (see [161]). In summary, all these studies provided a convincing numerical evidence that the heat conductivity is finite in this model. In particular, it was shown that the energy transport is diffusive for large values of ω (typically, larger than 10), while the linear response estimate was found to agree with numerical simulations. Much later, these results were essentially confirmed by a detailed series of simulations with longer chains and in a wider parameter range of the "ding-a-ling" model [162]. However, the temperature profile exhibits a nonlinear shape (see Fig. 2.32) different from the typical linear temperature profile predicted by the Fourier law.

Careful numerical studies performed on the original Dawson model [161] have confirmed the validity of Fourier's law in a wide temperature range. Besides direct nonequilibrium simulations with Maxwellian thermostats and the Green–Kubo formula, these studies implemented an efficient transient method that allowed them to explore the high-temperature regime ($T > 3$), where, because of the nearly integrable dynamics, a slow convergence of the averages with time and/or size is observed. In the low-temperature limit ($T < 0.1$) it was also proved that the heat conductivity vanishes as $\exp(-1/4T)$, in agreement with numerical results. Numerical evidence of finite thermal conductivity

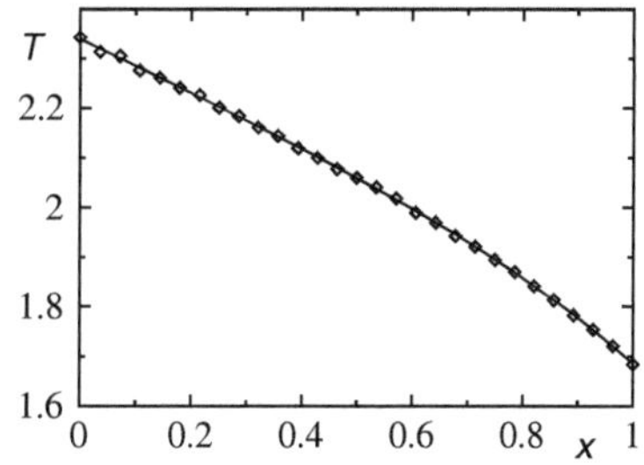

Fig. 2.32. Temperature profile for the ding-a-ling model, $\omega = 2$ (see [163])

was also found in a modified ding–a–ling model where the harmonic potential is replaced by a gravitational one [163].

The Klein–Gordon lattices are another important class of models with a substrate potential, where the interparticle potential is harmonic:

$$\mathcal{H} = \sum_{l=1}^{N} \left[\frac{p_l}{2m}^2 + U(q_l) + \frac{1}{2} C (q_{l+1} - q_l)^2 \right] . \tag{2.133}$$

The first and most complete study of the transport problem in this class of models has been carried out for the Frenkel–Kontorova potential [164].

$$U(x) \;=\; -U_0 \cos\left(\frac{2\pi x}{a}\right) . \tag{2.134}$$

The model can be interpreted as a chain of either coupled particles in an external periodic field or torsion pendula subject to gravity. In the latter case $a = 2\pi$ and q_l represents the angle with respect to the vertical direction: it can be read as the discretized (and nonintegrable) version of the well-known sine-Gordon field equation.

Besides energy, the dynamics admits a further conserved quantity, the winding number $\mathcal{P}$, which is an integer defined by the boundary condition $q_{l+N} = q_l + a\mathcal{P}$. In the particle interpretation, $\mathcal{P}$ represents the number of potential wells, while for the pendula it can be viewed as the degree of built-in twist in the system. Thermal conductivity was computed numerically in the general case of nonvanishing winding number, with three different methods: (i) attaching two heat baths; (ii) through the Green–Kubo formula; (iii) by adding an external field . All the methods give consistent results and clearly indicate that the thermal conductivity is finite. These results were later confirmed by further numerical studies [165], which investigated the dependence of the transport coefficient on the lattice length (for $\mathcal{P} = 0$). Similar conclusions were drawn for a more general version of the Frenkel–Kontorova model with an anharmonic inter-site potential [166].

In order to illustrate the type of behavior observed in this class of models Fig. 2.33 shows some data for the ϕ^4 chain

$$U(x) \;=\; \frac{a}{2} x^2 + \frac{b}{4} x^4 . \tag{2.135}$$

In panel (a), we present a case of fast convergence to a small κ value for a single-well potential; panel (b) refers instead to a low-temperature regime characterized by large thermal conductivity.

Evidence of a finite conductivity for the case $a = 0$ has been reported in [167, 168]. Two further examples were analyzed, the sine-Gordon and bounded single-well potentials:

$$U(x) \;=\; \cosh x \,-\, 1 \quad , \quad U(x) \;=\; \frac{1}{2}\left(1 - \operatorname{sech}^2 x\right) \tag{2.136}$$

representatives of the classes of hard and soft anharmonicity, respectively. In both cases there is numerical evidence for finite thermal conductivity [169].

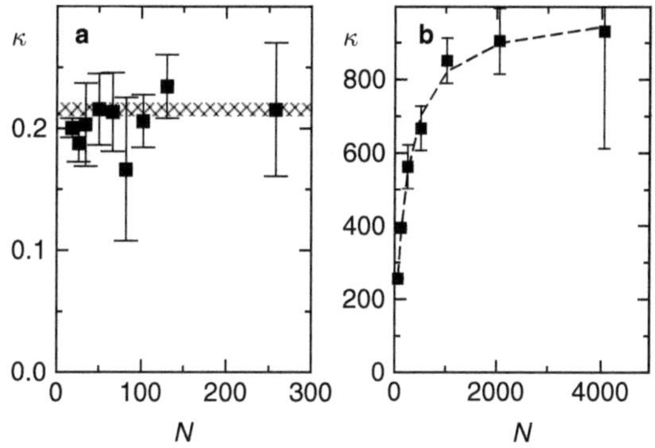

Fig. 2.33. Thermal conductivity vs. chain length in ϕ^4 chains with Nosé–Hoover thermostats. *Panel* **a** refers to the single well case [$a = b = 1$ in (2.135)] the results have been obtained for $C = 1$, $T_+ = 8$, and $T_- = 6$. The *shaded region* represents the value obtained from the Green–Kubo formula with its statistical uncertainty. *Panel* **b** refers to the double-well case ($a = -1$, $b = 1$) for an average temperature $T = 0.37$ and a temperature difference 0.002. The *dashed line* is just a guide for the eyes

2.8 Analytical Calculations and Estimates: Scaling Estimates for λ and for T_{eq} from Low Frequencies

2.8.1 Random Matrix Approximation in the Strong Mode Overlap Regime

The numerical computation of the energy density ($\varepsilon = E/N$) scaling of the largest Lyapunov exponent for the FPU-β model has put in evidence the existence of a *crossover* between two scaling laws: $\lambda(\varepsilon) \propto \varepsilon^2$ at low-energy density, and $\lambda(\varepsilon) \propto \varepsilon^{2/3}$ at larger ε values [60], reaching on an asymptotic value at large energy of $\lambda \propto \varepsilon^{1/4}$.

This transition has been called [64] the Strong Stochasticity Threshold (SST), and it has been ascribed to the (smooth) transition between a regime of *weak* chaos and a regime of *strong* chaos. Even if the system is prepared far from energy equipartition, corresponding to the crossover energy density ε_{c} there is a transition between slow and fast relaxation to equipartition; the relaxation time approximately constant for energy densities greater than ε_{c}, but steeply growing with decreasing the energy density below this value. Although the relaxation phenomenology depends on the details of the initial conditions, they always group into two families separated by ε_{c}. For example, below ε_{c} the initial excitation of high-frequency normal modes results in a slower relaxation to equipartition with respect to the initial excitation of low-frequency modes, but this situation is reversed above ε_{c} [64]. We can estimate the ε-scaling of λ to a transition from weak to strong chaos from a random matrix approximation (RMA) for the tangent dynamics, which approximately accounts for the high-energy scaling of λ, whereas the extrapolation of such a scaling law to lower energy increasingly overestimates the actual values of λ [53, 64].

The RMA assumes the absence of correlations: for a flow this can be assumed if the sampling time is not too short and if the dynamics itself mimics a true random process. It then appears reasonable to consider as strongly chaotic the dynamical regime where the RMA accounts for the $\lambda(\varepsilon)$ scaling and to consider weakly chaotic the dynamical regime where the RMA largely overestimates $\lambda(\varepsilon)$. In [53, 64] it has been suggested that in the strongly chaotic regime phase space diffusion occurs *across* resonances (and is therefore a fast diffusion entailing a fast mixing) thanks to the coalescence of the stochastic layers, generated near the resonant surfaces. At variance, below the SST, the widths of the stochastic layers could be such that the resonance overlap ceases—or is considerably reduced—so that phase space diffusion should be constrained to occur along tortuous paths, *along* resonances, in a definitely less efficient fashion; this qualitative picture has been given an interesting quantitative confirmation in [170]. Here the relaxation times of the weakly chaotic regime measured in [53] have been analytically estimated with a very good accuracy by means of a theoretical approach closely resembling the so-called quasi-linear theory of diffusion in plasma physics, a theory of slow diffusion applicable when chaos is not fully developed.

The numerical integration of a Hamiltonian flow consists in recursively computing the symplectic coordinate transformation $[q_i(t), p_i(t)] \to [q_i(t+\tau), p_i(t+\tau)]$

$$\begin{aligned} q_i(t+\tau) &= q_i(t) + \tau p_i(t) \\ p_i(t+\tau) &= p_i(t) + \tau F_i[q(t+\tau)] \end{aligned} \tag{2.137}$$

where $F_i = -\partial V(q)/\partial q^i$ are the forces and $q = (q_1, \ldots, q_N)$. The Jacobian matrix

$$\underline{J}(\Omega) = \begin{pmatrix} \mathbb{1} & \tau\mathbb{1} \\ \tau\Omega & \mathbb{1} + \tau^2\Omega \end{pmatrix} \tag{2.138}$$

of the discretized flow is symplectic, i.e

$$\underline{J}^{\mathrm{T}}\ \underline{E}\ \underline{J} = \underline{E}$$

where

$$\underline{E} = \begin{pmatrix} 0 & \mathbb{1} \\ -\mathbb{1} & 0 \end{pmatrix}.$$

Here $\mathbb{1}$ is the $N \times N$ identity matrix and Ω is the Hessian of the potential part of the Hamiltonian: $\Omega_{ij} = -\partial^2 V(q)/\partial q_i \partial q_j$ and τ is a discretization time (for instance the time integration step). $\underline{J}$ is a $2N \times 2N$ symplectic matrix which maps a vector $\xi(t)$ tangent to the flow at time t into a vector $\xi(t+\tau)$, that is: $\xi(t+\tau) = \underline{J}(t)\xi(t)$. Using Oseledec's multiplicative theorem [171], an approximation of λ is given by

$$\lambda = \lim_{n\to\infty} \frac{1}{n\tau} \ln \left\langle \left[\frac{\xi^{\mathrm{T}}(0)\left[\prod_{j=1}^{n} \underline{J}^{\mathrm{T}}(q(j\tau))\right]\cdot\left[\prod_{k=1}^{n} \underline{J}(q(k\tau))\right]\xi(0)}{\xi^{\mathrm{T}}(0)\cdot\xi(0)} \right]^{1/2} \right\rangle . \tag{2.139}$$

Now, if with a suitable time sampling the dynamics mimics a good random process then both Ω and $\underline{J}$ can be considered as random matrices. For the FPU-β model, the matrix Ω can be expressed as the sum of a tridiagonal constant matrix Ω_0 and of an implicitly time-dependent tridiagonal matrix $\widetilde{\Omega}$. The matrix elements of Ω contain combinations of terms like $(q_{i+1} - q_i)^2$. In the random matrix approximation, the hypothesis of δ-correlation in time is made for the fluctuating part $\widetilde{\Omega}$ of Ω, i.e. $\langle \widetilde{\Omega}_{ij}(k\tau)\,\widetilde{\Omega}_{ij}(l\tau)\rangle = (\gamma_{ij}/\tau)\delta_{kl}$. The average $\langle\cdot\rangle$ in (2.139) is carried over different realizations of the random matrix process. The theoretical computation of (2.139) yields [53]

$$\lambda \sim \tau^{2/3}\gamma^{1/3}, \tag{2.140}$$

where γ is defined as

$$\gamma = \frac{1}{N}\sum_{i=1}^{N} \left\langle [(q_{i\pm1} - q_i)^2 - \langle (q_{i\pm1} - q_i)^2\rangle]^2 \right\rangle . \tag{2.141}$$

By computing a statistical ensemble average of $\gamma(\varepsilon)$, one finds $\gamma(\varepsilon) \sim \varepsilon^2$ in an *intermediate* ε-range, as outlined below. Thus, from (2.140), $\lambda \sim \varepsilon^{2/3}$ if ε is not too large and if τ is *independent* of ε. We remark that this result is *not asymptotic*. The microcanonical measure should be used, but the canonical measure can equivalently work at large N. The canonical configurational partition function

$$Z_{\mathrm{C}}^{(N)} = \left[\Gamma\left(\frac{1}{2}\right)\left(\frac{\eta\beta}{2}\right)^{-1/4} \exp\left(\frac{\alpha^2\eta}{8\beta}\right) D_{-1/2}\left(\alpha\sqrt{\frac{\eta}{2\beta}}\right)\right]^{N},$$

where α is a dummy parameter multiplying the harmonic part of the FPU potential, $D_{-1/2}$ is a parabolic cylinder function, $\eta = 1/T$ ($k_{\mathrm{B}} = 1$) is the inverse of the average kinetic energy per particle which is, within a good approximation, proportional to ε thus $\eta \simeq 1/\varepsilon$. The ensemble average $\langle\gamma\rangle$ is then given by

$$\langle\gamma\rangle = Z_{\alpha=1}^{-1}\left[-\frac{4}{\eta}\left(\frac{\partial Z}{\partial\beta}\right)_{\alpha=1}\right] - \left[Z_{\alpha=1}^{-1}\left(-\frac{2}{\eta}\right)\left(\frac{\partial Z}{\partial\alpha}\right)_{\alpha=1}\right]^2,$$

and using the asymptotic approximation $D_{-1/2}(x) \sim \exp(-x^2/4)x^{-1/2}(1 - 3x^2/8 + \cdots)$ that holds good at $x \gg 0$ one immediately gets

$$\langle\gamma\rangle(\varepsilon) \sim \varepsilon^2 .$$

At high ε, where the above expansion for $D_{-1/2}$ worsens, the exponent 2 is lowered and consequently also the 2/3 exponent of $\lambda(\varepsilon)$ is lowered. In fact, the numerical results for $\lambda(\varepsilon)$ and the analytic results worked out by means of the geometric approach, outlined in Sect. 2.4.7 and applied to the FPU-β model in Sect. 2.8.2, give at high ε the asymptotic scaling $\lambda(\varepsilon) \sim \varepsilon^{1/4}$.

In conclusion, the transition from the low-energy density scaling $\lambda(\varepsilon) \sim \varepsilon^2$ to the scaling $\lambda(\varepsilon) \sim \varepsilon^{2/3}$ is attributed to the transition from two different regimes of chaoticity because the $\lambda(\varepsilon) \sim \varepsilon^{2/3}$ scaling is accounted for by the RMA for the tangent dynamics at moderately high ε values.

2.8.2 Geometric Calculation of λ for FPU-β

A completely analytical computation of $\lambda(\varepsilon)$ has been performed—in the $N \to \infty$ limit—for the FPU β-model using the geometric method sketched in Sect. 2.4.7. As we shall see below, the agreement is strikingly good. Particularly noticeable is the fact that the analytic values of λ check with the numerical ones within errors of few percent in a range of six orders of magnitude both in ε and λ, with no adjustable parameters. The analysis follows [60] and has been reviewed in [42].

The geometric quantities appearing in (2.125), that is Ω_0 and σ_Ω, written in the Eisenhart metric assume the simple form

$$\Omega_0 = \langle k_R \rangle_{\Sigma_E} = \frac{1}{N}\langle \triangle V \rangle_{\Sigma_E} \,, \tag{2.142}$$

$$\sigma_\Omega^2 = \langle \delta^2 k_R \rangle_{\Sigma_E} = \frac{1}{N}\left(\langle (\triangle V)^2 \rangle_{\Sigma_E} - \langle \triangle V \rangle_{\Sigma_E}^2\right) \,, \tag{2.143}$$

where $\triangle V$ is the euclidean Laplacian of the potential function $V(q)$.

Now, the microcanonical ensemble averages of $K_\mathrm{R}(q)$ and of its variance can be computed in terms of the corresponding quantities in the canonical ensemble as follows. The canonical configurational partition function $Z(\eta)$ is given by

$$Z(\eta) = \int \mathrm{d}q\, \mathrm{e}^{-\eta V(q)}, \tag{2.144}$$

where $\mathrm{d}q = \prod_{i=1}^N \mathrm{d}q_i$. The canonical average $\langle K_\mathrm{R} \rangle_\mathrm{can}$ of the Ricci curvature K_R follows as

$$\langle K_\mathrm{R} \rangle_\mathrm{can} = [Z(\eta)]^{-1} \int \mathrm{d}q\, K_\mathrm{R}(q)\, \mathrm{e}^{-\eta V(q)} \;. \tag{2.145}$$

From this average, we can obtain the microcanonical average of K_R, $\langle K_\mathrm{R} \rangle_{\Sigma_E}$, in the following (implicit) parametric form [172]

$$\left.\begin{aligned} \langle K_\mathrm{R} \rangle_{\Sigma_E}(\eta) &= \langle K_\mathrm{R} \rangle_\mathrm{can}(\eta) \\ \varepsilon(\eta) &= \frac{1}{2\eta} - \frac{1}{N}\frac{\partial}{\partial \eta}[\log Z(\eta)] \end{aligned}\right\} \to \langle K_\mathrm{R} \rangle_{\Sigma_E}(\varepsilon) \tag{2.146}$$

Note that (2.146) is strictly valid only in the thermodynamic limit; at finite N, $\langle K_{\rm R}\rangle_{\Sigma_E}(\eta) = \langle K_{\rm R}\rangle_{\rm can}(\eta) + {\rm O}(\frac{1}{N})$.

Contrary to the computation of $\langle K_{\rm R}\rangle$, which is insensitive to the choice of the probability measure in the $N \to \infty$ limit, computing the fluctuations of $K_{\rm R}$, i.e., of $\langle \delta^2 K_{\rm R}\rangle = \frac{1}{N}\langle (K_{\rm R} - \langle K_{\rm R}\rangle)^2\rangle$, by means of the canonical or microcanonical ensembles yields different results. The relationship between the canonical—i.e. computed with the Gibbsian weight $e^{-\beta H}$—and the microcanonical fluctuations, is given by the Lebowitz–Percus–Verlet formula [172]

$$\langle \delta^2 K_{\rm R}\rangle_{\Sigma_E}(\varepsilon) = \langle \delta^2 K_{\rm R}\rangle_{\rm can}(\eta) - \frac{\eta^2}{c_V}\left[\frac{\partial \langle K_{\rm R}\rangle_{\rm can}(\eta)}{\partial \eta}\right]^2 , \tag{2.147}$$

where

$$c_V = -\frac{\eta^2}{N}\frac{\partial \langle H\rangle_{\rm can}}{\partial \eta} \tag{2.148}$$

is the specific heat at constant volume and $\eta = \eta(\varepsilon)$ is given in implicit form by the second equation in (2.146).

For the FPU-β model, the Ricci curvature, written in the Eisenhart metric, simply reads

$$K_{\rm R} = 2N + 6\beta \sum_{i=1}^{N} (q_{i+1} - q_i)^2 . \tag{2.149}$$

Note that $K_{\rm R}$ is always positive. By taking advantage of the analytically known form of $Z(\eta)$ for the one-dimensional FPU-β model, one can exactly compute the analytic expressions of the microcanonical averages of (2.146) and (2.147) in the $N \to \infty$ limit. With Ω_0 and σ_Ω one computes τ and, substituting $\Omega_0(\varepsilon)$, $\sigma_\Omega(\varepsilon)$, and $\tau(\varepsilon)$ into (2.125), we finally get $\lambda(\varepsilon)$, which is reported in Fig. 2.34. As anticipated at the beginning of the present Section, the analytic values of λ are in excellent agreement with the numerical ones.

Other systems for which good results have been obtained are coupled rotators [60], as shown in Fig. 2.35, classical XY Heisenberg models [173], φ^4 models [141, 174], and the "mean-field" XY model [175], although some adjustments are necessary in these cases, mainly related with the nontrivial topology of the mechanical manifolds.

2.8.3 Estimates of Time to Equipartition with Strong Arnold Diffusion

In the energy regime in which the diffusion from a chaotic low-frequency driving resonance can efficiently couple to a high-frequency resonance, i.e. when the ratio of the driving to the driven frequencies $\omega_{\rm B}/\delta\omega_h \sim 1$, an analytic estimate of the equipartition time can be made [58]. The estimate predicts a scaling $T_{\rm eq} \propto \varepsilon^{-3}$, $\varepsilon \equiv E/N$, independent of E and N separately, for the FPU-β chain. The method has also been applied to estimate the equipartition time for energy placed initially in short-wavelength (high-frequency) modes,

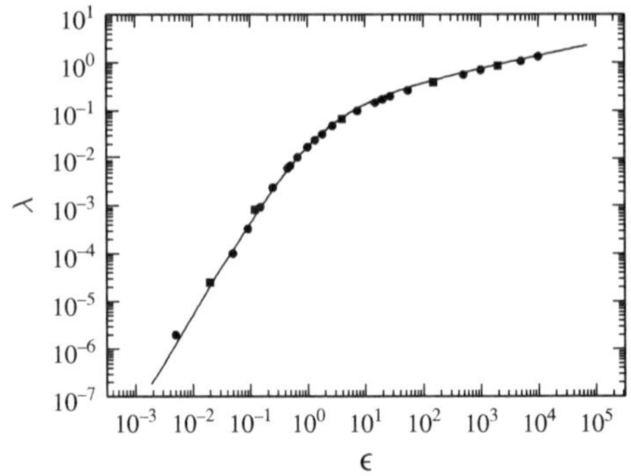

Fig. 2.34. Lyapunov exponent λ vs. energy density ε for the FPU-β model with $\beta = 0.1$. The *continuous line* is the theoretical computation according to (2.125), while the *circles* and *squares* are the results of numerical simulations with N respectively equal to 256 and 2000

obtaining a scaling $T_{\mathrm{eq}} \sim \varepsilon^{-2}$ [92]. These scalings agree with those found numerically, and furthermore give values of T_{eq} well within a factor of 10 of the numerical values. The basic method of making the estimation is outlined below.

The assumption is that there is an effective number of driving modes δk, dependent on the total energy, that transfer energy efficiently through the Arnold diffusion mechanism, i.e. satisfying (2.50). Considering that $E_\gamma \simeq E/\delta k$, then the number of interacting modes δk is, from (2.50),

$$\delta k = \mu \beta E \tag{2.150}$$

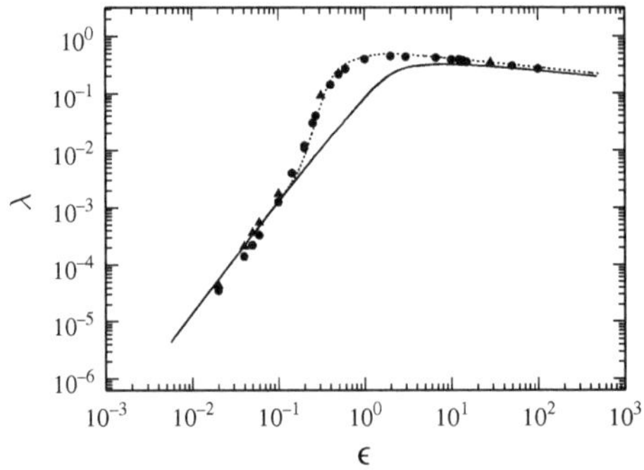

Fig. 2.35. Lyapunov exponent λ vs. energy density ε for the one-dimensional XY model. The *continuous line* is the theoretical computation according to (2.125), while *full circles, squares*, and *triangles* are the results of numerical simulations with N, respectively, equal to 150, 1000, and 1500. The *dotted line* is the theoretical result where the value of Ω_0 entering (2.125) has been corrected in order to empirically account for nontrivial configuration space topology

where μ is of order unity. The procedure is then to calculate the rate that energy is extracted from the modes containing energy, to be distributed among all modes, with an integration over time until all modes have equal energy (at equipartition). Since the energy is spread among many modes, we can take the energy to be primarily in the linear part of the Hamiltonian. To be specific we consider the case of transfer from long wavelengths to short wavelengths.

The change in the linear energy $E_i = (1/2)\left(P_i^2 + \omega_i^2 Q_i^2\right)$ of a driving low-frequency mode i, can be calculated from the action-angle form of the Hamiltonian by taking the derivative with respect to the angle θ_i

$$\frac{\mathrm{d}E_i}{\mathrm{d}t} = \left(\frac{-2\beta}{N+1}\right)\omega_i \sum_{j,h_1,h_2} B\left(i,j,h_1,h_2\right) \sqrt{\omega_i\omega_j\omega_{h1}\omega_{h2}I_iI_jI_hI_{h2}}\sin\theta_i \mathrm{ang}\left(j,h_1,h_2\right), \tag{2.151}$$

where $\mathrm{ang}(j,h_1,h_2)$ is a product of cosines of the angle variables. The notation $h1$ and $h2$ explicitly indicates that the energy transfer occurs between a low-frequency beat oscillation and a high-frequency mode difference oscillation through the Arnold diffusion mechanism. In the above equation, the summation is over indices j, h_1, and h_2 for a given i. The only terms to transfer energy to high-frequency modes are the ones where $j = i$, since then the product of the two low-frequency angles does not have a fast phase associated with it. Additionally, the selection rule requires that $B = 0$ unless

$$2i + h_1 + h_2 = 2N + 2\ , \tag{2.152}$$

which reduces the index to a single sum. This result for a single low-frequency mode i is an estimate for the average energy decay rate, which, from (2.151) is

$$\frac{\mathrm{d}E_i}{\mathrm{d}t} = -\left(\frac{2\beta}{N}\right)\omega_i\frac{\beta E}{2\pi}E_iE_h\left(t\right), \tag{2.153}$$

where $\omega_i = \pi i/N$. Since δk low-frequency modes, assumed to have energy, couple to δh high-frequency modes with $\delta k = \delta h$, the cross couplings imply each high-frequency mode is coupled on average to $\delta k/4$ low-frequency modes. The effect of the interaction of the phases between driving modes, when more than one driving term exists, has not been studied. The simplest assumption is that the effect from each low-frequency driving resonance is independent. Setting $\omega_i = \beta E/N$ ($i = \delta h/2 = \beta E/\pi$), and dividing by E_i, we obtain, an average, for each mode in the package

$$\frac{\mathrm{d}E_i}{E_i} = -\frac{\beta}{\pi}\left(\frac{\beta E}{N}\right)^2 E_h\left(t\right)\mathrm{d}t\ . \tag{2.154}$$

Integrating (2.154) in time, with $E_i(t)$ varying from $E/\delta k$ at $t = 0$ to the equipartition value E/N at the final time $T_{\rm eq}$ one obtains

$$\ln\left(\frac{N}{\delta k}\right) \simeq \left(\frac{\beta}{\pi}\right)\left(\frac{\beta E}{N}\right)^2 \int_0^{\rm T} E_h(t')\mathrm{d}t' , \qquad (2.155)$$

The final step in the approximation is to estimate the value of $\int_0^{\rm T} E_h(t')\mathrm{d}t'$ at $T = T_{\rm eq}$, a time of "near-equipartition." The quantity $E_h(t)$ appears in an integral, so that its exact form is not required. For a diffusive process in which the amplitudes of the modes increase with $t^{1/2}$, we might expect the mode energies to increase linearly with $t, E_h(t) \simeq (t/T)(E/N)$, such that the time dependence is independent of N. This is found to be approximately true, numerically, over most of the evolution to near-equipartition. Other forms of the time dependence of E_h lead to only small numerical differences. Evaluating the integral with the assumption of linear time dependence of $E_h(t)$ we obtain

$$T_{\rm eq} \simeq \frac{2\pi}{(\beta E/N)^3} \ln\left(\frac{\pi}{2\beta E/N}\right) . \qquad (2.156)$$

The numerical coefficient is only a rough estimate. Equation (2.156) exhibits a basic scaling of $T \propto (N/E)^3$. The scaling has been checked numerically, by rescaling the time as shown in Fig. 2.13. This scaling was also found by plotting the time to reach $n_{\rm eff}/N = 0.4$ against E/N. However, it was also pointed out that a "best fit" stretched exponential also fit the data (see Fig. 2.14). One can also compare the magnitude of $T_{\rm eq}$ in (2.156) with the numerics. From Fig. 2.15, taking $E/N = 0.05$ we find $T_{\rm eq} \sim 10^7$. Considering the many approximations, this value is reasonably close to the value of $T_{\rm eq} \simeq 3 \times 10^7$ obtained from (2.156).

The same calculation method can be modified to determine the scaling and estimate the value of $T_{\rm eq}$ from high-frequency mode initial conditions. The beat frequency in this case is the difference between the breather nonlinear frequency and the linear-mode frequencies, which are again proportional to $\beta\varepsilon$ as in the above analysis. However the high frequency is $\omega_i \simeq 2$ and thus a factor of $\beta\varepsilon$ is removed from the drive, which gives the basic scaling of $T_{\rm eq} \propto (\beta\varepsilon)^{-2}$. This scaling was confirmed numerically, and the theoretic and numerical values were again in reasonable agreement for high frequencies [92, 93, 94], as summarized in Sect. 2.9.2.

As previously noted, on-site potentials, as in the ϕ^4 chain, involve additional complications, and the comparisons were correspondingly less definitive. In particular, in [65] the m-scaling was estimated from the calculation to be related to the ε scaling by the factor ε/m^2, in contrast to the numerical result found in Fig. 2.18 which suggested $\varepsilon/m^{2.75} =$ const. to produce the best scaling.

2.9 Analytical Calculations and Estimates: Scaling Estimates from High Frequencies

2.9.1 Estimate of Coalescence into Single Chaotic Breather

It is not possible to make a complete quantitative calculation for the very complicated process of chaotic breather coalescence; however, arguments have been made for the FPU chain [91, 92] and for the ϕ^4 chain [94] that predict the scaling of the coalescence time with energy density. We review the main arguments without details. The coalescence time-constant τ_{B} is estimated from the standard description

$$\tau_{\mathrm{B}} = \frac{1}{n_{\mathrm{B}} \sigma_{\mathrm{B}} v_{\mathrm{B}}} , \tag{2.157}$$

where $n_{\mathrm{B}} = k_m N/2\pi$ is the initial number of breathers formed earlier in time by the instability, which scales with ψ_m , v_{B} is scaled by application of the virial theorem to a moving breather and σ_{B} is scaled from the Born approximation. For the FPU chain, the asymptotic scalings at small and large amplitude for each factor, are

$$n_{\mathrm{B}} \propto \begin{cases} \psi_m \\ \text{const.} \end{cases}, \qquad v_{\mathrm{B}} \propto \begin{cases} \psi_m \\ \text{const.} \end{cases}, \qquad \sigma_{\mathrm{B}} \propto \begin{cases} \psi_m^2, \text{ small amplitude} \\ \psi_m^4, \text{ large amplitude} \end{cases} \tag{2.158}$$

such that, taking the product in (2.157), we obtain a coalescence rate scaling with amplitude

$$\tau_{\mathrm{B}} \propto \psi_m^4 \tag{2.159}$$

essentially independent of the amplitude. Since numerically one can use energy as our independent variable, ψ_m can be related to E through the breather equilibria, as given in the above references, which vary from low to high energy, giving $\psi_m \propto E^{1/2}$ and $\psi_m \propto E^{1/4}$, for the FPU$-\beta$ at low and high energies, respectively. (The ϕ^4 proportionalities are similar but not exactly the same). Using the proportionalities above, for the FPU-β in (2.159), we obtain

$$\tau_{\mathrm{B}} \propto \begin{cases} E^{-2} \text{ small amplitude} \\ E^{-1} \text{ large amplitude} \end{cases} . \tag{2.160}$$

The results in (2.160) have been nummerically verified at the higher amplitude in [88] with some indication also of the steeper low amplitude scaling. The ϕ^4 chain has slightly different scaling and also some additional complications as discussed in the original paper [94]. It was also noted in that paper that n_{B} is not necessarily determined by the fastest growing unstable mode, for all initial conditions. In fact, most of the FPU numerics were performed with an initial mode $\gamma = 120$, for which $n_{\mathrm{B}} \simeq 9$, initially, independent of ψ_m. However, since most initial conditions were taken at sufficiently large ψ_m such that n_{B} is approximately constant, as calculated from the fastest growing mode, the scaling was the same for the two cases.

2.9.2 Estimate of Time to Equipartition

As already mentioned in Sect. 2.8.3, an estimate of the $T_{\rm eq}$ can be made, starting from high frequencies, using the same general procedure, as used there, for initial low frequencies. The main differences are
(i) That the stochastic beat frequency $\Omega_{\rm B}$ is given by

$$\Omega_{\rm B} = \omega_{\rm B} - \omega_h \, , \tag{2.161}$$

the difference between the breather frequency $\omega_{\rm B}$ and the high freequencies ω_h, where $\omega_h \simeq 2$, and from (2.89), $\omega_{\rm B} \simeq (4+6\beta\psi_m^2)^{1/2}$. The beat frequencies can be calculated analytically, but are quite complicated over the range of energies explored numerically. A good approximation was found to be

$$\Omega_{\rm B} \simeq 0.2\beta E_{\rm B} \, , \tag{2.162}$$

which was used in the calculation [92].
(ii) For faster than exponential Arnold diffusion the CB energy $E_{\rm B}$ needs to be sufficiently large that

$$\Omega_{\rm B}/\delta\omega_l > 1 \tag{2.163}$$

with the minimum $\delta\omega_l = \pi/N$ at low frequencies. Since this value is much larger than the minimum $\delta\omega_h$, the energies required to obtain stochastic diffusion from high to low frequencies is also larger, as seen numerically.
(iii) However, this is partly cancelled by the fact that $\omega_{\rm B}$ in (2.162) does not depend on a mode frequency that varies as N^{-1}. Because of the absence of the additional factor the resultant scaling is $T_{\rm eq} \propto \varepsilon^{-2}$, rather than the $T_{\rm eq} \propto \varepsilon^{-3}$ scaling for low energies. We do not repeat the calculation, which is, after these modifications, similar to that presented in Sect. 2.8.3, with the final result for the FPU-β chain [92]

$$T_{\rm eq} \simeq \frac{80\pi}{5}\left(\frac{\beta E}{N}\right)^{-2} . \tag{2.164}$$

The scaling of ε^{-2} has been checked numerically [88] with the absolute numerical time approximately a factor of 5 longer that that predicted by (2.164).

A similar calculation has been performed for the ϕ^4 chain [94]. There are some additional complications, and the resultant scaling is not precisely ε^{-2} but varies with somewhat steeper scaling for the lowest values of ε. We have already seen these results numerically in Sect. 2.6.2. These difficulties have been considered, resulting in a more complicated estimate for the scaling of $T_{\rm eq}(\varepsilon)$ than that given in (2.164) for the FPU-β chain.

Another question that arises is whether for high-frequency initial conditions there is a clear transition, with increasing energy from $T_{\rm eq} \propto \varepsilon^{-2}$ to weaker ε scalings as found for low-frequency initial conditions. Clearly there is a change of scaling at higher specific energy, as observed numerically in an extreme case, shown in Fig. 2.24. There is also numerical evidence for such transitions for both the FPU-β and ϕ^4 chains in Fig. 2.25. The transition has not been investigated in detail, to determine whether it is relatively sharp, i.e. a SST, or more diffuse. Since $\tau_B \propto \varepsilon^{-1}$ and $T_{eq} \propto \varepsilon^{-2}$, one would expect some type of significant transition at an ε sufficiently high that the two time-scales cross, such that a CB is not formed. Extrapolating the data in, for the FPU-β chain with E varied and $N = 128$, the cross-over occurs at $\varepsilon \simeq 4.5$ ($\beta = 0.1$). But this range of ε has not been numerically investigated. For the more complicated ϕ^4 chain there is no numerical evidence for a SST, but again there have been no computations in the ε-range for which such a transition might be expected.

2.10 Conclusions and Final Comments

It often happens that the failure of numerical or experimental results to substantiate theoretical predictions leads to the productive development of new physics. So it was with the Fermi–Pasta–Ulam (FPU) problem in which a one-dimensional chain of masses was connected to its nearest neighbors by nonlinear restoring forces (2.1). Enrico Fermi initially suggested the problem as a method of confirming his prediction that the nonlinearity would lead to equipartition among the degrees of freedom, therefore leading to a dynamical underpinning for statistical mechanics. However, the coupled differential equations, numerically integrated on a first-generation large digital computer, gave the seemingly contrary result that the energy initially placed in the lowest of the harmonic normal modes resulted in periodic (or near periodic) energy oscillations among the first few modes (see Fig. 2.1). The oscillations were soon explained from perturbation theory as the nonlinear beating among neighboring modes, but the fundamental question of whether equipartition would eventually be reached was not answered. The early attempts to analyze the dynamics led to better ways of employing perturbation theory and to better understanding of nonlinear mode coupling.

The most celebrated result, following the lack of confirmation of statistical properties of the FPU dynamics, was the development of soliton theory. In an attempt to understand the apparent stability of recurrences Norman Zabusky and Martin Kruskal found a Taylor expansion of the discreteness, valid for long-wavelength modes, that recovered partial differential equations different from the original nonlinear spring which produced the discretized chain of oscillators. The resulting equations are the Korteveg–de Vries (KdV) equation for the FPU-α chain with cubic nonlinearity, and the modified Korteveg–de Vries (mKdV) equation for the FPU-β chain with quartic nonlinearity. Nonlinear equations of this and related types had been known to have stable traveling solutions where the dispersion and nonlinearity balance to produce constant amplitude and propagation velocity. An arbitrary initial condition, such as the lowest linear mode on the FPU-β chain breaks up initially into a set of structures each having a steady traveling solution with its own velocity. Remarkably, these structures are sufficiently stable that they pass through one another without breaking up, and the observed recurrences can be interpreted in terms of their superpositions. But these results did not improve on the best perturbation calculations and are clearly limited to long-wavelength (low-frequency) modes by the approximations which led to (2.6). Partial differential equations have an infinite number of freedoms, such that general integrability from arbitrary initial conditions requires an infinite number of invariants of the motion. The real excitement came when it was shown that such an infinite set of invariants exists and the new field of soliton theory and applications was born. A single initial nonlinear mode solution of the mKdV equation is found to become unstable as the energy is increased. A linearization around the nonlinear structure predicted the unstable wave numbers and growth rates, and showed that the values correspond to the observed mode growth for the same discretized structure on the FPU-β chain (see Sect. 2.4.3). The result in which one soliton decomposes into a finite number is not inconsistent with general soliton theory, but the instability gives us insight into the dynamics that leads to chaotic motion and, ultimately, to equipartition.

Another important consequence of the numerical and theoretical work, which attempted to both explain the original results and extend them to other energies, initial conditions, and force laws, was that of relating high-dimensional oscillator chain dynamics to low-dimensional chaos theory. The development of KAM theory by Kolmogorov, Arnold, and Moser showed that despite the lack of global integrals of the motion in generic systems of two or more degrees of freedom (which in fact, motivated Fermi's initial FPU study), for small perturbations from integrable systems most of the phase space could still be regular. But countering this, another result by V. I. Arnold showed that generic systems of three or more degrees of freedom had stochastic resonance channels in the phase space that could reach close to any portion of the phase space. Furthermore, a heuristic understanding of high-dimensional systems indicated that the fraction of the phase volume that is stochastic increases with increasing number of freedoms. Investigation of the rates of diffusion through the stochastic web indicated that the "Arnold diffusion" was normally exponentially slow in the perturbation parameter, but could become large if the perturbation became large (see [15], Chap. 6). The results suggested that, for N freedoms and with fixed perturbation strength (fixed energy density), near equipartition would probably be obtained with increasing N (the thermodynamic limit), but the time-scales might be exponentially long and unapproachable computationally.

In addition to the low-dimensional theory that contributed to the understanding of high-dimensional Hamiltonian dynamics, the development of statistical measures were essential to the numerical investigations that elucidated the system behavior. Although various methods have been useful, the most important have been the calculation of Lyapunov exponents for the separation of neighboring trajectories, and the information entropy which qualitatively computes the number of modes taking part in the dynamics (see Sects. 2.1.6 and 2.3). To obtain equipartition it is necessary for the dynamics to be ergodic.

Mixing, which implies ergodicity, is a property that holds if all trajectories are on the average exponentially separating (positive KS entropy). This cannot be exactly true for a divided phase space, and thus it might be qualitatively stated that, if almost all of the space has positive KS entropy, then for practical purposes equipartition will be reached. Because the criterion depends on exponentially diverging trajectories, an obvious set of quantities to be examined are the Lyapunov exponents, giving the separation of trajectories, with positive values indicating exponential separation of trajectories, i.e. chaos. The numerical calculation of Lyapunov exponents has been used extensively to test for chaotic motion, as we have reviewed in Sect. 2.5.

Although the calculation of Lyapunov exponents is important in finding a necessary condition for obtaining equipartition, a more direct quantity for determining if equipartition is actually reached, and also for determining the time-scale to reach it, is the information entropy (see Sects. 2.1.6 and 2.3). Along with other statistical tools, computations of the information entropy have been the backbone of the numerical observations, as we have reviewed in Sects. 2.5 and 2.6. As described in Sect. 2.3.1, the information entropy can be calculated for either modes or oscillators, with the mode description being most useful if the initial conditions are from long wavelengths (low frequencies); the numerical results are given in Sect. 2.5. The oscillator description is most useful from short-wavelength (high-frequency) initial conditions, with numerical results in Sect. 2.6.

An important early breakthrough in understanding the onset with increasing energy of observable chaos, and the approach to equipartition on numerically observable time-scales, was made by the application of Chirikov's "overlap criterion" to the FPU system by Izrailev and Chirikov. The somewhat heuristic criterion was very useful in finding transitions from localized to extensive chaos in low-dimensional chaos. Estimates in the FPU system were made both for long wavelengths and for short wavelengths to determine the "overlap" of neighboring modes. However, numerical results demonstrated that mode overlap of long-wavelength modes is not a necessary condition for equipartition but approximates a transition between weak and strong stochasticity (the SST). The result for short-wavelength modes is neither necessary nor sufficient. It predicts easy overlap at short wavelengths due to mode crowding, while numerics show consistently that equipartition is more readily obtained from long-wavelength than from short-wavelength initial conditions.

A partial resolution of the discrepancy between numerical observations of transitions to observable equipartition and the analytic "overlap" estimates came with the recognition that resonances among groups of modes would satisfy an overlap criterion at much lower energies than overlap of the neighboring modes directly. We have described the method of calculating this behavior, both for the FPU-β chain with quartic nonlinearity, and the FPU-α chain with cubic nonlinearity, in Sect. 2.4.1. The results agree with numerical observations, predicting that packets of long-wavelength modes can overlap, once the energy is above a threshold, and determining scaling of the packet with energy and semi-quantitively determining its size. The scaling of the packet size can also be predicted by applying dimensional arguments to the partial differential equation approximation.

However, local interaction does not give a complete picture of the processes by which equipartition can be reached on numerically observable time-scales. To know how energy escapes from a mode packet at low frequencies requires the understanding of the Arnold diffusion mechanism in which energy can be transferred along resonances from long-wavelength to short-wavelength modes. We have briefly described this condition in Sect. 2.4.1, but a complete understanding requires a careful reading of the original work (see [55]). Once the criterion for strong (nonexponentially slow) Arnold diffusion is well satisfied, an estimate of the scaling of the time to equipartition can be made, valid in the thermodynamic limit. We gave a summary of the estimation method in Sect. 2.8.3 which agrees well with numerical results for the FPU-β oscillator chain, as presented in Sect. 2.5. The scaling of the equipartition time with the inverse third power of the energy density, $T_{\mathrm{eq}} \propto \varepsilon^{-3}$ also agrees well with numerical results from the FPU-α chain and for the ϕ^4 chain which has an onsite quartic potential.

A calculation of the Lyapunov exponent as a function of the energy density can also be made using concepts from the geometrization of Hamiltonian mechanics. We have reviewed the theory in Sect. 2.4.7 and presented the specific calculation for the FPU-β oscillator chain in Sect. 2.8.2. A complete exposition of the geometric method and its specific application to the FPU-β has been given in a review [42]. From the definition in (2.5), the Lyapunov exponent is determined by a long-time average, and is thus applicable from any initial condition. The comparison is most easily made with numerics starting from near-equipartition, which minimizes the effect of transients. By using this method, the scaling of $\lambda \propto \varepsilon^2$ was determined for the FPU-β in the region of energy density for which $T_{\mathrm{eq}} \propto \varepsilon^{-3}$ and the transition to the strong stochasticity regime, where the diffusion is across resonances, was also determined. Remarkably, the theoretically determined absolute values of λ are in excellent agreement with the numerical values.

The dynamics of the evolution from short wavelengths (high frequencies) is quite different than from the long wavelengths. The evolution, starting from a high-frequency mode initial condition, occurs at higher energy and on a slower time-scale than from energy initially in a low-frequency mode. A partial understanding of the increased stability of high-frequency modes comes from the analysis of breather-like structures on discrete systems that admit exact breather solutions. High-frequency mode initial conditions have symmetry of neighboring oscillators close to that of localized exact breathers. The resulting dynamics consists of three stages. First, there is an initial stage in which the mode breaks up into a number of breather-like structures. Second, on a slower time-scale, these structures coalesce into one large unstable structure, called a chaotic breather (CB). Third, the CB slowly disintegrates, resulting in equipartition. Since a single large CB closely approximates a stable breather, the final decay stage, toward equipartition, can be very slow. If the energy was placed in the highest frequency mode, with a periodic boundary condition that has strict alternation of the amplitudes from one oscillator to the next, the configuration is stable up to a particular energy at which a parametric instability occurs, leading to the events described above. However, the nonlinear evolution does not depend on special initial conditions, but generically evolves from any high-frequency mode initial condition that has predominately the alternating amplitude symmetry. One does not know, in this generic situation, whether there exists any true energy threshold to achieving equipartition. However, as discussed extensively with respect to low-frequency mode initial conditions, the practical thresholds refer to observable time-scales. The scaling with energy density of the time to equipartition is estimated for high-frequency initial conditions from the beating of the breather with the background, using the procedure developed to calculate the equipartition time from low-frequency initial conditions. The result gives the numerically observed scaling to equipartition of $T_{\mathrm{eq}} \propto \varepsilon^{-2}$, and reasonable quantitative agreement with the numerically determined values.

Considerable insight into the behavior of a nonlinear oscillator chain, starting from high-frequency mode initial conditions, can be obtained by introduction of an envelope function for the displacements of the oscillators. The initial conditions for the envelope function only contain significant long-wavelength perturbations. For the envelope function, an expansion is then possible to obtain a nonlinear partial differential equation (PDE) which approximates the behavior of the discrete system. Low-order expansions of this type produce PDEs that have integrable solutions in the form of envelope breathers, analogous to the solutions produced from low-frequency initial conditions. Higher order terms destroy the integrability, but we find, as expected, that discretized oscillator chains form localized structures that approximate the breathers on continuous systems, but are weakly unstable. The process by which the CBs coalesce into a single CB has also been estimated theoretically. On average, the large structures absorb energy from the smaller ones, as expected from general theoretical considerations. The time constant for coalescence into a single CB is estimated from the relation $\tau_{\mathrm{B}} \approx (n_{\mathrm{B}}\sigma v_{\mathrm{B}})^{-1}$ where n_{B} is the breather number, σ is a collision cross-section for absorption, and v_{B} is a characteristic breather velocity. Using this procedure in the numerically investigated energy range, reasonable agreement with the numerical scalings of $\tau_{\mathrm{B}} \propto \varepsilon^{-1}$ is obtained. The numerical results are presented in Sect. 2.6 and the analytic estimate for comparison to the numerics in Sect. 2.9.

To explore which processes are generic and which model dependent, it is necessary to investigate other oscillator chains which are related to the FPU chain but have significantly different parameters. The discretized Klein–Gordon equation with quartic nonlinearity (the ϕ^4 model) is such a chain. The FPU-β chain, with only an intersite potential, is translationally invariant (except for boundaries) and it is energy-rescalable on a single parameter. In contrast, the ϕ^4 chain has an additional on-site potential which adds a parameter [compare (2.1) with (2.9)], and is therefore more complicated to analyze and explores a larger parameter space. Nevertheless, a comparison of the ϕ^4 chain to the FPU-β chain reveals similarities and explainable differences. For a small-value linear on-site restoring force $m = 0.1$, of the individuals oscillators, the linear frequencies of most modes are similar in the ϕ^4 and FPU chains. Starting from long-wavelength (low-frequency) modes the numerical calculations of $T_{\text{eq}}(\varepsilon)$ show similar behavior of the two chains, having power-law behavior, depending only on ε over a wide range of ε, and N-dependent transitions at low ε to values of T_{eq} that may increase exponentially with decreasing ε. For larger m (but $m < 1$) the value of ε required to achieve a given T_{eq} is much larger with strong m-scaling, which can be qualitatively predicted from a heuristic argument. For short wavelengths (high frequencies) the behavior of the two chains is similar and a simple argument predicts that the same value of T_{eq} can be obtained for ϕ^4 as for the FPU if the ϕ^4 value of ε is a factor of 4 larger. The reasoning is that since the phases of neighboring oscillators alternate for the FPU, the quadratic term is factor of 16 larger, which is larger by a factor of 4 in energy; thus the energy in ϕ^4 should be a factor of 4 larger to bring T_{eq} into correspondence. The predicted factor of 4 separation for the higher values of ε (higher E) is found when the scaling follows ε^{-2} for both potentials, but the separation becomes larger as the ϕ^4 potential scaling becomes steeper at low energy densities, due to additional correlations.

The knowledge that nonlinear oscillator chains, such as the FPU, produced stochastic dynamics, encouraged the idea that they would reproduce the Fourier heat law. But this was not the case. The configuration for studying heat conductivity is different from that for studying equipartition, in that the former is not a closed conservative system, but must be connected to heat baths which emit and absorb energy. Since the steady state is not an equilibrium, the dynamics involves short-time effects. The lack of normal thermal conductivity for the FPU-β chain was found to be due to the excitation of nonlinear waves, which were not diffusive. Momentum conservation in the FPU chains is a key ingredient in preventing the stochasticity from producing the required diffusive energy flow. This led to the exploration of models which included substrate potentials, such as various Klein–Gordon models, which do produce normal heat conduction when their dynamics is primarily stochastic. The history of these developments is given in Sect. 2.1.10, and the basic theoretical ideas and numerical results are given in Sect. 2.7.

So what do we know about the generic properties of oscillator chains and their implications for physics; and what are the outstanding problems that have not been fully addressed? First, we can say that Fermi's original intuition about the role of nonlinear dynamics in underpinning statistical physics has essentially been proved correct. For large systems (approximating the thermodynamic limit) and at reasonably large energy densities, generic nonlinear oscillator chains with nearest neighbor coupling dynamically exhibit stochastic diffusion leading to equipartition among oscillators and modes in isolated systems. Over a wide range of energy densities the Lyapunov exponents, measuring the exponential separation of trajectories, and the resulting times to achieve equipartition are both proportional to inverse power laws of the energy density, but with different exponents. For the most thoroughly studied FPU-β chain, these power laws have been calculated in various approximations that have yielded good agreement with numerics in the ranges of oscillator number and energy density for which computers can yield numerical results. Geometrical methods have been used to accurately predict the scaling of the Lyapunov exponent with energy density for both weak and strong diffusion, and the transition between the two regimes.

The transient dynamics, starting from a variety of initial conditions at either long or short wavelengths, can be quite complicated, but is also reasonably well understood. Again, referring to the most-studied FPU-β chain from long-wavelength modes, the formation of mode packets on short time-scales is numerically observed and theoretically understood. The relationship with soliton formation and soliton instabilities has also been established. The understanding of the role played by the web of resonance channels in phase space, and the Arnold diffusion through the web to transport energy among modes, is qualitatively understood, and the three resonance model for diffusion has been successfully used to predict the observed numerics. From short-wavelength modes, parametric instability coalescence into chaotic breathers, which ultimately dissipate, is seen numerically and the scaling of its time-scale with energy density (or energy) can be predicted. For other oscillator chains there are differences and additional complications which are generally qualitatively understood. The FPU-α chain is a lowest order approximation to the integral Toda chain and is therefore more stable, having a plateau in n_{eff} during the time that the FPU-α Lyapunov exponent tracks the decay of the Lyapunov exponent for the Toda chain (whose asymptotic value is zero). The similarities and differences between the FPU-β and ϕ^4 chains are also seen numerically and qualitatively understood. These detailed dynamical processes are very interesting from the general perspective of nonlinear dynamics. However, their significance for physical problems has not been explored in any detail.

Less understood are the transitions at fixed N as ε is decreased. For large values of N, and starting from generic long-wavelength initial conditions, one expects transitions from inverse power-law scaling to exponential scaling, when the diffusion is through decreasingly thin channels. Some numerical hints of this behavior have been seen in the FPU-β and ϕ^4 chains, but the time-scales become exponentially long so that numerical results become increasingly difficult to obtain. There are also special initial conditions for which no diffusion occurs, an example being the π-mode with periodic boundary conditions for an energy below the border of parametric instability. Other situations are not so clear, particularly for small N, where numerical observations do not distinguish between initial conditions lying on regular orbits and lying on stochastic orbits for which the stochasticity is unobservable.

Another area of uncertainly is the relation between the inverse power-law of $T_{\mathrm{eq}} \propto \varepsilon^{-3}$ for various oscillator chains, and the diffusion through the stochastic channels. For the FPU-β system, starting from long wavelengths, the universal s-shaped curve of n_{eff} vs. $\log(\varepsilon^3 t)$ (see Fig. 2.13) must consist of a complicated averaging over many increasingly fine resonance channels as equipartition is approached. No theory exists to explain the shape of this curve. The lack of explanation of the detailed evolution of this macroscopic parameter points to the difficulty in understanding detailed microscopic dynamics in problems with many degrees of freedom.

For what concerns the problem of heat conduction, the most puzzling question to be answered is the universality of the power-law divergence of heat conductivity for FPU-like chains. Since numerics is unable to yield conclusive results, one can only hope that a theoretical approach, such as rigorous hydrodynamic theory of transport in FPU-like chains, could provide the explanation.

Finally, we remark that even in the simplest many-degree-of-freedom systems, with the FPU-α and FPU-β systems being prime examples, rigorous results that are also useful in making quantitative calculations are difficult to obtain. From the core material of this review, it is evident that numerical calculations underpin our understanding of the dynamics and validate our analytic calculations. For regions of the parameter space in which numerical calculations are not practical, extrapolation can be useful, but the uncertainty of relying on answers from calculations grows with increasing extrapolation and our understanding becomes less secure. Clearly, there remain challenges for theorists to obtain results that are both rigorous and useful for calculations, and for numerics to be extended into the areas of uncertainty.

Acknowledgments

We thank the University of Firenze for financially supporting the visit of AJL to Florence. SR thanks The Isaac Newton Institute in Cambridge UK for hospitality and financial support. SR thanks also Steve Greenham at Newton Institute for help in producing some of the figures. Antonio Giorgilli is acknowledged for providing us a copy of Fig. 2.14.

References

1. E. Fermi, J.R. Pasta and S. Ulam, in *Collected papers of Enrico Fermi*, E. Segre (ed.). University of Chicago Press, Chicago, 2, 978, 1965.
2. H. Poincaré, *Les méthodes Nouvelles de la Mécanique Celeste.* Gauthier-Villars, Paris, 1892.
3. E. Fermi, Zeit. Phys. **24**, 261 (1923).
4. E. Fermi, Nuovo Cimento **25**, 267; **26**, 105 (1923).

5. J. Ford and J. Waters, J. Math. Phys. **4**, 1293 (1963).
6. E.A. Jackson, J. Math. Phys. **4**, 551, 686 (1963).
7. R.L. Bivins, N. Metropolis and J.R. Pasta, J. Comp. Phys. **12**, 62 (1972).
8. A.N. Kolmogorov, Dokl. Akad. Nauk. SSSR **98**, 527 (1954).
9. V.I. Arnold, Soviet Math. Dokl. **2**, 501 (1961).
10. J. Moser, Nachr. Akad Wiss. Gottingen, **K1**, 1 (1962).
11. M. Henon and C. Heiles, Astr. J. **69**, 73 (1964).
12. B.V. Chirikov, Plasma Phys. (J.N.E. Pt C) **1**, 253 (1960).
13. B.V. Chirikov, Phys. Rep. **52**, 265 (1979).
14. V.I. Arnold, Russian Math. Surveys **18**, 85 (1964).
15. A.J. Lichtenberg and M.A. Lieberman, Regular and chaotic dynamics, 2nd edition Springer, New York, 1992.
16. M. Falcioni, U.M.B. Marconi and A. Vulpiani, Phys. Rev. A **44**, 2263 (1991).
17. N.N. Nekhorochev, Usp. Mat. Nauk. (USSR) **32**, 6 (1977).
18. A. Giorgilli, Ann. Inst. H. Poincare Phys. Theor. **48**, 423 (1988).
19. P. Lochak, Phys. Lett. A **143**, 39 (1990).
20. M. Toda, Prog. Theor. Phys. Suppl. **45**, 174 (1970).
21. J. Ford, S.D. Stoddard and J.S. Turner, Prog. Theor. Phys. **50**, 1547 (1973).
22. M. Henon, Phys. Rev. B **9**, 1925 (1974).
23. A.N. Kolmogorov, Dokl. Akad. Nauk. SSSR **124**, 754 (1959).
24. Ya. G. Sinai, Dokl. Akad. Nauk. SSSR **124**, 768 (1959).
25. G. Benettin, L. Galgani and J.M. Strelcyn, Phys. Rev. A **14**, 2338 (1976).
26. V.I. Arnold and A. Avez, *Ergodic problems of statistical mechanics.* Benjamin, New York, 1968.
27. V.I. Arnold, *Mathematical methods of classical mechanics.* Springer-Verlag, Berlin, 1978.
28. N.S. Krylov, *Works on the foundations of statistical physics*, Princeton University Press, Princeton, 1979.
29. J. Hadamard, J. Math. Pur. Appl. **4**, 27 (1898).
30. G.A. Hedlund, Bull. Am. Math. Soc. **45**, 241 (1939).
31. E. Hopf, Proc. Nat. Acad. Sci. **18**, 263 (1932).
32. V.I. Anosov, Proc. Steklov Math. Inst. **90**, 1 (1967).
33. Ya.G. Sinai, Dokl. Akad. Nauk SSSR **153**, 1261 (1963).
34. C.P. Ong, Adv. Math. **15**, 269 (1975).
35. M.C. Gutzwiller, J. Math. Phys. **18**, 806 (1977).
36. A. Knauf, Comm. Math. Phys. **110**, 89 (1987).
37. M. Szydlowski, M. Heller and W. Sasin, J. Math. Phys. **37**, 346 (1996).
38. H.E. Kandrup, Astrophys. J. **364**, 420 (1990).
39. H.E. Kandrup, Physica A **169**, 73 (1990).
40. H.E. Kandrup, Phys. Rev. E **56**, 2722 (1997).
41. T.J. Hunt and MacKay, R.S., Nonlinearity **16**, 1499 (2003).
42. L. Casetti, M. Pettini and E.G.D. Cohen, Phys. Rep. **337**, 237 (2000).
43. M.D. Kruskal and N.J. Zabusky, J. Math. Phys. **5**, 231 (1964).
44. N.J. Zabusky and M.D. Kruskal Phys. Rev. Lett. **15**, 240 (1965).
45. C.F. Driscoll and T.M. O'Neil, Phys. Rev. Lett. **37**, 69 (1976).
46. C.F. Driscoll and T.M. O'Neil, Rocky Mt. J. Math **5**, 211 (1978).
47. F.M. Izrailev and B.V. Chirikov, Sov. Phys. Dokl. **11**, 30 (1966).
48. G. Benettin, L. Galgani and A. Giorgilli, Phys. Lett. A **120**, 23 (1987).
49. G. Benettin, L. Galgani and A. Giorgilli, Commun. Math. Phys. **113**, 87 (1987).

50. R. Livi, M. Pettini, S. Ruffo, M. Sparpaglione and A. Vulpiani, Phys. Rev. A **28**, 3544 (1983).
51. R. Livi, M. Pettini, S. Ruffo, M. Sparpaglione and A. Vulpiani, Phys. Rev. A **31**, 1039 (1985).
52. R. Livi, M. Pettini, S. Ruffo and A. Vulpiani, Phys. Rev. A **31**, 2740 (1985).
53. M. Pettini and M. Landolfi, Phys. Rev. A **41**, 768 (1990).
54. C.G. Goedde, A.J. Lichtenberg and M.A. Lieberman, Physica D **59**, 200 (1992).
55. J. DeLuca, A.J. Lichtenberg and M.A. Lieberman, Chaos **5**, 283 (1995).
56. H. Kantz, R. Livi and S. Ruffo, J. Stat. Phys. **76**, 627 (1994).
57. J. DeLuca, A.J. Lichtenberg and S. Ruffo, Phys. Rev. E **51**, 2877 (1995).
58. J. DeLuca, A.J. Lichtenberg and S. Ruffo, Phys. Rev. E **60**, 3781 (1999).
59. L. Berchialla, A. Giorgilli and S. Paleari, Phys. Lett. A **321**, 167 (2004).
60. L. Casetti, C. Clementi and M. Pettini, Phys. Rev. E **54**, 5969 (1996).
61. L. Casetti, M. Cerruti-Sola, M. Pettini and E.G.D. Cohen, Phys. Rev. E **55**, 6566 (1997).
62. D.L. Shepelyansky, Nonlinearity **10**, 1331 (1997).
63. M. Cerruti-Sola, M. Pettini and E.G.D. Cohen, Phys. Rev. E **62**, 6078 (2000).
64. M. Pettini and M. Cerruti-Sola, Phys. Rev. A **41**, 44, 975 (1991).
65. J. DeLuca and A.J. Lichtenberg, Phys. Rev. E **66**, 026206 (2002).
66. A. Ponno, L. Galgani and F. Guerra, Phys. Rev. E **61**, 7081 (2000).
67. N.J. Zabusky and G.S. Deem, J. Comp. Phys. **2**, 126 (1967).
68. N. Budinsky and T. Bountis, Physica D **8**, 445 (1983).
69. S. Flach, Physica D **91**, 223 (1996).
70. P. Poggi and S. Ruffo, Physica D **103**, 251 (1997).
71. G.M. Chechin, N.V. Novikova and A.A. Abramenko, Physica D **166**, 208 (2002).
72. G.M. Chechin, D.S. Ryabov and K.G. Zhukov, Physica D **203**, 121 (2005).
73. B. Rink, Physica D **175**, 31 (2003).
74. T. Dauxois, S. Ruffo and A.Torcini, Phys. Rev. E **56**, R 6229 (1997).
75. V.M. Burlakov, S.A. Kiselev and V.I. Rupasov Phys. Lett. A **147**, 130 (1990).
76. K.W. Sandusky and J.B. Page, Phys. Rev. B **50**, 866 (1994).
77. T. Dauxois, S. Ruffo and A.Torcini, J. Phys. IV **8**, 147 (1998).
78. Yu.S. Kivshar and M. Peyrard, Phys. Rev. A **46**, 3198 (1992).
79. T.B. Benjamin and J.E. Feir, J. Fluid Mech. **27**, 417 (1967).
80. V.E. Zakharov and A.B. Shabat Zhurnal Eksperimentalnoi I Teoreticheskoi Fiziki **64**, 1627 (1973).
81. G.P. Berman and A.R. Kolovskii, Zh. Eksp. Teor. Fiz. **87**, 1938 and Sov. Phys. JETP **60**, 1116 (1984).
82. A.J. Sievers and S. Takeno, Phys. Rev. Lett. **61**, 970 (1988).
83. V.M. Burlakov and S.A. Kiselev, Sov. Phys. JETP **72**, 854 (1991).
84. S. Flach and C.R. Willis, Phys. Rep. **295**, 181 (1998).
85. A.I. D'yachenko, V.E. Zackharov, A.N. Pushkarev, V.E. Shvets and V.V. Yan'kov, Sov. Phys. JETP **69**, 1144 (1989).
86. O. Bang and M. Peyrard, Phys. Rev. E **53**, 4143 (1996).
87. T. Cretegny, T. Dauxois, S. Ruffo and A. Torcini, Physica D **121**, 109 (1998).
88. K. Ullman, A.J. Litchenberg and E. Corso, Phys. Rev E **61**, 2471 (2000).
89. S. Flach, C.R. Willis and E. Olbrich, Phys. Rev. E **49**, 836, (1994).
90. Yu.A. Kosevich, Phys. Rev. B **47**, 3138 (1993).
91. Yu.A. Kosevich and S. Lepri, Phys. Rev. B **61**, 299 (2000).

92. V.V. Mirnov, A.J. Lichtenberg and H. Guclu, Physica D **157**, 251 (2001).
93. A.J. Lichtenberg and V.V. Mirnov, Physica D, **202**, 116 (2005).
94. A.J. Lichtenberg, V.V. Mirnov and C. Day, Chaos, **15**, 015109 (2005).
95. M.C. Forrest, C.G. Goedde and S. Sinha, Phys. Rev. Lett. **68**, 2722 (1992).
96. J.L. Marin and S. Aubry, Physica D, **119**, 163 (1998).
97. S.R. de Groot and P. Mazur, *Non-equilibrium thermodynamics*, Dover, NY, 1984.
98. P. Debye, Communication at the congress on the kinetic theory of matter. Gottingen, 1913.
99. R.E. Peierls, *Quantum theory of solids.* Oxford University Press (1955).
100. Z. Rieder, J.L. Lebowitz and E. Lieb, J. Math. Phys. **8**, 1073 (1967).
101. S. Lepri, R. Livi and A. Politi, Phys. Rep. **377**, 1 (2003).
102. D.N. Payton, M. Rich and W.M. Visscher, Phys. Rev. **160**, 706 (1967).
103. E.A. Jackson, J.R. Pasta and J.F. Waters, J. Comput. Phys. **2**, 207 (1968).
104. N. Nakazawa, Progr. Theor. Phys. (Suppl. 45), 231 (1970).
105. W.M. Visscher, *Methods in computational physics* Vol. **15**, p. 371. Academic Press, New York, 1976.
106. H. Kaburaki and M. Machida, Phys. Lett. A **181**, 85 (1993).
107. Y. Ohtsuboi, N. Nishiguchi and T. Sakuma, J. Phys. Condens. Matter **6**, 3013 (1994).
108. F. Mokross and H. Buttner, J. Phys. C **16**, 4539 (1983).
109. E.A. Jackson and A.D. Mistriotis, J. Phys. Cond. Mat. **1**, 1223 (1984).
110. N. Nishiguchi and T. Sakuma, J. Phys. Condens. Matter **2**, 7575 (1990).
111. O.V. Gendelman and L.I. Manevich, Sov. Phys. JETP **75**, 271 (1992).
112. S. Lepri, R. Livi and A. Politi, Phys. Rev. Lett. **78**, 1896 (1997).
113. S. Lepri, R. Livi and A. Politi, Physica D **119**, 140 (1998).
114. S. Lepri, R. Livi and A. Politi, Europhys. Lett. **43**, 271 (1998).
115. S. Lepri, Eur. Phys. J. B **18**, 441 (2000).
116. T. Hatano, Phys. Rev. E **59**, R1 (1999).
117. M. Vassalli, Diploma Thesis, University of Florence, 1999.
118. L. Galgani, A. Giorgilli, A. Martinoli and S. Vanzini, Physics D **59**, 334 (1992).
119. G. Casati, J. Ford , F. Vivaldi and W.M. Visscher, Phys. Rev. Lett. **52**, 1861 (1984).
120. T. Prozen and D.K. Campbell, Chaos **15**, D15117 (2005).
121. J. De Luca and A.J. Lichtenberg, Phys. Rev. E. **66**, 026206 (2002).
122. D. Sholl, Phys. Lett. A **149**, 253 (1990).
123. R. Livi and A. Vulpiani, (eds), *L' Heritage de Kolmogorov en Physique*, Editions Belin, Paris; *The Kolmogorv Legacy in Physics* Lectures Notes in Physics, Springer, Berlin and New York, 2003.
124. J.P. Eckmann and D. Ruelle, Rev. Mod. Phys. **57**, 617 (1985).
125. M. Tabor, *Chaos and integrability in nonlinear dynamics*, John Wiley & Sons. Inc., New York, 1989.
126. E. Ott, *Chaos in dynamical systems.* Cambridge University Press, Cambridge 1993.
127. Ya. B. Pesin, Russ. Math. Surveys **32**, 55 (1977).
128. R. Livi, A. Politi and S. Ruffo, J. Phys. A **19**, 2033 (1986).
129. J.P. Eckmann and E. Wayne, J. Stat. Phys. **50**, 853 (1988).
130. Ya.G. Sinai, Int. J. Bifur. Chaos **6**, 1137 (1996).
131. L. Casetti, R. Livi and M. Pettini, Phys. Rev. Lett. **74**, 375 (1995).

132. S. Isola, R. Livi, S. Ruffo and A. Vulpiani, Phys. Rev. A **33**, 1163 (1986).
133. S. Flach, M.V. Ivanchenko and O.I. Kanakov, Phys. Rev. Lett. **95**, 064102 (2005).
134. J. DeLuca, A.J. Lichtenberg and S. Ruffo, Phys. Rev. E **54**, 2329 (1996).
135. A. Ponno and Bambusi D., Energy cascade in Fermi–Pasta–Ulam models, Proceedings of the International Conference on Symmetry and Perturbation Theory 2004, G. Gaeta et al. (eds), World Scientific Publishing, 263–270 and private communication, 2005.
136. J.A. Biello, P.R. Kramer, Y.V. L'vov, Discr. Cont. Dynam. Sys. (Suppl. **113**) (2003).
137. L. Berchialla, Galgani L. and Giorgilli A., Discr. Cont. Dynam. Sys. (Suppl., **855**), (2005).
138. T. Dauxois, R. Khomeriki, F. Piazza and S. Ruffo, Chaos **15**, 015110 (2005).
139. A. Cafarella, M. Leo and R.A. Leo, Phys. Rev. E **69**, 046604 (2004).
140. G.P. Berman, F.M. Izrailev, Chaos, **15**, 015104 (2005).
141. L. Caiani, L. Casetti and M. Pettini, J. Phys. A **31**, 3357 (1998).
142. M. Pettini, Phys. Rev. E **47**, 828 (1993).
143. M. Cerruti-Sola and M. Pettini, Phys. Rev. E **51**, 53 (1995).
144. M. Cerruti-Sola and M. Pettini, Phys. Rev. E **53**, 179 (1996).
145. A. Abraham and J.E. Marsden, *Foundations of mechanics*. Addison-Wesley, Redwood City, 1987.
146. M.P. do Carmo, *Riemannian geometry*. Birkhäuser, Boston-Basel, 1993.
147. L.P. Eisenhart, Ann. of Math. **30**, 591 (1929).
148. J. Guckenheimer and P. Holmes, *Nonlinear oscillations, dynamical systems and bifurcations of vector fields*, ed. Springer, New York, 1983.
149. M. Pettini and R. Valdettaro, Chaos **5**, 646 (1995).
150. N.G. Van Kampen, Phys. Rep. **24**, 71 (1976).
151. R. Livi, M. Pettini, S. Ruffo and A. Vulpiani, J. Stat. Phys. **48**, 539 (1987).
152. G. Benettin, L. Galgani, A. Giorgilli and J.M. Strelcyn, Meccanica **15**, 9 and 21 (1980).
153. G. Benettin, L. Galgani and A. Giorgilli, Il Nuovo Cimento B **89**, 89 (1985).
154. R. Kubo, M. Toda and N. Hashitsume, *Statistical physics II*. Springer Series in Solid State Sciences, Vol. 31, Springer, Berlin, 1991.
155. K. Aoki and D. Kusnezov, Phys. Rev. Lett. **86**, 4029 (2001).
156. D. Escande, H. Kantz, R. Livi and S. Ruffo, J. Stat. Phys. **76**, 605 (1994).
157. C. Giardinà, R. Livi, A. Politi and M. Vassalli, Phys. Rev. Lett. **84**, 2144 (2000).
158. O.V. Gendelman and A.V. Savin, Phys. Rev. Lett. **84**, 2381 (2000).
159. J. Dawson, Phys. Fluids **5**, 445 1962.
160. G. Casati, Found. Phys. **16**, 51 (1986).
161. T. Prozen and M. Robnik, J. Phys. A **25**, 3449 (1992).
162. D.J.R. Mimnagh and L.E. Ballentine, Phys. Rev. E **56**, 5332 (1997).
163. H.A. Posch and W.G. Hoover, Phys. Rev. E **58**, 4344 (1998).
164. M.J. Gillan and R.W. Holloway, J. Phys. C **18**, 5705 (1985).
165. B. Hu, B. Li and H. Zhao, Phys. Rev. E **57**, 2992 (1998).
166. A. Filippov, B. Hu, B. Li and A. Zelser, J. Phys. A **31**, 7719 (1998).
167. B. Hu, B. Li and H. Zhao, Phys. Rev. E **61**, 3828 (1999).
168. K. Aoki and D. Kusnezov, Phys. Lett. A **265**, 250 (2000).
169. G.P. Tsironis, A.R. Bishop, A.V. Savin and A.V. Zolotaryuk, Phys. Rev. E **60**, 6610 (1999).

170. G. Tsaur and J. Wang, Phys. Rev. E **54**, 4657 (1996).
171. V.I. Oseledec, Trans. Moscow Math. Soc. **19**, 197 (1968).
172. J.L. Lebowitz, Percus J.K. and Verlet L., Phys. Rev. **153**, 250 (1967).
173. L. Caiani, L. Casetti, C. Clementi and M. Pettini, Phys. Rev. Lett. **79**, 4361 (1997).
174. L. Caiani, L. Casetti, C. Clementi, G. Pettini, M. Pettini and R. Gatto, Phys. Rev. E **57**, 3886 (1998).
175. M.-C. Firpo, Phys. Rev. E **57**, 6599 (1998).

3

Role of Chaos for the Validity of Statistical Mechanics Laws: Diffusion and Conduction

Massimo Cencini[1], Fabio Cecconi[1], Massimo Falcioni[2] and Angelo Vulpiani[2]

[1] INFM and Istituto dei Sistemi Complessi (ISC-CNR) Via dei Taurini 19, I-00185 Roma (Italy)
Massimo.Cencini@roma1.infn.it,
Fabio.Cecconi@roma1.infn.it

[2] INFM and Dipartimento di Fisica Università "La Sapienza" P.le Aldo Moro 2, I-00185 Roma (Italy)
Massimo.Falcioni@roma1.infn.it,
Angelo.Vulpiani@roma1.infn.it

Abstract. Several years after the pioneering work by Fermi, Pasta and Ulam, fundamental questions about the link between dynamical and statistical properties remain still open in modern statistical mechanics. Particularly controversial is the role of deterministic chaos for the validity and consistency of statistical approaches. This contribution reexamines such a debated issue taking inspiration from the problem of diffusion and heat conduction in deterministic systems. Is microscopic chaos a necessary ingredient to observe such macroscopic phenomena?

3.1 Introduction

Statistical mechanics, founded by Maxwell, Boltzmann and Gibbs, aims to explain the macroscopic properties of systems with a huge number of degrees of freedom without specific assumptions on the microscopic dynamics, a part from ergodicity [1, 2]. The discovery of deterministic chaos [3], beyond its undoubted important implications on many natural phenomena, enforced us to reconsider some basic problems standing at the foundations of statistical mechanics such as, for instance, the applicability of a statistical description to low-dimensional systems. However, even after many years, the experts do not agree yet on the basic conditions which should ensure the validity of statistical mechanics.

The spectrum of viewpoints found in literature is rather wide, ranging from the Landau (and Khinchin [4]) earlier belief on the key role of the many degrees of freedom and the (almost) complete irrelevance of ergodicity, to the opinion of those who, as Prigogine and his school [5] consider chaos as the crucial requirement to develop consistent statistical approaches. Recently some authors (e.g. Lebowitz [6] and Bricmont [7]) have given new life to the

M. Cencini et al.: *Role of Chaos for the Validity of Statistical Mechanics Laws: Diffusion and Conduction*, Lect. Notes Phys. **728**, 123–149 (2008)
DOI 10.1007/978-3-540-72995-2_3

debate [10, 5], renewing the intuition of Boltzmann [8] and Maxwell [9] on the relevance of the huge number of particles in macroscopic systems.

This volume offers the opportunity to celebrate the 100th and 50th anniversaries of two of the most influential works in statistical physics: Einstein's work on Brownian motion (1905) [11] and Fermi's one (1955) on the non-linear chain of oscillators (*al secolo* the FPU work, from the authors Fermi, Pasta and Ulam [12]). We shall discuss some aspects related to diffusion problems and heat conduction focusing on the role of (microscopic) chaos for the occurrence and robustness of these (macroscopic) phenomena. Transport phenomena, despite their ubiquity in everyday life, are still subject of debate among theoretic physicists.

Because of the variety of specific interactions and technical difficulties in realistic systems, simplified microscopic models are unavoidable tools for the study of transport mechanism. Several simulations and theoretical works have shown that, in systems with very strong chaos (namely hyperbolic systems), there exists a close relationship between transport coefficients (e.g. viscosity, diffusivity, thermal and electrical conductivity) and indicators of chaos (Lyapunov exponents, KS entropy, escape rates) [13, 14]. At a first glance, the existence of such relations would support the point of view of who considers chaos as the basic ingredient for the applicability of statistical mechanics. However, it is not possible to extend those results to generic systems. In fact, we shall see that many counterexamples prove that chaos is not a necessary condition for the emergence of robust statistical behaviors [15, 16]. In particular, we shall see that phenomena such as diffusion [17] and heat conduction [18] may take place also in non-chaotic systems. These and many other examples provide indication that microscopic chaos is not the unique possible origin of macroscopic transport in dynamical systems.

The material is organized into two, almost self-contained, parts. In the first, after a brief historical introduction to the different microscopic models proposed to explain macroscopic diffusion, we discuss a recent experiment (and the consequent debate it stimulated) aimed to prove that microscopic chaos is at the origin of Brownian motion. This gives us the possibility to introduce and discuss the problem of diffusion in non-chaotic deterministic systems, and to point out the necessary microscopic conditions to observe diffusion. The second part is mostly devoted to a discussion of the celebrated FPU numerical experiments and its consequences for the ergodic problem and heat conduction. We shall see that there are non-chaotic models displaying (macroscopic) heat conduction, confirming the non-essential role of chaos on transport.

3.2 On the Microscopic Origin of Macroscopic Diffusion

At the beginning of the twentieth century, the atomistic theory of matter was not yet fully accepted by the scientific community. While searching for phenomena that would prove, beyond any doubt, the existence of atoms, Einstein realized that "...according to the molecular-kinetic theory of heat, bodies of microscopically-visible size suspended in a liquid will perform movements of such magnitude that they can be easily observed in a microscope...," as he wrote in his celebrated paper in 1905 [11]. In this work, devoted to compare the different predictions that classical thermodynamics and molecular-kinetic theory of heat make about those small bodies, Einstein argued that their motion has a diffusive character. Moreover, he discovered an important relation among the diffusion coefficient D, the fluid viscosity η, the particles radius a (having assumed spherical particles), Avogadro's number N_{A}, the temperature T and the gas constant R:

$$D = \frac{1}{N_{\mathrm{A}}} \frac{RT}{6\pi\eta a} . \tag{3.1}$$

Einstein relation (3.1), which may be seen as the first example of the fluctuation–dissipation theorem [19], allowed for the determination of Avogadro's number [20] and gave one of the ultimate evidences of the existence of atoms.

Einstein's work on Brownian motion (BM) is based on statistical mechanics and thermodynamical considerations applied to suspended particles, with the assumption of velocity decorrelation.

One of the first successful attempts to develop a purely dynamical theory of BM dates back to Langevin [21] that, as himself wrote, gave "... a demonstration [of Einstein results] that is infinitely more simple by means of a method that is entirely different." Langevin considered the Newton equation for a small spherical particle in a fluid, taking into account that the Stokes viscous force it experiences is only a mean force. In one direction, e.g. the x-direction, one has

$$m \, \frac{\mathrm{d}^2 x}{\mathrm{d}t^2} = -6\pi\eta a \, \frac{\mathrm{d}x}{\mathrm{d}t} + F , \tag{3.2}$$

where m is the mass of the particle. The first term of the r.h.s. is the Stokes viscous force. The second one $F(t)$ is a fluctuating random force, independent of $v = \mathrm{d}x/\mathrm{d}t$, modeling the effects of the huge number of impacts with the surrounding fluid molecules, which is taken as a zero-mean, Gaussian process with covariance $\langle F(t)F(t')\rangle = c\delta(t-t')$. The constant c is determined by the equipartition condition $\langle(\mathrm{d}x/\mathrm{d}t)^2\rangle = RT/(mN_{\mathrm{A}})$, i.e. $c = 12\pi\eta a RT/N_{\mathrm{A}}$.

Langevin's work along with that of Ornstein and Uhlenbeck [22] are at the foundation of the theory of stochastic differential equations. The stochastic approach is however unsatisfactory being a phenomenological description.

The next theoretical challenge toward the building of a dynamical theory of Brownian motion is to understand its microscopic origin from first principles. Almost contemporarily to Einstein's efforts, Smoluchowski tried to derive the large-scale diffusion of Brownian particles from the similar physical assumptions about their collisions with the fluid molecules [23].

A renewed interest on the subject appeared some years later, when it was realized that even purely deterministic systems composed of a large number of particles give rise to macroscopic diffusion, at least on finite time scales. These models had an important impact in justifying Brownian motion theory and, more in general, in deriving a consistent microscopic theory of irreversibility.

Some of these works considered chains of harmonic oscillators of equal masses [24, 25, 26, 27], while others [28, 29] analyzed the motion of a heavy impurity linearly coupled to a chain of equal mass oscillators. When the number of oscillators goes to infinity, the momentum of the heavy particle was proved to behave as a genuine stochastic process described by the Langevin equation (3.2). When their number is finite, diffusion remains an effective phenomenon lasting for a (long but) finite time.

Soon after the discovery of dynamical chaos [30], it was realized that simple low-dimensional deterministic systems may also exhibit a diffusive behavior. In this framework, the two-dimensional Lorentz gas [31], describing the motion of a free particle through a lattice of hard round obstacles, provided the most valuable example. As a consequence of the obstacle convexity, particle trajectories are chaotic, i.e. aside from a set initial conditions of zero measure, exhibit a positive and finite Lyapunov exponent,. At long times, for the case of billiards, the mean squared displacement from the particle initial position grows linearly in time. A Lorentz system with periodically arranged scatterers is closely related to the Sinai billiard [32, 33], which can be obtained from the former by folding the trajectories into the unitary lattice cell. The extensive study on billiards has shown that chaotic behavior might usually be associated with diffusion in simple low-dimensional models, supporting the idea that chaos was at the very origin of diffusion. However, more recently (see, e.g. [17]) it has been shown that even non-chaotic deterministic systems, such as a bouncing particle in a two-dimensional billiard with polygonal but randomly distributed obstacles (wind-tree Ehrenfest model), may exhibit a diffusion-like properties. This example can lead to think that the external source of randomness may play a role similar to chaos (for a more detailed discussion about this point see Sect. 3.2.2).

Deterministic diffusion is a generic phenomenon present also in simple chaotic maps on the line. Among the many contributions we mention the work by Fujisaka, Grossmann [34, 35] and Geisel [36, 37]. A typical example is the one-dimensional discrete-time dynamical system:

$$x(t+1) = [x(t)] + F(x(t) - [x(t)]) \, , \tag{3.3}$$

where $x(t)$ (the position of a point-like particle) performs diffusion in the real axis. The bracket $[\cdot]$ denotes the integer part of the argument. $F(u)$ is a map defined on the interval $[0, 1]$ that fulfills the following requirements:

(i) The map, $u(t+1) = F(u(t))$ (mod 1) is chaotic.

(ii) $F(u)$ must be larger than 1 and smaller than 0 for some values of u, so to have a non-vanishing probability to escape from each unit cell (a unit cell of real axis is every interval $C_\ell \equiv [\ell, \ell+1]$, with $\ell \in Z$).

(iii) $F_r(u) = 1 - F_l(1-u)$, where F_l and F_r define the map in $u \in [0, 1/2[$ and $u \in [1/2, 1]$ respectively. This anti-symmetry condition with respect to $u = 1/2$ is introduced to avoid a net drift.

A very simple and much studied example of F is

$$F(u) = \begin{cases} 2(1+a)u & \text{if } u \in [0, 1/2[\\ 2(1+a)(u-1)+1 & \text{if } u \in [1/2, 1] , \end{cases} \tag{3.4}$$

where $a > 0$ is the control parameter. It is useful to remind the link between diffusion and velocity correlation, i.e. the Taylor–Kubo formula, that helps to unravel how diffusion can be realized in different ways. The velocity correlation function is defined as $C(\tau) = \langle v(\tau)v(0)\rangle$, where $v(t)$ is the velocity of the particle at time t. It is easy to see that for continuous time systems [e.g. (3.2)]

$$\langle (x(t) - x(0))^2 \rangle \simeq 2\, t \int_0^t \mathrm{d}\tau\, C(\tau)\ . \tag{3.5}$$

Standard diffusion, with $D = \int_0^\infty \mathrm{d}\tau\, C(\tau)$, is always obtained whenever the hypotheses for the validity of the central limit theorem are verified:

(i) finite variance of the velocity: $\langle v^2 \rangle < \infty$;
(ii) faster than τ^{-1} decay of the velocity correlation function $C(\tau)$.[1]

The first condition, independently of the microscopic dynamics under consideration (stochastic, deterministic chaotic or regular), excludes unphysical models, i.e. with infinite variance for the velocity. The second requirement corresponds to a rapid memory loss of initial conditions. It is surely verified for the Langevin dynamics where the presence of the stochastic force entails a rapid decay of $C(\tau)$. In deterministic regular systems, such as the model of many oscillators, the velocity decorrelation (i.e. the small fluctuations of $C(\tau)$ around zero, for almost all the time) is the result of the huge number of degrees of freedom that act as a heat bath on a single oscillator. In the (non-chaotic) Ehrenfest wind-tree model decorrelation originates from the disorder in the obstacle positions. Deterministic chaotic systems, in spite of the fact that nonlinear instabilities generically lead to a memory loss, are more subtle. Indeed, there are many examples, namely intermittent systems [38], characterized by a slow decay of the velocity correlation function.

We end this section by asking whether it is possible to determine, by the analysis of a Brownian particle, if the microscopic dynamics underlying the observed macroscopic diffusion is stochastic, deterministic chaotic or regular.

[1] In discrete-time systems, the velocity $v(t)$ and the integral $\int \mathrm{d}\tau C(\tau)$ are replaced by the finite difference $x(t+1) - x(t)$ and by the quantity $\langle v(0)^2\rangle/2 + \sum_{\tau \geq 1} C(\tau)$, respectively.

3.2.1 Chaos or Noise? A Difficult Dilemma

Inferring the microscopic deterministic character of Brownian motion on an experimental basis would be attractive from a fundamental viewpoint. Moreover it could provide further evidence to some recent theoretical and numerical studies [39, 40]. Before discussing a recent experiment [41] in this direction, we must open the "Pandora box" of the longstanding and controversial problem of distinguishing chaos from noise in signal analysis [42] (see also [45, 46, 47, 48, 49, 50]). For the sake of clearness on the terminology used here, we specify that "chaos" refers to the motions originating from a deterministic system with at least one positive but finite Lyapunov exponent, and therefore a positive and finite Kolmogorov–Sinai entropy; "noise" instead denotes the outcomes of a continuous valued stochastic process with infinite value of Kolmogorov–Sinai entropy.

The first observation concerning the chaos/noise distinction is that, very often in the analysis of experimental time series, there is not a unique model of the "system" that produced the data. Moreover, even the knowledge of the "true" model might not be an adequate answer about the character of the signal. From this point of view, BM is a paradigmatic example: in fact it can be modeled by a stochastic as well as by a deterministic chaotic or regular process.

In principle, a definite answer exists. If we were able to determine the maximum Lyapunov exponent (λ) or the Kolmogorov–Sinai entropy (h_{KS}) of a data sequence, we would know without uncertainty whether the sequence was generated by a deterministic law ($\lambda, h_{\mathrm{KS}} < \infty$) or by a stochastic one ($h_{\mathrm{KS}} \to \infty$). Nevertheless, there are unavoidable practical limitations in computing such quantities. They are indeed defined as infinite time averages taken in the limit of arbitrary fine resolution. But, in experiments, we have access only to a finite, and often very limited, range of scales and times.

However, there are measurable quantities that are appropriate for extracting meaningful information from the signal. In particular, we shall consider the (ϵ, τ)-entropy per unit time [51, 52, 53] $h(\epsilon, \tau)$ that generalizes the Kolmogorov–Sinai entropy (for details see next section (3.8)). In a nutshell, while for evaluating h_{KS} one has to detect the properties of a system with infinite resolution, for $h(\epsilon, \tau)$ a finite scale (resolution) ϵ is involved. The Kolmogorov–Sinai entropy is recovered in the limit $\epsilon \to 0$, i.e. $h(\epsilon, \tau) \to h_{\mathrm{KS}}$. This means that if we had access to arbitrarily small scales, we could answer the original question about the character of the law that generated the recorded signal. Even if this limit is unattainable, still the behavior of $h(\epsilon, \tau)$ provides a very useful scale-dependent description of the signal character [42, 54].

For instance, chaotic systems ($0 < h_{\mathrm{KS}} < \infty$) are typically characterized by $h(\epsilon, \tau)$ attaining a plateau $\approx h_{\mathrm{KS}}$, below a resolution threshold, ϵ_c, associated with the smallest characteristic length scale of the system. Instead, for $\epsilon > \epsilon_c$ $h(\epsilon, \tau) < h_{\mathrm{KS}}$, and in this range the details of the ϵ-dependence may be informative on the large scale (slow) dynamics of the system (see, e.g. [42, 54]). Indeed, at large scales typically chaotic systems give rise to behaviors rather similar to stochastic processes (e.g. the diffusive behavior discussed in the previous subsection) with characteristic ϵ-entropy. In stochastic signals, although $h_{\rm KS} = \infty$, for any $\epsilon > 0$, $h(\epsilon, \tau)$ is a finite function of ϵ and τ. The dependence of $h(\epsilon, \tau)$ on ϵ and τ, when known, provides a characterization of the underlying stochastic process (see [51, 53]). For instance, stationary Gaussian processes with a power spectrum[2] $S(\omega) \propto \omega^{-(2\alpha+1)}$ (being $0 < \alpha < 1$) are characterized by a power-law ϵ-entropy [51]:

$$\lim_{\tau \to 0} h(\epsilon, \tau) \sim \epsilon^{-1/\alpha} \, . \tag{3.6}$$

The case $\alpha = 1/2$, corresponding to the power spectrum of a Brownian signal, would give $h(\epsilon) \sim \epsilon^{-2}$. Other stochastic processes, such as time uncorrelated and bounded ones, are characterized by a logarithmic divergence below a critical scale, ϵ_c, which may depend on τ.

Definition and Computation of the ϵ-Entropy

For the sake of self-consistency in this subsection we provide some basic information on the definition and measurement (from experimental data) of the ϵ-entropy, which was originally introduced in the context of information theory by Shannon [52] and, later, by Kolmogorov [51] in the theory of stochastic processes. The interested reader may find more details in [53] and [55].

An operative definition of $h(\epsilon, \tau)$ is as follows. Given the time evolution of a continuous variable $\mathbf{x}(t) \in \Re^d$, that represents the state of a d-dimensional system, one introduces the vector in $\Re^{md}$

$$\mathbf{X}^{(m)}(t) = (\mathbf{x}(t), \ldots, \mathbf{x}(t + m\tau - \tau)) \, , \tag{3.7}$$

which represents a portion of the trajectory, sampled at a discrete time interval τ. After partitioning the phase space $\Re^d$ using hyper-cubic cells of side ϵ, $\mathbf{X}^{(m)}(t)$ is coded into an m-word: $W^m(\epsilon, t) = [i(\epsilon, t), \ldots, i(\epsilon, t + m\tau - \tau)]$, where $i(\epsilon, t + j\tau)$ labels the cell in $\Re^d$ containing $\mathbf{x}(t + j\tau)$. For bounded motions, the number of available cells (i.e. the alphabet) is finite. Under the hypothesis of stationarity, the probabilities $P(W^m(\epsilon))$ of the admissible words $\{W^m(\epsilon)\}$ are obtained from the time evolution of $\mathbf{X}^{(m)}(t)$. Then one introduces the m-block entropy, $H_m(\epsilon, \tau) = -\sum_{\{W^m(\epsilon)\}} P(W^m(\epsilon)) \ln P(W^m(\epsilon))$, and the quantity $h_m(\epsilon, \tau) = [H_{m+1}(\epsilon, \tau) - H_m(\epsilon, \tau)]/\tau$. The (ϵ, τ)-entropy per unit time, $h(\epsilon, \tau)$ is defined by [52]

$$h(\epsilon, \tau) = \lim_{m \to \infty} h_m(\epsilon, \tau) \, . \tag{3.8}$$

The Kolmogorov–Sinai entropy is obtained in the limit of small ϵ:

[2] The power spectrum $S(\omega)$ is the Fourier transform of $\langle (x(t) - x(0))^2 \rangle$.

$$h_{\mathrm{KS}} = \lim_{\epsilon \to 0} h(\epsilon, \tau) \,. \tag{3.9}$$

In principle, in deterministic systems $h(\epsilon)$, and henceforth h_{KS}, depend neither on the sampling time τ [3] nor on the chosen partition because its rigorous definition[53] would require the infimum to be taken over all possible partitions with elements of size smaller than ϵ. However, in practical computations, the specific value of τ is important, and the impossibility to take the infimum over all the partitions implies that, at finite ϵ, $h(\epsilon)$ may depend on the chosen partition. Nevertheless, for small ϵ, the correct value of the Kolmogorov–Sinai entropy is usually recovered independently of the partition [3].

Let us stress that partitioning the phase space does not mean a discretization of the states of the dynamical system, which still evolves on a continuum. The partitioning procedure corresponds to a coarse-grained description (due, for instance, to measurements performed with a finite resolution), that does not change the dynamics. On the contrary, discretizing the states would change the dynamics, implying periodic motions in any deterministic systems. This happens, for instance, in any floating point computer simulations; however such periods are, apart from trivial cases, very long and practically undetectable.

In experimental signals, usually, only a scalar variable $u(t)$ can be measured, and moreover the dimensionality of the phase space is not known a priori. In these cases, one uses delay-embedding techniques [45, 46], where the vector $\mathbf{X}^{(m)}(t)$ is build as $(u(t), u(t+\tau), \ldots, u(t+m\tau-\tau))$, now in $\Re^m$. This is a special instance of (3.7). Then to determine the entropies $H_m(\epsilon)$, very efficient numerical methods are available (the reader may find an exhaustive review in [45]). The delay-embedding procedure can be applied to compute the ϵ-entropy of deterministic and stochastic signals as well. The dependence of the ϵ-entropy on the observation scale ϵ can be used to characterize the process underlying the signal [53].

In the following, we exemplify the typical difficulties by analyzing the map:

$$x(t+1) = f(x(t)) = x(t) + p\ \sin(2\pi x(t)) \,. \tag{3.10}$$

As soon as $p > 0.7326\ldots$, $f(x)$ is such that $f(x) > 1$ and $f(x) < 0$ for some $x \in]0,1[$. This implies that the trajectory can travel across different unitary cells giving rise to large-scale diffusion, i.e. asymptotically:

$$\langle [x(t) - x(0)]^2 \rangle \simeq 2Dt \,, \tag{3.11}$$

where D is the diffusion coefficient. We note that $p = \mathrm{O}(1)$ sets the intrinsic scale of the displacements to be $\mathrm{O}(1)$. Therefore, as far as the ϵ-entropy is concerned, for $\epsilon \ll 1$ (small-scale observations) one should be able to recognize that the system is chaotic, i.e. $h(\epsilon)$ displays a plateau at $h_{\mathrm{KS}} = \lambda$. For $\epsilon \gg 1$ (large scale observations), due to the diffusive behavior, $h(\epsilon)$ is characterized by the scaling (3.6) with $\alpha = 1/2$, therefore

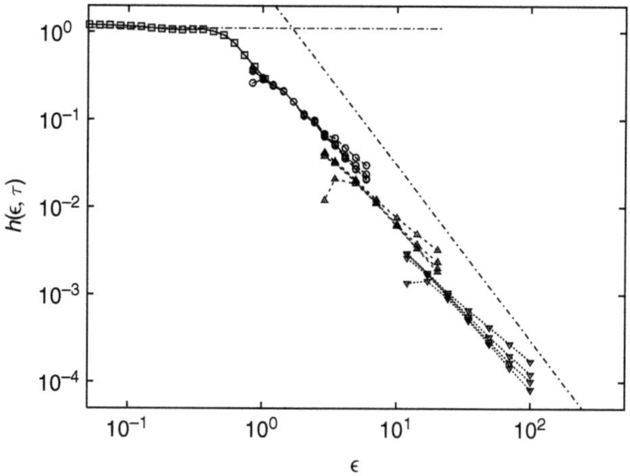

Fig. 3.1. Numerically evaluated (ϵ, τ)-entropy for the map (3.10) with $p = 0.8$ computed by the standard techniques [45] at $\tau = 1$ ($\circ$), $\tau = 10$ ($\triangle$) and $\tau = 100$ ($\bigtriangledown$) and different *block length* ($m = 4, 8, 12, 20$). The *boxes* refer to the entropy computed with $\tau = 1$ but by using periodic boundary condition over 40 cells. The use of periodic boundary conditions is necessary to probe scales small enough to recover the Lyapunov exponent. The *straight lines* correspond to the two asymptotic behaviors, $h(\epsilon) = h_{\mathrm{KS}} \simeq 1.15$ and $h(\epsilon) \sim \epsilon^{-2}$

$$h(\epsilon) \simeq \begin{cases} \lambda & \text{for } \epsilon \ll 1 \\ D/\epsilon^2 & \text{for } \epsilon \gg 1 \end{cases}, \tag{3.12}$$

where λ is the Lyapunov exponent and D is the diffusion coefficient. The typical problems encountered in numerically computing $h(\epsilon)$ can be appreciated in Fig. 3.1. First notice that the deterministic character (i.e. $h(\epsilon, \tau) \approx h_{\mathrm{KS}}$) appears only at $\epsilon < \epsilon_{\mathrm{c}} \approx 1$. However, the finiteness of the data set imposes a lower cut-off scale ϵ_d below which no information can be extracted from the data (see [56]). As for the importance of the choice of τ note that if τ is much larger or much shorter than the characteristic time-scale of the system at the scale ϵ, then the correct behavior of the ϵ-entropy [42] cannot be properly recovered. Indeed the diffusive behavior $h(\epsilon) \sim \epsilon^{-2}$ is roughly obtained only by considering the envelope of $h_m(\epsilon, \tau)$ evaluated at different values of τ. The reason for this is that the characteristic time of the system is determined by its diffusive behavior $T_\epsilon \approx \epsilon^2/D$. On the other hand, the plateau at the value h_{KS} can be recovered only for $\tau \approx 1$, even if, in principle, any value of τ could be used.

We also mention that if the system is deterministic, to have a meaningful measure of the entropy, the embedding dimension m has to be larger than information dimension of the attractor [3].

Experiments on the Microscopic Origin of Brownian Motion

We are now ready to discuss the experiment and its analysis reported in [41]. In this experiment, a long-time record (about 1.5×10^5 data points) of the motion of a small colloidal particle in water was sampled at regular time intervals ($\Delta t = 1/60$ s) with a remarkable high spatial resolution (25 nm). To our knowledge, this is the most accurate measurement of a BM. The data were then processed by means of standard methods of non-linear time-series analysis [45] to compute the ϵ-entropy.[3] This computation shows a power-law dependence $h(\epsilon) \sim \epsilon^{-2}$. Actually, similarly to what is displayed in Fig. 3.1, this behavior is recovered only by considering the envelope of the $h(\epsilon, \tau)$-curves, for different τs. However, unlike to Fig. 3.1, no saturation $h(\epsilon, \tau) \approx$ const. is observed for small ϵ. Nevertheless, the authors *assume* from the outset that the system dynamics is deterministic and, since in deterministic systems $h(\epsilon, \tau) \leq h_{\mathrm{KS}} \leq \sum_i^+ \lambda_i$, deduce from the positivity of $h(\epsilon)$ the existence of positive Lyapunov exponents. Their conclusion is thus that microscopic chaos is at the origin of the macroscopic diffusive behavior.

However, as several works pointed out (see [57, 58]), the huge amount of involved degrees of freedom (Brownian particle and the fluid molecules), the impossibility to reach a (spatial and temporal) resolution high enough, and the limited amount of data points do not allow for such optimistic conclusions. Avoiding a technical discussion on these three points we simply notice that the limitation induced by the finite resolution is particularly relevant to the experiment. For example, if the analysis of Fig. 3.1 would be restricted to the region with $\epsilon > 1$ only, then discerning whether the data were originated by a chaotic system or by a stochastic process would be impossible.

Particularly interesting is the fact that, as shown by Dettman et al. [17, 57], the finite amount of data severely limits our ability to distinguish not only if the signal is deterministic, chaotic or stochastic but also if it is deterministic regular, i.e. of zero entropy. The following example serves as a clue to better understand the way in which a deterministic non-chaotic systems may give rise (at least on certain temporal and spatial scales) to a stochastic behavior.

Let us consider two signals, the first generated by a continuous random walk:

$$\dot{x}(t) = \sqrt{2D}\eta(t) , \tag{3.13}$$

where η is a zero mean Gaussian variable with $\langle \eta(t)\eta(t') \rangle = \delta(t - t')$, and the second obtained as a superpositions of Fourier modes:

---

[3] Of course in data analysis, only scalar time series are available and the dimensionality of the space of state vectors is a priori unknown. However, one can use the delay embedding technique to reconstruct the phase-space. In this way, the ϵ-entropy can be evaluated as discussed in the previous section. It is worth stressing that this procedure can be applied even though the equations of motion of the system, which generated the signal, are unknown. Moreover, this approach is meaningful independently of the stochastic or deterministic nature of the considered signal.

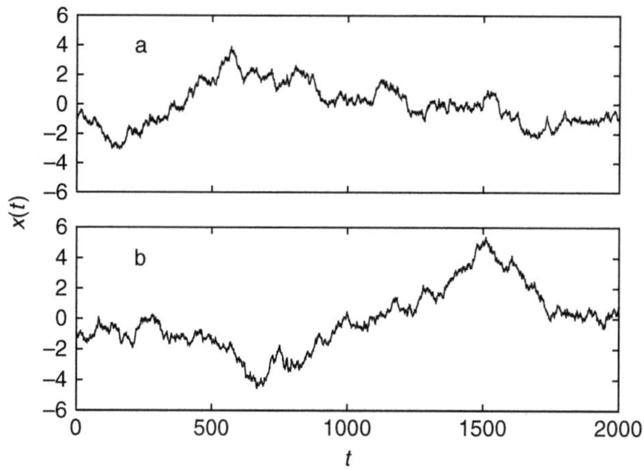

Fig. 3.2. **a** Signals obtained from (3.14) with $M = 10^4$ and random phases uniformly distributed in $[0, 2\pi]$. The numerically computed diffusion constant is $D \approx 0.007$. **b** Time record obtained with a continuous random walk (3.13) with the same value of the diffusion constant as in **a**. In both cases, data are sampled with $\tau = 0.02$, i.e. 10^5 data points

$$x(t) = \sum_{i=1}^{M} X_{0i} \sin(\Omega_i t + \phi_i) \ . \tag{3.14}$$

The coordinate $x(t)$ in (3.14), upon properly choosing the frequencies [29, 42] and the amplitudes (e.g. $X_{0i} \propto \Omega_i^{-1}$), describes the motion of a heavy impurity in a chain of M linearly coupled harmonic oscillators. We know [29] that $x(t)$ performs a genuine BM in the limit $M \to \infty$. For $M < \infty$ the motion is quasi-periodic and regular, nevertheless for large but finite times it is impossible to distinguish signals obtained by (3.13) and (3.14) (see Fig. 3.2). This is even more striking looking at the computed ϵ-entropy of both signals (see Fig. 3.3).

The results of Fig. 3.3 along with those by Dettman et al. [57] suggest that, by assuming also the deterministic character of the system, we are in the practical impossibility of discerning chaotic from regular motion.

It is worth mentioning that recently some interesting works [43, 44] applied the entropy analysis to the motion of a heavy impurity embedded in an FPU-chain (see Sect. 3.3.1), which is a chaotic variant of the above example. The purpose was again to infer the chaotic character of the whole FPU-chain by observations on the impurity motion only. It was found that the impurity does not alter the behavior of the FPU-chain so it can be considered as a true probe of the dynamics. The impurity performs a motion that, when observed at small but finite resolutions, closely resembles a Brownian motion. Time series (ϵ, τ)-entropy analysis both in momentum and position allows for detecting the chaotic nature of the FPU unperturbed system, and clearly locating the stochasticity threshold.

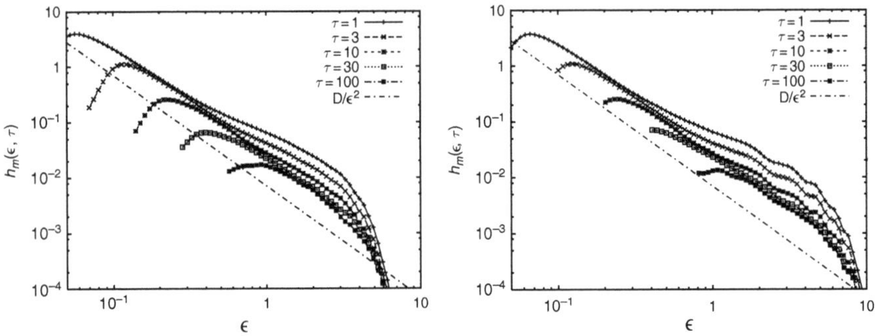

Fig. 3.3. Numerically evaluated (ϵ, τ)-entropy using using 10^5 points from the time series of Fig. 3.2. We show the results for embedding dimension $m = 50$. The *straight lines* show the D/ϵ^2 behavior

From the above discussion, one reaches a pessimistic view on the possibility to detect the "true" nature of a signal by means of data analysis only. However, the situation is not so bad if the question about the character of a signal is asked only relatively to a certain interval of scales. In this case, in fact, it is possible to give an unambiguous classification of the signal character based solely on the entropy analysis and free from any prior knowledge of the system/model that generated the data. Moreover the behavior of $h(\epsilon, \tau)$ as a function of (ϵ, τ) provides a very useful "dynamical" classification of stochastic processes [53, 59]. One has then a practical tool to classify the character of a signal as deterministic or stochastic, on a given scale, without referring to a specific model, and is no longer obliged to answer the metaphysical question, whether the system that produced the data was a deterministic or a stochastic [42, 60] one.

3.2.2 Diffusion in Deterministic Non-chaotic Systems

With all the proviso on its interpretation, Gaspard et al. [41] experiment had a very positive role not only in stimulating the discussion about the chaos/noise distinction but also in focusing the attention on deep conceptual aspects of diffusion. From a theoretical point of view, the study of chaotic models exhibiting diffusion and their non-chaotic counterpart is indeed important to better understand the role of microscopic chaos on macroscopic diffusion.

In Lorentz gases, the diffusion coefficient is related, by means of periodic orbits expansion methods [13, 14, 61], to chaotic indicators such as the Lyapunov exponents. This suggested that chaos was or might have been the basic ingredient for diffusion. However, as argued by Dettman and Cohen [17], even an accurate numerical analysis based on the ϵ-entropy, being limited by the finiteness of the data points, has no chance to detect differences in the diffusive behavior between a chaotic Lorentz gas and its non-chaotic counterpart, such as the wind-tree Ehrenfest's model. In the latter model, particles (wind) scatter against square obstacles (trees) randomly distributed in the plane but with fixed orientation. Since the reflection by the flat edges of the obstacles cannot produce exponential separation of trajectories, the maximal Lyapunov exponent is zero. The result of [17] implies thus that chaos may be not indispensable for having deterministic diffusion. The question may be now posed on what are the necessary microscopic ingredients to observe deterministic diffusion at large scales.

We would like to remark that, in the wind-tree Ehrenfest's model, the external randomness amounting to the disordered distribution of the obstacles is crucial. Hence, one may conjecture that a finite spatial entropy density h_S is necessary for observing diffusion. In this case, deterministic diffusion might be a consequence either of a non-zero "dynamical" entropy ($h_{KS} > 0$) in chaotic systems or of a non-zero "static" entropy ($h_S > 0$) in non-chaotic systems. This is a key-point, because someone can argue that a deterministic infinite system with spatial randomness can be interpreted as an effective stochastic system.[4]

With the aim of clarifying this point, we consider here a spatially disordered non-chaotic model [62], which is the one-dimensional analog of a two-dimensional non-chaotic Lorentz system with polygonal obstacles. Let us start with the map defined by (3.3) and (3.4), and introduce some modifications to make it non-chaotic. One can proceed as exemplified in Fig. 3.4, that is by replacing the function (3.4) on each unit cell by its step-wise approximation generated as follows. The first-half of C_ℓ is partitioned in N micro-intervals

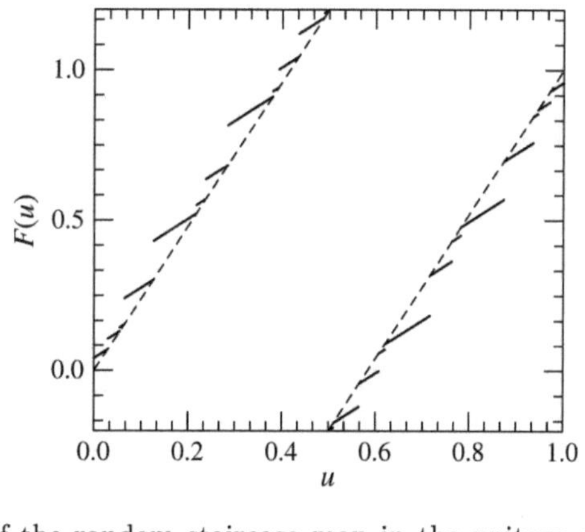

Fig. 3.4. Sketch of the random staircase map in the unitary cell. The parameter a defining the macroscopic slope is set to 0.23. Half domain $[0, 1/2]$ is divided into $N = 12$ micro-intervals of random size. The map on $[1/2, 1]$ is obtained by applying the antisymmetric transformation with respect to the center of the cell $(1/2, 1/2)$

[4] This is probably a "matter of taste."

$[\ell+\xi_{n-1}, \ell+\xi_n[$, $n = 1, \ldots, N$, with $\xi_0 = 0 < \xi_1 < \xi_2 < \cdots < \xi_{N-1} < \xi_N = 1/2$. In each interval, the map is defined by its linear approximation

$$F_\Delta(u) = u - \xi_n + F(\xi_n) \qquad \text{if } u \in [\xi_{n-1}, \xi_n[\;, \tag{3.15}$$

where $F(\xi_n)$ is (3.4) evaluated at ξ_n. The map in the second half of the unit cell is then determined by the anti-symmetry condition with respect to the middle of the cell. The quenched random variables $\{\xi_k\}_{k=1}^{N-1}$ are uniformly distributed in the interval $[0, 1/2]$, i.e. the micro-intervals have a *random* extension. Further they are chosen independently in each cell C_ℓ (so one should properly write $\xi_n^{(\ell)}$). All cells are partitioned into the same number N of randomly chosen micro-intervals (of mean size $\Delta = 1/N$). This modification of the continuous chaotic system is conceptually equivalent to replacing circular by polygonal obstacles in the Lorentz system [17].

Since F_Δ has slope 1 almost everywhere, the map is no longer chaotic, violating the condition *i)* (see Sect. 3.2). For $\Delta \to 0$ (i.e. $N \to \infty$) the continuous chaotic map (3.3) is recovered. However, this limit is singular and as soon as the number of intervals is finite, even if extremely large, chaos is absent. It has been found [62] that this model still exhibits diffusion in the presence of both quenched disorder and a quasi-periodic external perturbation

$$x(t+1) = [x(t)] + F_\Delta(x(t) - [x(t)]) + \gamma \cos(\alpha t) \;. \tag{3.16}$$

The strength of the external forcing is controlled by γ and α defines its frequency, while Δ indicates a specific quenched disorder realization. The sign of γ is irrelevant; without lack of generality we study the case $\gamma > 0$.

The diffusion coefficient D is then numerically computed from the linear asymptotic behavior of the mean quadratic displacement, see (3.11). The results, summarized in Fig. 3.5, show that D is significantly different from zero only for values $\gamma > \gamma_c$. For $\gamma > \gamma_c$, D exhibits a saturation close to the value of the chaotic system (horizontal line) defined by (3.3) and (3.4). The existence of a threshold γ_c is not surprising. Due to the staircase nature of the system, the perturbation has to exceed the typical discontinuity of F_Δ to activate the "macroscopic" instability which is the first step toward the diffusion. Data collapsing, obtained by plotting D versus γN, in Fig. 3.5 confirms this argument. These findings are robust and do not depend on the details of forcing. Therefore, we have an example of a non-chaotic model in the Lyapunov sense by construction, which performs diffusion.

Now the question concerns the possibility that the diffusive behavior arises from the presence of a quenched randomness with non-zero spatial entropy per unit length. To clarify this point, similarly to [17], the model can be modified in such a way that the spatial entropy per unit cell is forced to be zero, and see if the diffusion still persists.

Zero spatial entropy per unit length may be obtained by repeating the same disorder configuration every M cells (i.e. $\xi_n^{(\ell)} = \xi_n^{(\ell+M)}$). Looking at the diffusion of an ensemble of walkers it was observed that diffusion is still present

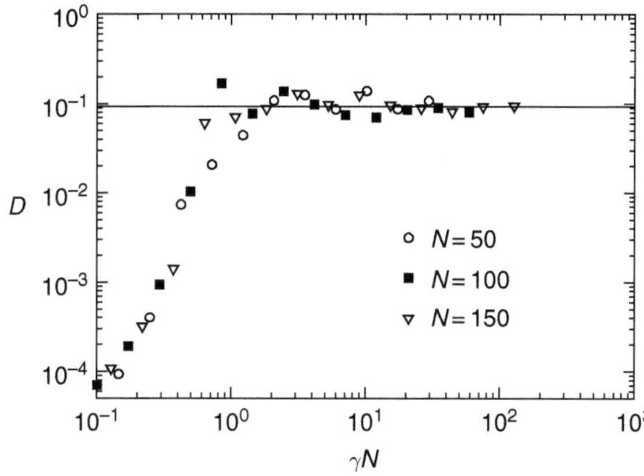

Fig. 3.5. Log–Log plot of the dependence of the diffusion coefficient D on the external forcing strength γ. Different data relative to a number of cell micro-intervals $N = 50$, 100 and 150 are plotted vs the natural scaling variable γN to obtain a collapse of the curves. Horizontal line represents the result for chaotic system (3.3, 3.4)

with D very close to the expected value (as in Fig. 3.5). A careful analysis (see [62] for details) showed that the system displays genuine diffusion for a very long times even with a vanishing (spatial) entropy density, at least for sufficiently large M.

These results along with those by Dettman and Cohen [17] allow us to draw some conclusions on the fundamental ingredients for observing deterministic diffusion (both in chaotic and non-chaotic systems).

- An instability mechanism is necessary to ensure particle dispersion at small scales (here small means inside the cells). In chaotic systems, this is realized by the sensitivity to the initial condition. In non-chaotic systems, this may be induced by a finite size instability mechanisms. Also, with zero maximal Lyapunov exponent one can have a fast increase of the distance between two trajectories initially close [63]. In the wind-tree Ehrenfest model this stems from the edges of the obstacles, in the "stepwise" system of Fig. 3.4 from the jumps.
- A mechanisms able to suppress periodic orbits and therefore to allow for a diffusion at large scale.

It is clear that the first requirement is not very strong while the second is more subtle. In systems with "strong chaos," all periodic orbits are unstable and, so, it is automatically fulfilled. In non-chaotic systems, such as the non-chaotic billiards studied by Dettman and Cohen and the map (3.16), the stable periodic orbits seem to be suppressed or, at least, strongly depressed, by the quenched randomness (also in the limit of zero spatial entropy). However, unlike the two-dimensional non-chaotic billiards, in the one-dimensional system (3.4,3.15,3.16), the periodic orbits may survive to the presence of disorder, so we need the aid of a quasi-periodic perturbation to obtain their destruction and the consequent diffusion.

3.3 The Heritage of the Fermi–Pasta–Ulam Problem for the Statistical Mechanics

The ergodic theory begun with Boltzmann's effort to justify the determination of average values in kinetic theory. Ergodic hypothesis states that time averages of observables of an isolated system at the equilibrium can be computed as phase averages over the constant-energy hyper-surface. This statement can be regarded as the first attempt to establish a link between statistical mechanics and the dynamics of the underlying system. One can say that proving the validity of ergodic hypothesis provides a "dynamical justification" of statistical ensembles.

The ergodic problem, at an abstract level, had been attacked by Birkhoff and von Neumann who proved their fundamental theorems on the existence of time averages and established a necessary and sufficient condition for the ergodicity. In spite of their mathematical importance, on a practical ground such theorems do not help very much to really solve the ergodic problem in statistical physics.

There exists a point of view according to which the effectiveness of a statistical mechanics approach resides mainly on the presence of many degrees of freedom rather than on the underlying (chaotic or regular) dynamics. Khinchin in his celebrated book *Mathematical Foundation of the Statistical Mechanics* [4] presents some important results on the ergodic problem that need no metrical transitivity. The main point of his approach relies on the concept of relevant physical observables in systems with a huge number of degrees of freedom. Since physical observables are non-generic functions (in mathematical sense), the equivalence between time and ensemble averages should be proved only for a restricted class of relevant observables. Moreover for physical purposes, it is "fair" to accept the failure of ergodicity for few (in the sense of sets of small measure) initial conditions.

In plain words, Khinchin's formulation, coinciding with Boltzmann's point of view (see, e.g., Chap. 1 of [2]), asserts that statistical mechanics works, independently of ergodicity, because the (most meaningful) physical observables are practically constant, a part in regions of very small measure, on the constant energy surface. Within this approach, dynamics have a marginal role, and the existence of "good statistical properties" is granted by the large number of degrees of freedom. However, the validity of Khinchin's statement restricts to a special class of observables not covering all the physically interesting possibilities. Therefore for each case, a detailed study of the specific dynamics is generally needed to assess the statistical properties of a given system.

The issue of ergodicity is naturally entangled with the problem of the existence of non-trivial conserved quantities (first integrals) in Hamiltonian systems. Consider a system governed by the Hamiltonian

$$H(\mathrm{I},\, \phi) = H_0(\mathrm{I}) + \epsilon H_1(\mathrm{I},\, \phi)\ , \tag{3.17}$$

where $\mathrm{I} = (I_1, \ldots, I_M)$ are the action variables and $\phi = (\phi_1, \ldots, \phi_M)$ are the phase variables. If $\epsilon = 0$ the system is integrable, there are M independent first integrals (the actions I_i) and the motion evolves on M-dimensional tori. Two questions arise naturally. Do the trajectories of the system (3.17) remain "close" to those of the integrable one? Do some conserved quantity, besides energy, survive in the presence of a generic (small) perturbation $\epsilon H_1(\mathrm{I},\, \phi)$? Of course whenever other first integrals exist the system cannot be ergodic.

In a seminal work, H. Poincaré [64] showed that generally a system like (3.17) with $\epsilon \neq 0$ does not possess analytic first integrals other than energy. This result sounds rather positive for the statistical mechanics approach. In 1923, Fermi [65], generalizing Poincaré's result, proved that for generic perturbations H_1 and $M > 2$, there cannot exist, on the $2M - 1$ dimensional constant-energy surface, even a single smooth[5] surface of dimension $2M - 2$ that is analytical in the variables (I, ϕ) and ϵ. From this result, Fermi argued that generic (non-integrable) Hamiltonian systems are ergodic.

At least in the physicists' community, this conclusion was generally accepted and, even in the absence of a rigorous demonstration, there was a vast consensus that the non-existence theorems of regular first integrals implied ergodicity.

3.3.1 FPU: Relaxation to Equilibrium and Ergodicity Violation

Thirty-two years later Fermi itself, together with Pasta and Ulam, with one of the first numerical experiments, in the celebrated paper *Studies of Nonlinear Problems* [12] (often referred with the acronym FPU) showed that the ergodic problem was still far from being solved. The FPU model studies the time evolution of a chain of N particles, interacting by means of non-linear springs:

$$H = \sum_{n=0}^{N} \left[\frac{p_n^2}{2m} + \frac{K}{2}(q_{n+1} - q_n)^2 + \frac{\epsilon}{\alpha}(q_{n+1} - q_n)^\alpha \right]\ , \tag{3.18}$$

with boundary conditions $q_0 = q_{N+1} = p_0 = p_{N+1} = 0$, $\alpha = 3$ or 4 and $K > 0$. The Hamiltonian is of the form (3.17) with a harmonic (integrable) part and a non-integrable (anharmonic) term $O(\epsilon)$. For $\epsilon = 0$, one has a collection of N non-interacting harmonic modes of energies E_ks, which remain constant. What happens if an initial condition is chosen in such a way that all the energy is concentrated in a few normal modes, for instance $E_1(0) \neq 0$ and $E_k(0) = 0$ for $k = 2, \ldots, N$? Before the FPU work, the general expectation would have

[5] For instance, analytic or differentiable enough.

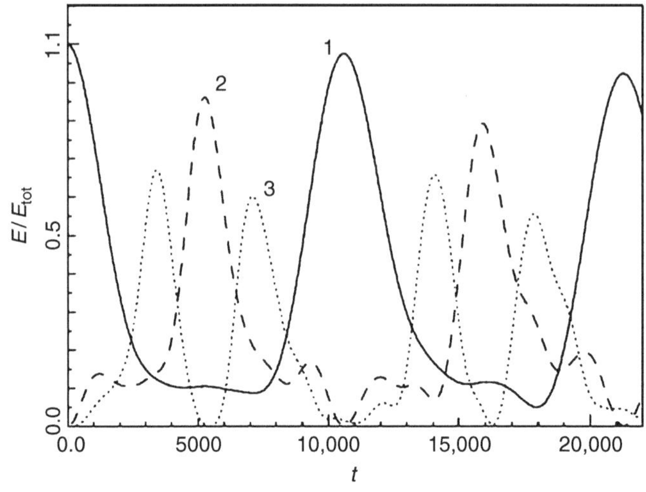

Fig. 3.6. $E_1(t)/E_{\rm tot}$, $E_2(t)/E_{\rm tot}$, $E_3(t)/E_{\rm tot}$ for the FPU system, with $N = 32$, $\alpha = 3$, $\epsilon = 0.1$ and energy density $\mathcal{E} = E_{\rm tot}/N = 0.07$. (Courtesy of G. Benettin [66])

been that the first normal mode would have progressively transferred energy to the others and that, after some relaxation time, every $E_k(t)$ would fluctuate around the common value. Therefore, it came as a surprise the fact that no tendency toward equipartition was observed, even for long times. In other words, a violation of ergodicity and mixing was found. Figure 3.6 shows the time evolution of the fraction of energy contained in three modes ($k = 1, 2, 3$), in a system with $N = 32$.

At the beginning all the energy is contained in mode 1. Instead of a distribution of the energy among all the available modes, with a loss of memory of the initial state, the system exhibits a close to periodic behavior. The absence of equipartition can be well appreciated looking at Fig. 3.7, where the quantities

$$E_{(\mathrm{av})k}(T) = \frac{1}{T}\int_0^T E_k(t)\mathrm{d}t\,, \qquad \text{with} \quad k = 1, \ldots, N\,, \tag{3.19}$$

i.e. the energies in the modes, averaged along the observation time T, are displayed. As one can see, almost all of the energy remains confined in the first four modes.

The existence of non-ergodic behavior in non-integrable Hamiltonian systems is actually a consequence of the so-called KAM theorem [67, 68, 69], whose first formulation, due to A. N. Kolmogorov, dates back to the year before the FPU paper. This was surely unbeknown to Fermi and his colleagues. The FPU result can be seen (a posteriori) as a numerical "verification" of the KAM theorem and, above all, of its physical relevance, i.e. the tori survival for physically significant values of the non-linearity. After Kolmogorov and

FPU, it is now well established that ergodicity is a non-generic property of mechanical systems.

Concerning the FPU problem, in terms of the KAM theorem, the following scenario, at least for large but finite times, can be outlined [70, 71, 72]. For N particles and for a given energy density $\mathcal{E} = E/N$ there is a threshold ϵ_c for the strength of the perturbation such that

(a) if $\epsilon < \epsilon_c$ the KAM tori are dominant and the system is essentially regular;
(b) if $\epsilon > \epsilon_c$ the KAM tori are negligible and the system is essentially chaotic.

However, the long-time evolution of very large chains with small ϵ is hindered by the presence of metastable states. To probe such an asymptotics by numerical simulations is extremely hard, for a discussion on the subject see the contributions by Benettin et al. and Lichtenberg et al. in this volume.

In most of the physical situations where the strength of the perturbation (i.e. the Hamiltonian) is fixed, the control parameter is $\mathcal{E}$. There exists a critical energy density, separating regular from chaotic behaviors. This is evident by comparing Fig. 3.8 with Fig. 3.7. In Fig. 3.8 the same quantities of Fig. 3.7 are plotted, but now they refer to a system where the energy density is much greater than before: $\mathcal{E} = 1.2$; the system has entered the chaotic region and equipartition is established.

However, also when most KAM tori are destroyed, and the system turns out to be chaotic, the validity of ordinary statistical mechanics is not automatically granted. Indeed the relaxation time for reaching equipartition may

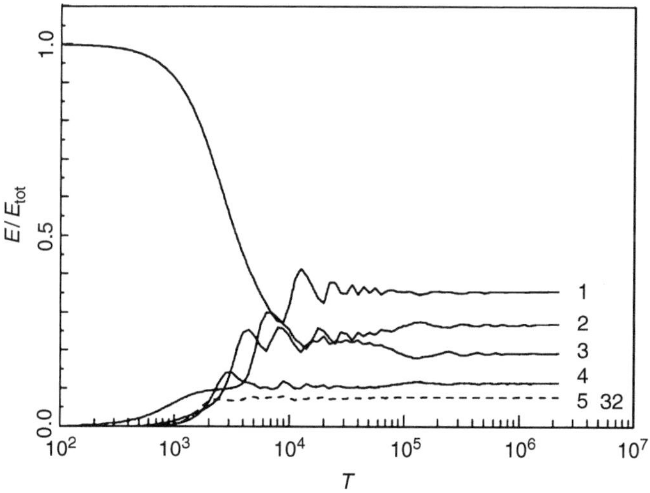

Fig. 3.7. Time averaged fraction of energy, in modes $k = 1, 2, 3, 4$ (*bold lines*, from top to below) and $\sum_{k=5}^{32} E_{(\mathrm{av})k}(T)/E_{\mathrm{tot}}$ (*dashed line*). The parameters of the system are the same as in Fig. 3.6 (Courtesy of G. Benettin [66].)

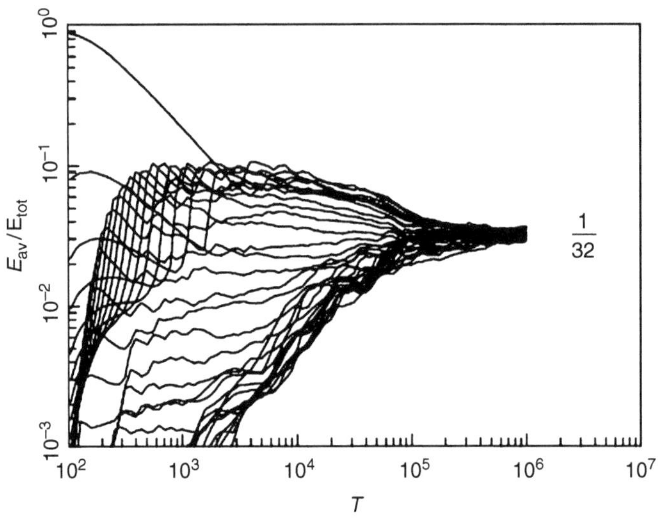

Fig. 3.8. Time-averaged fraction of energy, in all the modes $k = 1, \ldots, 32$. The parameters of the system are $N = 32$, $\alpha = 3$, $\epsilon = 0.1$ and energy density $\mathcal{E} = E_{\text{tot}}/N = 1.2$ (Courtesy of G. Benettin [66])

become very large (see [73, 74, 75, 76, 77, 78, 79] for a detailed discussion about this point).

The problem of slow relaxation is rather common in high-dimensional Hamiltonian systems, where [80, 81] though the phase-space volume occupied by KAM tori decreases exponentially with the number of degrees of freedom (which sounds like a good news for statistical mechanics) nonetheless very long time-scales are involved. This means that it takes an extremely long time for the individual trajectories to forget their initial conditions and to invade a non-negligible part of the phase space. Indeed, even for very large systems, Arnol'd diffusion is very weak and different trajectories, although with a high value of the Lyapunov exponent, maintain some of their own features for a very long time.

We conclude this part emphasizing that also in high-dimensional systems the actual role of chaos is not yet well understood. For instance, in [82] detailed numerical computations on the FPU system show that both the internal energy and the specific heat, computed with a time average, as functions of the temperature are rather close to the prediction of the canonical ensemble. This is true also in the low-energy region (i.e. low temperature) where the system behaves in a regular way (the KAM tori are dominant). This supports Khinchin approach (though the observables are not in the class of the sum functions[6]) on the poor role of dynamics. Indeed strong chaos seems to be

[6] Khinchin defines sum functions as any function of the form $\sum_{n=1}^{N} f_n(q_n, p_n)$, f_n assuming order 1 values. Such observables, in the large N limit, are self-averaging,

unnecessary for the prediction of the statistical mechanics to hold. However, this is not the end of the story because in other non-linear systems (such as a chain of coupled rotators) the situation is different: even in the presence of strong chaos one can observe disagreement between time average and ensemble average [82].

In the following, we discuss the problem of heat transport that allows us to discuss the role of chaos for the validity of transport properties.

3.3.2 Heat Transport in Chaotic and Non-chaotic Systems

As stated in the introduction, a part of the statistical mechanics community accepts the picture according to which the instabilities of microscopic particle dynamics are the basic requirement for the onset of macroscopic transport. In this framework, several works [13, 14] have shown that, in some systems, there exists a relationship between transport coefficients (thermal or electrical conductivity, viscosity, diffusivity, etc.) and Lyapunov exponents. Such a link is of remarkable importance because it establishes a straightforward connection between the microscopic dynamical properties of a system and its macroscopic behavior, which is the main goal of statistical mechanics. However, as exemplified in the previous sections, chaotic dynamics does not seem to be a necessary condition to both equilibrium and out-of-equilibrium statistical mechanics approaches. In fact, we have seen that transport may occur even in the absence of deterministic chaos. These counterexamples pose some doubts on the generality and so on the conceptual relevance of the links found between chaotic indicators and macroscopic transport coefficients.

Heat conduction is a typical phenomenon that needs a microscopic mechanism leading to normal diffusion that distributes particles and their energy across the whole system. Since a chaotic motion has the same statistical properties of a "random walk," when observed at finite resolution, this mechanism can be found in the presence of either exponential instability in deterministic dynamics or intrinsic disorder and non-linearities.

In the context of the conduction problem, FPU chains have recently played an important role in further clarifying the transport properties of low spatial dimension systems. FPU models represent simple but non-trivial candidates to study heat transport by phonons in solids whenever their boundaries are kept at different temperatures. This issue becomes even more interesting at low spatial dimensions where the constraints set by the geometry may induce anomalous transport properties characterized by the presence of divergent transport coefficients in the thermodynamic limit [83]. Thermal conductivity χ, defined via the Fourier's Law

$$J = -\chi \nabla T \ ,$$

i.e. they are practically constant on the constant-energy surface, aside a region of small measure.

relates the heat (energy) flux J to a temperature gradient. When a small temperature difference $\delta T = T_1 - T_2$ is applied to the ends of a system of linear size L, the heat current across the system is expected to be

$$J = -\frac{\chi \delta T}{L} .$$

For some one- and even two-dimensional systems, theoretical arguments, confirmed by several simulations, predict a scaling behavior $J \sim L^{\alpha - 1}$ implying a size dependent conductivity

$$\chi(L) = L^{\alpha} . \tag{3.20}$$

As a consequence, χ diverges in the limit $L \to \infty$ with a power law whose exponent $\alpha > 0$ depends on the specific system considered. The presence of this divergence is referred to as *anomalous heat conduction* in contrast with normal conduction which, according to dimensional analysis of Fourier's Law, prescribes a finite limit for χ. FPU chains are systems where the anomaly in the heat transport is clearly observed. Its origin can be traced back to the existence of low-energy modes which survive long enough to propagate freely before scattering with other modes. Such modes can carry much energy and since their motion is mainly ballistic rather than diffusive, the overall heat transport results to be anomalous. Models other than FPU indeed presents this peculiar conduction, as widely shown in the literature [18, 84, 85]. Then the issue is the general understanding of the conditions leading to this phenomenon and more specifically the role of microscopic dynamical instabilities. A well-known chaotic system, such as the Lorentz Gas in a channel [86] configuration, provides an example of a system with normal heat conduction. This model consists of a series of semicircular obstacles with radius R arranged in a lattice along a slab of size $L \times h$ ($h << L$) see Fig. 3.9. As in a Lorentz system, particles scatter against obstacles but do not interact with each other.

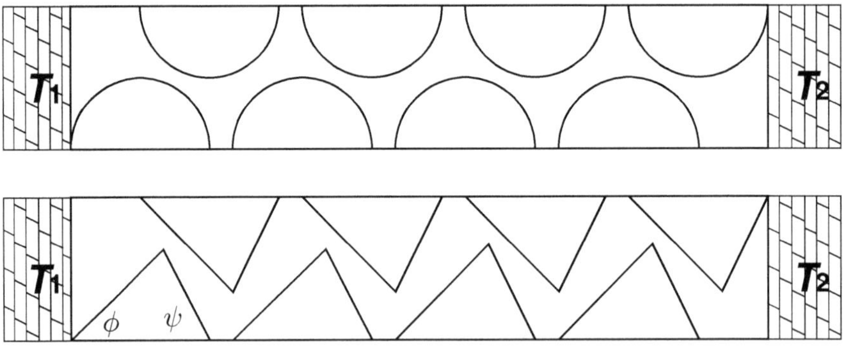

Fig. 3.9. Example of channel geometry used in [84, 86] to study heat transport in low-dimensional chaotic (*upper panel*) and non chaotic billiards (*lower panel*)

Two thermostats at temperatures T_1 and T_2 respectively are placed at each end of the slab to induce transport. They reinject into the system those particles reaching the ends with a velocity drawn from a Gaussian velocity distribution with variance proportional to the temperatures T_1 and T_2. In the case of semicircular obstacles, the system is chaotic and one observes a standard Fourier's Law [86].

In [18] and [84], some non-chaotic variants of the Lorentz channel have been proposed in order to unravel the role of exponential instabilities in the heat conduction. In those models, called the Ehrenfest Channel, the semicircular obstacles were replaced with triangular ones, so that the system is trivially non-chaotic since collisions with flat edges of the obstacles cannot separate trajectories more than algebraically. The results show that when two angles (e.g. ϕ and ψ) of the triangles are irrational multiple of π, the system exhibits a normal heat conduction. On the contrary, for rational ratio, such as isosceles right triangles, the conduction becomes anomalous. The single particle heat flux across N cells $J_1(N)$ scales as $J_1(N) \sim N^\alpha$, while the temperature gradient behaves as $1/N$ implying that $\chi(N)$ diverges as $N \to \infty$. The explanation of such a divergence can be found in the single-particle diffusivity along the channel direction which occurs in a non-standard way. Indeed, the evolution of a large set of particles has a mean squared displacements from initial conditions which grows in time with a power-law behavior

$$\langle [x(t) - x(0)]^2 \rangle \sim t^b$$

with an exponent $1 < b < 2$. This super-diffusion is the unique responsible for a divergent thermal conductivity independently of Lyapunov instabilities, since the model has a zero Lyapunov exponents.

When an Ehrenfest Channel with anomalous thermal conductivity is disordered, for instance, by randomly modulating the height of triangular obstacles or their positions along the channel, the conduction follows Fourier's law, becoming normal [18]. This scenario is rather similar to that one discussed in Sect. 3.2.2 for diffusion on non chaotic maps.

The works in [83, 87, 88] suggest that the anomalous conduction is associated with the presence of a mean free path of energy carriers that can behave abnormally in the thermodynamic limit. For FPU the long mean free path is due to soliton-like ballistic modes. In the channels, the long free flights, between consecutive particle collisions, become relevant. The above considerations suggest a very week role of chaos for heat transport, and for transport in general, since also systems without exponential instability may show transport, even anomalous.

3.4 Concluding Remarks

The problem of distinguishing chaos from noise cannot receive an absolute answer in the framework of time series analysis. This is due to the finiteness of the observational data set and the impossibility to reach an arbitrary fine resolution and high embedding dimension. However, we can classify the signal behavior, without referring to any specific model, as stochastic or deterministic on a certain range of scales.

Diffusion may be realized in both stochastic and deterministic systems. In particular, as the analysis of polygonal billiards and non-chaotic maps (see Sect. 3.2.2) shows, chaos is not a prerequisite for observing diffusion and, more in general, nontrivial statistical behaviors.

In a similar way, we have that for the validity of heat conduction chaos is not a necessary ingredient. Also in systems with zero maximal Lyapunov exponent (see [84, 85, 86]) the Fourier's law (or its anomalous version) can hold.

We conclude by noticing that the poor role of exponential instability for the validity of statistical laws does not seem to be limited to transport problems. For instance it is worth mentioning the interesting results of Lepri et al. [89] showing that the Gallavotti–Cohen formula [90], originally proposed for chaotic systems, holds also in some non-chaotic model.

Acknowledgments

The authors express their gratitude to D. Del-Castillo-Negrete, O. Kantz and E. Olbrich who recently collaborated with them on the issues discussed in this chapter. A special thanks to G. Benettin for having provided us with Figs. 3.6, 3.7, 3.8.

References

1. P. Ehrenfest and T. Ehrenfest, *The conceptual foundation of the statistical approach in mechanics.* Cornell University Press, New York, 1956, original edition in German 1912.
2. G. Gallavotti, *Statistical mechanics. A short treatise.* Springer-Verlag, Berlin, 1999.
3. J.P. Eckmann and D. Ruelle, Rev. Mod. Phys. **57**, 617 (1985).
4. A.I. Khinchin, *Mathematical foundations of statistical mechanics.* Dover Publications, Inc., New York, 1949.
5. I. Prigogine, *Les Lois du Chaos.* Flammarion, Paris, 1994.
6. J.L. Lebowitz, Phys. Today **46**, 32 (1993).
7. J. Bricmont, Phys Mag. **17**, 159 (1995).
8. C. Cercignani, *Ludwig Boltzmann: the man who trusted atoms.* Oxford University Press, 1998.
9. E. Garber, S.G. Brush and C.W.F. Everitt, *Maxwell on heat and statistical mechanics.* Lehigh University Press, Bethlehem, PA, 1995.
10. D. Driebe, Phys. Today, Nov. 1994 (letter to editor), p. 13.
11. A. Einstein, Ann. Phys. **17**, 549 (1905) [English translation in: *Investigations on the theory of the Brownian movement.* Dover Publications, Inc., New York, 1956.

12. E. Fermi, J. Pasta and S. Ulam, *Studies of non linear problems*, Los Alamos Sci. Lab. Rep. LA-1940 (1955).
13. P. Gaspard *Chaos, scattering, and statistical mechanics.* Cambridge University Press, Cambridge, 1998.
14. J.R. Dorfman, *An introduction to chaos in nonequilibrium statistical mechanics.* Cambridge University Press, Cambridge, 1999.
15. J.L. Vega, T. Uzer and J. Ford, Phys. Rev. E **48**, 4314 (1993).
16. G.M. Zaslavsky, Phys. Rep. **371**, 461 (2002).
17. C.P. Dettmann and E.D.G. Cohen, J. Stat. Phys. **101**, 775 (2000).
18. B. Li, L. Wang and B. Hu, Phys. Rev. Lett. **88**, 223901 (2002).
19. R. Kubo, Science **233**, 330 (1986).
20. For a historical introduction to the Brownian motion see S. Chandrasekhar, Rev. Mod. Phys. **15**, 1 (1943).
21. P. Langevin, C. R. Acad. Sci. (Paris) **146**, 530 (1908) [English translation: Am. J. Phys. **65**, 1079 (1997)].
22. G.E. Uhlenbeck and L.S. Ornstein, Phys Rev. **36**, 823 (1930).
23. M. Smoluchowski, Ann. Phys. **21**, 756 (1906).
24. R.E. Turner, Physica (Amsterdam) **26**, 274 (1960).
25. P. Mazur and E. Montroll, J. Math. Phys. **1**, 70 (1960).
26. G.W. Ford, M. Kac and P. Mazur, J. Math. Phys. **6**, 504 (1965).
27. P.E. Phillipson, J. Math. Phys. **15**, 2127 (1974).
28. R.J. Rubin, J. Math. Phys. **1**, 309 (1960).
29. P. Mazur and E. Braun, Physica (Amsterdam) **30**, 1973 (1964).
30. E.N. Lorenz, J. Atmos. Sci. **20**, 130 (1963).
31. H.A. Lorentz, Proc. Amst. Acad. **7**, 438; 585; 604 (1905).
32. Ya.G. Sinai, Funkts. Anal. Ego Prilozh. **13**, 46 (1979).
33. L.A. Bunimovich and Ya.G. Sinai, Commun. Math. Phys. **78**, 479 (1981).
34. H. Fujisaka and S. Grossmann, Z. Phys. B **48**, 261 (1982).
35. S. Grossmann and S. Thomae, Phys. Lett. A **97**, 263 (1983).
36. T. Geisel and S. Thomae, Phys. Rev. Lett. **52**, 1936 (1984).
37. T. Geisel, J. Nierwetberg and A. Zacherl, Phys. Rev. Lett. **54**, 616 (1985).
38. G.M. Zaslavsky, D. Stevens and H. Weitzener, Phys. Rev. E **48**, 1683 (1993).
39. H.A. Posch and W.G. Hoover, Phys. Rev. A **38**, 473 (1988).
40. H. van Beijeren, J.R. Dorfman, H.A. Posch and Ch. Dellago, Phys. Rev. E **56**, 5272 (1997).
41. P. Gaspard, M.E. Briggs, M.K. Francis, J.V. Sengers, R.W. Gammon, J.R. Dorfman and R.V. Calabrese, Nature **394**, 865 (1998).
42. M. Cencini, M. Falcioni, H. Kantz, E. Olbrich and A. Vulpiani, Phys. Rev. E **62**, 427 (2000).
43. M. Romero-Bastida, Phys. Rev. E **69** 056204 (2004).
44. M. Romero-Bastida, D. Castañeda and E. Braun, Phys. Rev. E **71** 046207 (2005).
45. H. Kantz and T. Schreiber, *Nonlinear time series analysis.* Cambridge University Press, Cambridge, UK, 1997.
46. H.D.I. Abarnabel, *Analysis of observed chaotic data.* Springer-Verlag, New York, 1996.
47. G. Sugihara and R. May, Nature **344**, 734 (1990).
48. M. Casdagli, J. Roy. Statist. Soc. Ser. B **54**, 303 (1991).
49. D.T. Kaplan and L. Glass, Phys. Rev. Lett. **68**, 427 (1992).

50. D.T. Kaplan and L. Glass, Physica D **64**, 431 (1993).
51. A.N. Kolmogorov, IRE Trans. Inf. Theory **1**, 102 (1956).
52. C.E. Shannon, Bell Sys. Tech. J. **27**, 623 (1948); **27**, 379 (1948).
53. P. Gaspard and X.J. Wang, Phys. Rep. **235**, 291 (1993).
54. G. Boffetta, M. Cencini, M. Falcioni and A. Vulpiani, Phys. Rep. **356**, 367 (2002).
55. T. Berger, *Rate distortion theory.* Prentice-Hall, Englewood Cliffs, 1971.
56. E. Olbrich and H. Kantz, Phys. Lett. A **232**, 63, 1997.
57. C. Dettman, E. Cohen, and H. van Beijeren, Nature **401**, 875 (1999).
58. P. Grassbeger and T. Schreiber, Nature **401**, 875 (1999).
59. M. Abel, L. Biferale, M. Cencini, M. Falcioni, D. Vergni and A. Vulpiani, Physica D **147**, 12 (2000).
60. G. Kubin, *Workshop on nonlinear signal and image processing, Vol. 1*, IEEE. IEEE, Halkidiki, Greece, 1995, pp. 141–145.
61. G.P. Morris and L. Rondoni, J. Stat. Phys. **75**, 553 (1994).
62. F. Cecconi, D. del-Castillo-Negrete, M. Falcioni and A. Vulpiani, Physica D **180**, 129 (2003).
63. A. Torcini, P. Grassberger and A. Politi, J. Phys. A **27**, 4533 (1995).
64. H. Poincaré, Acta Math. **13**, 1 (1890).
65. E. Fermi, Phys. Zeits. **24**, 261 (1923).
66. G. Benettin, in *Introduzione alla teoria Ergodica*, from the web page http://www.math.unipd.it/~benettin/ .
67. A.N. Kolmogorov, Dokl. Akad. Nauk SSSR **98**, 527 (1954).
68. V.I. Arnold, Russ. Math. Surv. **18**, 9 (1963).
69. J.K. Moser, Nachr. Akad. Wiss. Göttingen Math. Phys. kl. **2**, 1 (1962).
70. F.M. Izrailev and B.V. Chirikov, Dokl. Akad. Nauk SSSR **166**, 57 (1966).
71. P. Bocchieri, A. Scotti, B. Bearzi and A. Loinger, Phys. Rev. A **2**, 2013 (1970); M. Casartelli, G. Casati, E. Diana, L. Galgani and A. Scotti, Theor. Math. Phys. **29**, 205 (1976);
R. Livi, M. Pettini, S. Ruffo, M. Sparpaglione and A. Vulpiani, Phys. Rev. A **31**, 1039 (1985).
72. G. Benettin, *Molecular-dynamics simulation of statistical–mechanical systems.* G. Ciccotti and W.G. Hoover, ed. North-Holland, Amsterdam, 1986.
73. G. Benettin and A. Tenenbaum, Phys. Rev. A **28**, 3020 (1983).
74. H. Kantz, Physica D **39**, 322 (1989);
H. Kantz, R. Livi and S. Ruffo, J. Stat. Phys. **76**, 627 (1994).
75. L. Casetti, M. Cerruti-Sola, M. Pettini and E.G.D. Cohen, Phys. Rev. E **55**, 6566 (1997).
76. J. De Luca, A.J. Lichtenberg and S. Ruffo, Phys. Rev. E **60**, 3781 (1999).
77. S. Ruffo, *Chance in physics: foundations and perspectives.* J. Bricmont et al., eds, Springer-Verlag, Berlin, 2000.
78. C. Alabiso and M. Casartelli, J. Phys. A **33**, 831 (2000).
79. R. Livi, M. Pettini, S. Ruffo and A. Vulpiani, J. Stat. Phys. **48**, 539 (1987).
80. M. Falcioni, U. Marini-Bettolo-Marconi and A. Vulpiani, Phys. Rev. A **44**, 2263 (1991).
81. L. Hurd, C. Grebogy and E. Ott, in *Hamiltonian mechanics* J. Siemenis, ed., Plenum, New York, 1994.
82. R. Livi, M. Pettini, S. Ruffo and A. Vulpiani, J. Stat. Phys. **48**, 539 (1987).
83. S. Lepri, R. Livi and A. Politi, Phys. Reports, **377**, 1–80 (2003).

84. B. Li, G. Casati and L. Wang, Phys. Rev. E **67**, 021204 (2003).
85. P. Grassberger, W. Nadler and L.Yang, Phys. Rev. Lett. **89**, 180601 (2002).
86. D. Alonso, R. Artuso, G. Casati and I. Guarneri, Phys. Rev. Lett. **82**, 1859 (1999).
87. B. Li and J. Wang, Phys. Rev. Lett. **91**, 044301 (2003).
88. B. Li, H. Zhao and B. Hu, Phys. Rev. Lett. **86**, 63 (2001).
89. S. Lepri, L. Rondoni and G. Benettin, J. Stat. Phys. **99**, 857 (2000).
90. G. Gallavotti and E.G.D. Cohen, Phys. Rev. Lett **74**, 2694 (1995).

4

The Fermi–Pasta–Ulam Problem and the Metastability Perspective

G. Benettin,[1] A. Carati,[2] L. Galgani[2] and A. Giorgilli[2]

[1] Università di Padova, Dipartimento di Matematica Pura e Applicata, Via G. Belzoni 7, 35131 Padova, Italy
[2] Università di Milano, Dipartimento di Matematica, Via Saldini 50, 20133 Milano, Italy

Abstract. A review is given of the works on the FPU problem that were particularly relevant in connection with the metastability perspective, proposed in the year 1982. The idea is that there exists a specific energy threshold above which the time-averages of the relevant quantities quickly agree with the predictions of classical equilibrium statistical mechanics, whereas below it there exist two time scales. First there is a quick formation of a packet of low-frequency modes which do share the energy, and this produces a metastable state that lasts for a long time; then the system attains the final equilibrium state. There are strong indications that the specific energy threshold does not vanish in the limit of infinitely many particles. The review is given for the case of a one-dimensional FPU chain.

4.1 Introduction

If one looks at the scientific literature on the FPU problem, 50 years after the original paper (or rather *report*) [1], one will find a rather large amount of papers (see, for example the recent special issue of the journal Chaos [2]). But if one tries to extract from them any clear conclusion about the mathematical status of the problem or the physical meaning of the results, one may remain rather perplexed and have the impression of a certain confusion. Or, even, one can find statements as if the problem had already been solved and there were nothing more to be said (see [3], p. 2). In this chapter, we will try to indicate, among the many papers on the subject, the ones which in our opinion played a significant role with respect to the main question we have in mind, namely that of establishing whether the FPU problem may have some relevant physical impact or not. We will try to show how the question is still completely open, although one may be confident that it may be solved in a near future.

We now give a preliminary summary of the history we are going to trace back in this chapter, in the above mentioned perspective. First of all, let us

G. Benettin et al.: *The Fermi–Pasta–Ulam Problem and the Metastability Perspective*, Lect. Notes Phys. **728**, 151–189 (2008)
DOI 10.1007/978-3-540-72995-2_4

recall that the essential result of the original FPU report was the exhibition of what we now call "the FPU paradox". Namely, numerical solutions of the equations of motion were performed for a model of a discretized string (or equivalently of a one-dimensional crystal, actually, a chain of N particles with nearest-neighbour nonlinear interactions), and it was observed that, starting from a long-wavelength initial datum (and thus very far from statistical equilibrium), there was quickly formed an apparently stationary state, extremely different from the one expected according to classical equilibrium statistical mechanics.

The FPU report had, 10 years later (1965), a great impact in mathematics, because it stimulated the well-known work [4] of Zabusky and Kruskal, in which the FPU result was interpreted in terms of solitons. In turn, this fact paved the way to the whole research on infinite-dimensional integrable systems, which quickly became a fashionable and extremely interesting mathematical field in itself, with the result that its relation to the FPU problem was somehow neglected. We will point out later how the relations between the two subjects, solitons and FPU problem, were reestablished in very recent times. In may be worth remarking that, as the soliton theory is essentially equivalent to integrability, by some naive transitivity some people may have been induced to associate FPU to integrability, which corresponds to even exalting the FPU paradox.

The next essential step, which by the way also eliminated the possible confusion just mentioned, was made one year later (1966) by Izrailev and Chirikov in the work [5] (see also [6]), with the discovery that the paradox disappears (i.e., a quick agreement with the predictions of classical equilibrium statistical mechanics is found) if initial data are taken of the FPU type (long-wavelength), but with a sufficiently high energy. In other words, there somehow exists a critical energy E_{c}, above which the paradox disappears. However, Izrailev and Chirikov appeared even to go beyond such a result, because they also advanced the additional conjecture (supported by some kind of analytical considerations, later adjusted by their pupil Shepelyansky in [7], with arguments subsequently critically discussed by Ponno in [8]) that the FPU paradox disappears at all in the thermodynamic limit (N tending to infinity, with positive specific energy $\epsilon = E/N$). On the other hand, a little later (1971) Bocchieri et al. (in [9]) reported numerical results that appeared to support the opposite conjecture. This fact was particularly emphasized by Galgani and Scotti and by Cercignani (see [10], [11] and the review [12]), who were pointing out that the FPU paradox, if it persists in the thermodynamic limit, may have a deep physical impact. With such papers ends the first phase of the history of the FPU problem, at least in our personal way of reconstructing it. At that moment the alternative seemed to be whether the paradox disappears in the thermodynamic limit or not, i.e., whether one has $\epsilon_{\mathrm{c}} \to 0$ for $N \to \infty$ or not, where $\epsilon_{\mathrm{c}} = E_{\mathrm{c}}/N$ is the specific energy threshold.

But in such an alternative the mathematical (and even the physical) setting of the problem was a rather "naive" one, because it appeared that one had to decide whether, in the terminology then used, the motions are of "ordered" or of "chaotic" type (below or above the threshold respectively), whereas the deep question of determining the "relaxation times" for the approach to equilibrium (which actually was the very question raised in the original FPU report itself) was completely overlooked. The breakthrough in this direction came from a paper of the year 1982 by Fucito et al. (see [13]), which clearly was conceived within a scientific frame, the theory of glasses (particularly studied by a group of people around Parisi in Roma), in which a special attention was naturally payed to the possibility that relaxation times of quite different orders of magnitude may show up. This actually constitutes what we now call *the metastability scenario*: The time-averages of the relevant physical quantities are expected to agree with the predictions of classical equilibrium statistical mechanics at any energy after a sufficiently long time-scale, the "final" one (which is the one described by the limit $t \to \infty$), and the paradox is interpreted as corresponding to the existence of another, shorter, time-scale (the fast scale), within which a relaxation is produced to some intermediate state. Such an intermediate state at first sight appears as an equilibrium one, although it is destined to subsequently relax, on a much longer time-scale, to the "final"equilibrium state. The way in which the existence of an energy threshold can be conceived within such a metastability scenario was understood much later (see [14]).

In an attempt to trace back, in the present days, a kind of historical review on the subject, one cannot but remain perplexed by remarking that the paper of Fucito et al. did not receive at that time the attention that would appear natural today. Indeed, apart from a bunch of papers written immediately later, for a long time the metastability scenario was essentially forgotten. In particular, no discussion was given of the relevant problem that was left open within such a scenario, namely to establish whether the formation of a metastable state is a phenomenon that persists in the thermodynamic limit or not. The idea of the metastability perspective actually reappeared only rather recently, under the stimulus of the work [15] by Carati and Galgani (see also [16]), devoted to the problem of estimating the specific heats in systems of FPU type. In such a paper, the existence of relaxation times of different orders of magnitude was reported, and a qualitative analogy with the problem of glasses was explicitly pointed out. Finally, a vivid numerical illustration of the metastability phenomenon, with a particularly impressive exhibition of two quite different relaxation times, was given by Berchialla et al. in [14]. In particular, it was found that the phenomenon of the two separated time-scales occurs only below a certain energy, which can thus be interpreted as the critical energy in the sense of Bocchieri et al.

Finally, a deep analytical understanding of the metastability scenario in the FPU problem was given in a paper by Bambusi and Ponno (see [17]), through a result holding in the thermodynamic limit. In such a paper, by the way, a bridge with the old Zabusky and Kruskal contribution was given. Indeed, a general mathematical frame (the method of resonant normal forms) was devised in order to approximate the FPU system for not too long times, and this actually amounts to justify the use of a pair of KdV equations for not too long times, thus explaining the quick formation of the metastable state, in a way that is essentially equivalent to that of Fucito et al. The result of Bambusi and Ponno actually holds only for an extremely special class of initial conditions, but a strong indication that significant results may be obtained also for a much broader set of initial data, is afforded by a very recent result of one of us (see [18]), where for the first time it was proved, in a concrete model, that the techniques of Hamiltonian perturbation theory can be extended to the thermodynamic limit (previous results uniform in N were given, by Bambusi and Giorgilli in [19], but only for a finite energy E, namely for vanishing specific energy $\epsilon = E/N$ in the limit $N \to \infty$).

On the basis of the successes obtained with such recent results, one may be tempted to conclude that the FPU paradox should persist in the thermodynamic limit. But a deep question still remains open, namely the "question of the dimensions". Indeed, all the results previously mentioned refer to the FPU problem in its original formulation, namely in the one-dimensional case (a chain of particles), and there remains the problem of establishing whether the phenomenon of the quick formation of a metastable state persists (still in the thermodynamic limit) when one passes to the case of dimension two and especially to the "physical case" of dimension three. At the moment, a few results in dimension larger than one are available in the literature, for example [20, 21], but in our opinion they do not allow one to draw a definite conclusion. So in the present review we shall not enter the question. However, by judging from the results that were recently obtained in the one-dimensional case, we are rather confident that the problem may find a solution in the near future.

4.2 The First Phase, 1955–1972: From FPU to Izrailev and Chirikov and to Bocchieri et al.; the Suggestion of a Possible Physical Interpretation

4.2.1 The Original FPU Paper and the FPU Paradox

Fermi, Pasta and Ulam considered the simplest model of a discretized nonlinear string, which can also be interpreted as a model of a one-dimensional crystal, namely, a chain of equal particles with nearest-neighbours nonlinear interactions (nonlinear springs), and fixed ends. Denoting by x_j the displacements of the particles from their equilibrium positions and by p_j the corresponding momenta, $j = 0, \ldots, N+1$, with the boundary conditions

$$x_0 = 0 \,, \quad x_{N+1} = 0 \,,$$

the Hamiltonian of the system is then

$$H(x_1, \cdots, x_N, p_1, \cdots, p_N) = \frac{1}{2} \sum_{j=1}^{N} p_j^2 + \sum_{j=1}^{N+1} V(x_{j+1} - x_j) \tag{4.1}$$

where the potential V actually chosen was

$$V(x) = \frac{1}{2}x^2 + \frac{\alpha}{2}x^3 + \frac{\beta}{3}x^4 .$$

Here, the mass of the particles and the harmonic constant of the springs have been set equal to 1, while α and β are positive parameters. It is well known that the corresponding linearized system ($\alpha = \beta = 0$) can be transformed, through a linear change of variables, to a system of uncoupled linear oscillators (normal modes) with a corresponding Hamiltonian H_2 which, in terms of action-angle variables I_k, φ_k, takes the form

$$H_2 = \sum_{k=1}^{N} E_k ,$$

where

$$E_k = \omega_k I_k$$

are the normal-mode energies, having angular frequencies ω_k given by

$$\omega_k = 2 \sin \frac{k\pi}{2(N+1)} .$$

According to classical equilibrium statistical mechanics, the statistical properties of an isolated system (such as the FPU one) at equilibrium at a given total energy E should be described by the microcanonical measure (the one naturally induced on the "energy surface" $H = E$ by the Lebesgue measure in the whole phase space) or equivalently (at least for sufficiently large N) by the "canonical" or Gibbs measure in the whole phase space with a suitable temperature $T = T(E)$. The fundamental result of classical equilibrium statistical mechanics is then the "equipartition theorem", according to which, in the harmonic limit ($\alpha = \beta = 0$), the expected values of the harmonic energies E_k at a given total energy E, which we denote by $< E_k >_E$, are all equal, independent of k,

$$< E_k >_E = E/N \equiv \epsilon , \quad k = 1, \ldots, N , \tag{4.2}$$

where $\epsilon = E/N$ is the specific energy, and for the common value ϵ one has the interpretation $\epsilon = k_{\rm B}T$, where $k_{\rm B}$ is the Boltzmann constant and T the absolute temperature. The result does not change qualitatively for a slightly anharmonic system (α and β small) and for a small temperature T (i.e., a small specific energy $\epsilon = E/N$), because the anharmonic corrections to the relations (4.2) do vanish in the limit $\alpha, \beta \to 0$ or $T \to 0$ (i.e., $\epsilon \to 0$).

Let us now come to the dynamics. In the harmonic case the system is "integrable", namely it has N integrals of motion (the harmonic actions I_k or equivalently the harmonic energies $E_k = \omega_k I_k$) which are independent and in involution (their mutual Poisson brackets vanish). Instead, the system is expected to be ergodic, i.e., to have no (measurable) integral of motion apart from the total Hamiltonian H itself, when the perturbation is present (for $\alpha \neq 0$ or $\beta \neq 0$), no matter how small the perturbation be. This was suggested by a famous theorem of Poincaré (see [22]), to which Fermi himself had contributed in one of the first works of his youth (see [23] and also [24]), and this was probably the main reason for him to come back again to such a problem near the end of his life.

Thus one meets with the problem of how is it possible to reconcile such a dichotomy (N integrals of motion in the harmonic case $\alpha = \beta = 0$, no integral of motion independent of the Hamiltonian in the perturbed case, no matter how small the perturbation be) with the continuity of the solutions of the equations of motion with respect to the parameters. A possible way of recovering continuity is by making reference to the notion of relaxation time. In order to make this point clear, let us recall what the ergodicity property is in our particular case of a Hamiltonian system with a phase space M coinciding with the energy surface Γ_E defined by $H = E$. As above, by $< f >_E$ we denote the corresponding microcanonical expectation of a dynamical variable $f : M \to \mathbb{R}$. Denoting by $\{g^t\}_{t \in \mathbb{R}}$, $g^t : M \to M$, the flow induced by the equations of motion, and by x a point in phase space, then the ergodicity of the microcanonical distribution amounts to the property

$$\overline{f}(t,x) \to < f >_E \quad \text{as} \quad t \to \infty$$

for all measurable dynamical variables f and for almost all initial data $x \in M$, where $\overline{f}(t,x)$ is the "time-average" of the function f up to time t with initial datum x:

$$\overline{f}(t,x) = \frac{1}{t}\int_0^t f(g^s x)\,\mathrm{d}s \ .$$

Now, as particularly pointed out by von Neumann (see [25]), for every significant dynamical variable f there should exist a typical relaxation time τ, defined as the first time such that the time-average essentially coincides with the "phase average" $< f >_E$ for all times larger than it.

Obviously, the elimination of the just mentioned dichotomy should correspond to the fact that the relaxation time τ does actually depend on the parameters (α, β and E), and should tend to infinity as they tend to zero. i.e., as the linear system is approached. So FPU had in mind to determine, through numerical solutions of the equations of motion, the relaxation times for the time-averages $\overline{E}_k(t,x)$ of the energies E_k, for initial data very far from equilibrium. As the equilibrium expectations of such energies are all equal (equipartition), the most significant initial datum corresponding to a situation out of equilibrium is the one in which the energy is given to just one mode, for example the "first one" (i.e., the one with lowest frequency), namely the initial datum with $E_1 = E$, $E_k = 0$ for $k = 2, \cdots, N$, and for example all particles in their equilibrium positions, $x_j = 0$, $j = 1, \ldots, N$.

The essence of their numerical computations is well summarized by the first and the last figures of their paper (corresponding to Figs. 4.1 and 4.2 here). They considered the case $N = 32$ with the first mode initially excited (in the way just mentioned) for a certain value of the total energy E and certain values of α and β. They were expecting that the energy would soon spread over all other modes $k = 2, \cdots, N$. Instead, they found that the values of the instantaneous mode energies E_k versus time t were as in Fig. 4.1. One sees that the energy, initially given to mode 1, passes to the modes 2, 3, 4 and 5 (each of such modes entering the sharing of energy at a proper characteristic

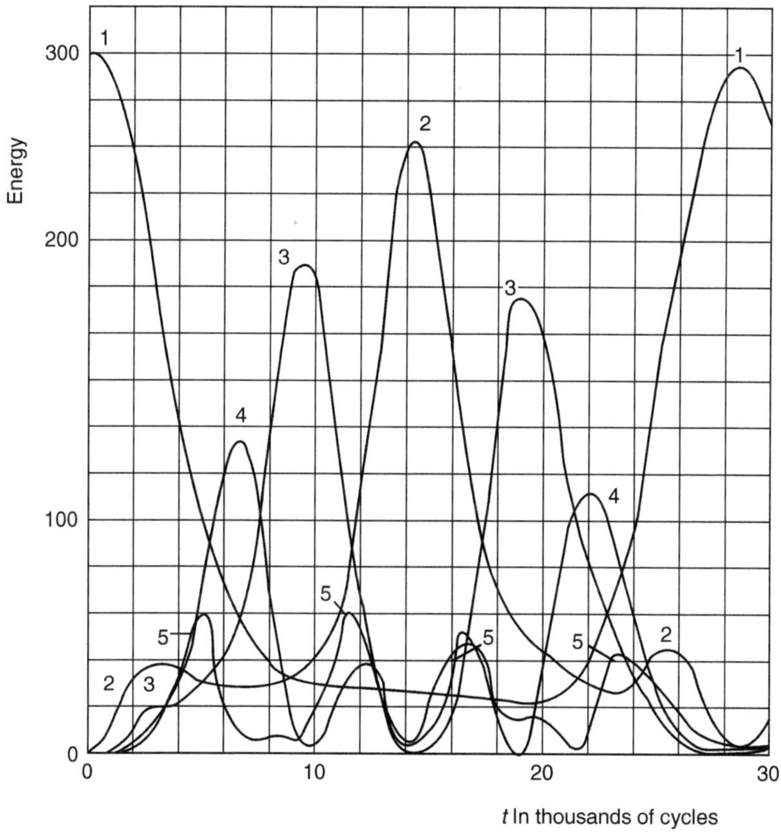

Fig. 4.1. The time evolution of the harmonic energies. The figure is a reproduction of the first one of the original FPU report. Here, $N = 32$ (with $\alpha = 1/4$, $\beta = 0$), and the energy was given initially just to the lowest frequency mode. One sees that the energy, instead of flowing to all the 32 modes, remains confined within a packet of low-frequency modes, namely modes 1 up to 5

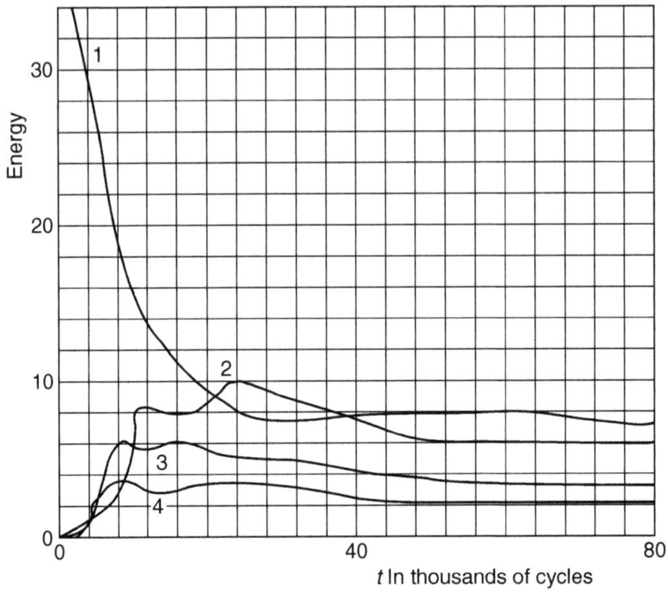

Fig. 4.2. Time-averaged harmonic energies $\overline{E}_k$ versus time. The figure is a reproduction of the last one of the original FPU report

time—this is a point to which we will come back later), and then flows back almost completely to the first mode (this is called the *recurrence property*). In any case, the energy does not appear to flow to the high-frequency modes at all (or almost at all). The most striking feature was however exhibited by the last figure of their paper (Fig. 4.2), where the time-averages $\overline{E}_k(t, x)$ of the energies E_k up to time t were plotted versus time. Indeed such a figure clearly shows not only that the final state predicted by classical equilibrium statistical mechanics was not attained, but also that a relaxation had indeed been attained to some other kind of (apparently stationary) state (after a certain time, the time-averages do not appear to change any more), which is completely different from the final expected one (equipartition). The stabilization of the averages is much more evident in Fig. 4.3, where the calculation has been pushed to a much longer time with respect to Fermi's one.

This is what we like to call *the FPU paradox*: Instead of a slow relaxation to the final equilibrium state, there is exhibited a rather quick relaxation to some kind of "nonstandard" state, in which the energy turns out to be shared only within a *packet* of low-frequency modes, having a certain well defined width, i.e., extending up to some characteristic frequency. One somehow has a kind of "partial thermalization" involving just such a packet, with the high-frequency modes essentially excluded, as if the system were composed only of an *effective number of degrees of freedom*, substantially smaller than N. This fact is well exhibited in Fig. 4.4, where we report the corresponding

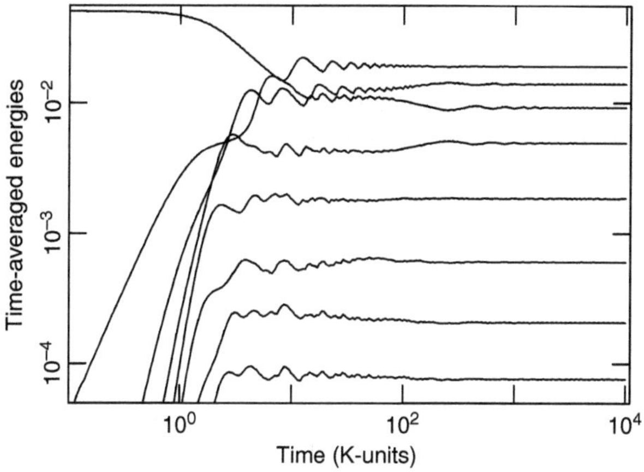

Fig. 4.3. The FPU phenomenon: exhibition of the apparent stabilization. Time-averaged harmonic energies $\overline{E}_k$ versus time in log–log scale, for a time interval much longer than in the original FPU report. The curves are drawn only for the first eight modes. Notice that the "final" value of $\overline{E}_k$ is a decreasing function of the wave-number k (which is not indicated on the corresponding curve in the figure), actually of exponential type (at least for $k > 3$). Here, $N = 32$ and $E = 0.05$ (and thus specific energy $\epsilon = E/N \simeq 0.0015$). Taken from [26]

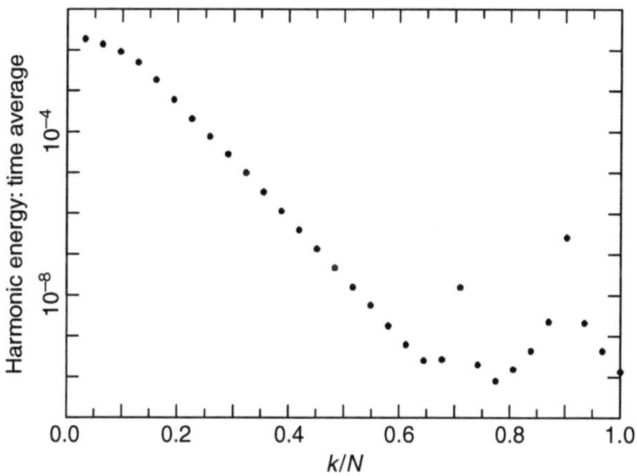

Fig. 4.4. The spectrum (namely the time-averaged energies $\overline{E}_k$ versus k/N), for the same orbit of Fig. 4.3, at the final time of the calculation. Notice the logarithmic scale for the energies. This exhibits how the modes involved in the energy sharing constitute a low-frequency packet with a tail presenting an exponential decay towards the high frequencies

"spectrum", namely the values of the time-averages $\overline{E}_k(t,x)$ versus the mode-number k (actually, versus k/N) at the final time of the calculation. Notice the exponential tail, on which we will come back later. The reaction of Fermi (who had passed away before the paper was written down) is reported by Ulam, in the preface to the reproduction of the paper in Fermi's Collected Papers, in the following terms: "The results of the calculations...were interesting and quite surprising to Fermi. He expressed to me the opinion that they really constituted a little discovery in providing intimations that the prevalent beliefs in the universality of mixing and thermalization in nonlinear systems may not be always justified".

4.2.2 The Paper of Zabusky and Kruskal, and the KdV Equation

With the paper [4] of Zabusky and Kruskal (1965), the Korteweg–de Vries (KdV) equation

$$u_t + uu_x + u_{xxx} = 0$$

entered the game. Here one thinks of a function $u = u(x,t)$ which gives, at time t, the profile of a continuous nonlinear string interpolating the FPU chain of particles. The fact that the interpolation of the FPU chain is rather well described by the KdV equation in certain situations is since then a well-known fact, and is proved in some standard way by multi-scale methods which are familiar in several fields of applied mathematics (see, for example the application given later in [27]).

From the way in which the KdV equation was associated by Kruskal and Zabusky to the FPU model, it is completely clear that the solutions of the KdV equation should provide a good approximation to those of the FPU model only for initial data corresponding to an excitation of low-frequency modes (i.e., for long-wavelength initial data). A relevant further point is however that the agreement should be expected to hold only for not too long times, as was particularly emphasized in the later "deduction" of the KdV equation that was given quite recently by Bambusi and Ponno, through a technique extending to Hamiltonian partial differential equations certain methods of perturbation theory (Birkhoff normal forms) well known in the case of a finite number N of degrees of freedom.

We give here a particular emphasis to the latter fact, because no explicit mention of it is made in the original Zabusky–Kruskal paper. Rather, just at the beginning of the paper, it is said that the KdV equation "can be used to describe the one-dimensional, long time, behavior of small, but finite amplitude,... long waves in the anharmonic crystal." Here, we are pointing out that this should be understood as meaning "long time" within the time-scale up to which the KdV equation provides a good approximation to the solutions of the FPU equations themselves. We will come back to this point later.

In any case, Zabusky and Kruskal studied the KdV equation, and were able to exhibit the existence of three time-scales (or time intervals, as they say), namely, in their words: "(I) Initially, the first two terms (of the KdV equation) dominate and the classical overtaking phenomenon occurs; that is, u steepens in regions where it has a negative slope. (II) Second, after u has steepened sufficiently, the third term becomes important and serves to prevent the formation of a discontinuity. Instead, oscillations of small wavelength...develop on the left of the front. The amplitudes of the oscillations grow and finally each oscillation achieves an almost steady amplitude...and has a shape almost identical to that of an individual solitary-wave solution (of the KdV equation). (III) Finally, each such 'solitary-wave pulse' or 'soliton' begins to move uniformly...". For a recent numerical illustration of this description, see [28] by Lorenzoni and Paleari.[1]

So the theory of solitons had come to its modern life, and started to be pursued in itself, giving rise to the whole theory of infinite-dimensional integrable systems, while its relation to the FPU problem was somehow neglected. To such a connection we will come back later.

4.2.3 The Izrailev–Chirikov Contribution

a) The discovery of a stochasticity threshold.

The next fundamental contribution was the discovery, by Izrailev and Chirikov, of the so-called stochasticity threshold (see [5, 6]). That is, the FPU paradox disappears if the initial energy is sufficiently large, i.e., there exists a critical energy $E_c = E_c(N)$ such that one has a quick equipartition for $E > E_c$. This is illustrated in Fig. 4.5, where the time evolution of the harmonic energies is calculates for a much larger energy than in Fig. 4.3 ($E = 1$).

Actually, Izrailev and Chirikov considered initial data of a certain broader class than FPU, in that they gave the energy to a packet of modes of nearby frequencies, considering the characteristic frequency of the packet as a parameter. The analog of the FPU paradox was found to occur for any frequency of the excited packet, and in all cases the paradox disappeared above a critical energy E_c depending on the frequency of the excited packet, and on N.

Concerning the theoretical motivation behind such a discovery of the energy threshold, a reading of the Izrailev–Chirikov paper clearly indicates that they had in mind the results on Hamiltonian perturbation theory (KAM theorem) that had just been obtained in Russia by the school of Kolmogorov (see [29]). If one considers a Hamiltonian perturbation of an integrable system, for small perturbations the system resembles very much the integrable one: there exist invariant tori, near the unperturbed ones, and the relative measure of the set of such perturbed invariant tori tends to 1 as the perturbation tends

[1] This paper should be compared with [27], where the first time-scale of Zabusky and Kruskal (namely, the characteristic one for the formation of the packet) was interpreted as the time-scale for equipartition.

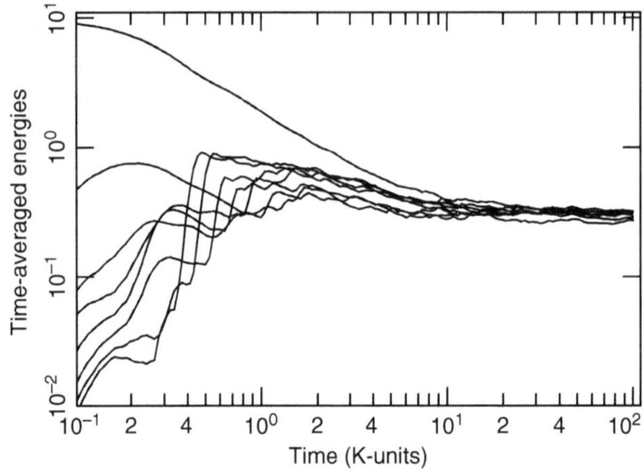

Fig. 4.5. The Izrailev–Chirikov discovery: equipartition of energy is quickly attained if energy is large enough. Time-averaged harmonic energies $\overline{E}_k$ versus time in log–log scale, still for $N = 32$ but now for $E = 1$ (i.e., $\epsilon \simeq 0.9$). Compare with Fig. 4.3, which refers to $E = 0.05$ (i.e., $\epsilon \simeq 0.0015$). Taken from [26]

to 0. Continuity is thus obtained in such a measure-theoretic sense. But the relative measure of the invariant tori is expected in general to decrease as the perturbation is increased so that, at a large enough perturbation, the resemblance of the system to the unperturbed one is essentially completely lost, and the motions are in general expected to present "chaotic" features. This is the reason why in the FPU problem one might expect that, for sufficiently large energies E, chaotic motions should dominate, and this actually led Izrailev and Chirikov to the conception of the existence of a stochasticity threshold E_c.

b) The conjecture of the disappearing of the FPU paradox in the thermodynamic limit.

The Izrailev–Chirikov discovery previously recalled, certainly constituted an extremely relevant contribution. The authors however added something more, by indicating a way in which the FPU paradox might be removed altogether, for the purposes of statistical mechanics.

Indeed, for the aims of statistical mechanics one has to consider the case of extremely large numbers N, i.e., formally, the limit $N \to \infty$. So they pointed out that one should estimate the value of the specific critical energy $\epsilon_c(N) = E_c(N)/N$ in the limit $N \to \infty$. Indeed, the FPU paradox would be completely removed if one could prove that $\epsilon_c(N) \to 0$ for $N \to \infty$. In such a way one would be guaranteed that the FPU phenomenon does not occur for large systems at any positive specific energy $\epsilon > 0$ (i.e., at any finite positive temperature $T > 0$).

The authors even indicated some kind of mathematical mechanism which should govern the vanishing of the limit-specific energy threshold. In this connection, a relevant role should be played by resonances (i.e., relations of the type $m\omega + n\bar{\omega} = 0$ for two frequencies ω, $\bar{\omega}$, with m, n integers), because the authors had in mind that resonances would lead to stochasticity, as was familiar to them through the so-called Chirikov criterion of the overlapping or resonances. On the other hand, they pointed out that, in the limit $N \to \infty$, the FPU system presents infinitely many resonances. So they worked out some estimates based on this idea, for the case of initial data with an excitation of a few high-frequency modes, and they interpreted their considerations as suggesting that $\epsilon_c(N) \to 0$ in that case. An analogous conclusion could not be drawn for the case of initial data with excitations of low-frequency modes (the case considered in the FPU work). Quite recently, their pupil Shepelyansky, elaborating on their methods, maintained to have extended such a result to that case too (see [7]).

Serious doubts on the significance of the criterion of the overlapping of resonances may actually be raised (see [8]). In any case, however, one can say that a physical conjecture had emerged, namely, that the FPU paradox may disappear entirely in the thermodynamic limit.

4.2.4 The Result of Bocchieri et al.

Five years later (1971), in [9] Bocchieri et al. gave numerical indications in the opposite direction: The FPU paradox should persist in the limit $N \to \infty$. They actually performed computations for a slight modification of the FPU model, inasmuch as they introduced a "realistic" potential of Lennard-Jones type, namely,

$$V(r) = 4V_0 \left[(\sigma/r)^{12} - (\sigma/r)^6 \right]$$

involving two parameters, the depth V_0 of the potential well and the typical distance σ, at which the potential passes from positive to negative values. They considered several types of initial data with a few nearby modes excited, of low, or of high, or of intermediate frequency, and found that in a short time equilibrium is attained (equipartition of the time-averages of the mode-energies was obtained), if the initial energy is sufficiently large, i.e., for $E > E_c(N)$ for some critical energy $E_c(N)$, in agreement with the discovery of Izrailev and Chirikov. For what concerns the dependence of the specific stochasticity threshold $\epsilon_c(N) = E_c(N)/N$ on N, they found a large dependence for small N, say for $2 < N < 10$, whereas $\epsilon_c(N)$ was found to be essentially constant for "large" N (concretely, in their computations, for $10 \leq N \leq 100$). They actually found for $\epsilon_c(N)$ the "limit" value $\simeq (3/100)V_0$. In their words: "When the energy of vibration per particle is equal or larger than 2 or 3 percent of the potential well and the number of particles is sufficiently large, one has, in time average, equipartition of energy among the normal modes".[2] Namely,

$$\varepsilon_c \simeq 0.03\, V_0 \; .$$

[2] It must however be added that the authors were completely aware of the possible relevance of the actual times of observation, because they also added: "We may

4.2.5 The Suggestion of a Possible Physical Interpretation

At this point the situation was as follows. Izrailev and Chirikov had with an extreme clarity indicated how one might eliminate the FPU paradox entirely, for systems of interest to statistical mechanics: One should prove that the specific energy threshold $\epsilon_c(N) = E_c(N)/N$ vanishes in the limit $N \to \infty$. Moreover, they believed to have shown that this is the case at least in the case of high-frequency excitations. On the other hand, Bocchieri et al. had given indications in the opposite direction. Thus, there was the problem of which could be a physical interpretation for the apparently stationary state exhibited by FPU, in case the indications of Bocchieri et al. were confirmed.

The idea that the relations between classical mechanics and quantum mechanics may be much subtler than usually believed was very much discussed within the group of theoretical physicists in Milano, particularly under the stimulus of Caldirola and Loinger. Thus, as one of the strongest and deepest manifestations of quantum mechanics in a statistical mechanics frame, actually the one that gave rise to quantum mechanics itself, is the lack of energy equipartition at low temperatures, the systems behaving as if the high-frequency modes were excluded from the energy sharing, quite naturally there arose the idea that the "nonstandard" apparently stationary FPU states may be a sort of classical analogs of quantum states. These are characterized by the Planck spectrum E_k^{P} given by

$$E_k^{\mathrm{P}} = \frac{\hbar\omega_k}{\exp(\beta\hbar\omega_k) - 1}$$

where $\hbar$ is Planck's constant and $\beta = 1/(k_{\mathrm{B}}T)$ the "inverse temperature".

In such a way, after many discussions with Bocchieri and Loinger, Galgani and Scotti started out an investigation in which the FPU spectrum below threshold was fitted to a Planck-like distribution E_k^{Plike}, namely,

$$E_k^{\mathrm{Plike}} = \frac{A\omega_k}{\exp(\beta A\omega_k) - 1} ,$$

with two free parameters A and β, having the dimensions of an action and of an inverse temperature, respectively. The fits were made to data obtained with the same computer program used by Bocchieri et al., in which the molecular parameters m (mass of the particles), V_0 and σ (the ones entering the

---

conclude by saying that, in the case of very low total energies, the relaxation mechanism towards the standard Boltzmann distribution of the normal modes may act so slowly that the coupling of the system with a thermal bath could be very important in determining the approach of the model towards such a distribution." This remark, by the way, actually opens another relevant problem, because it may happen that also the mechanisms of transfer of energy between a FPU system and a heat reservoir are slowed down when temperature is lowered. In fact, this actually seems to be the case.

Lennard-Jones potential) actually considered were those of Argon, as taken from standard available handbooks.

The result found was a rather striking one. Indeed, not only the qualitative fit to the Planck-like law was found to be rather good, with β behaving as expected (namely, as an inverse temperature depending only on the specific energy $\epsilon = E/N$), but it was also found that the parameter A on the one hand was pretty constant, i.e. independent of the specific energy, and on the other hand happened to have a value quite near to the Planck constant $\hbar$.

It took some time to understand how this could have happened. The simple reason is that A, being an action, has to be proportional to the natural action obtained from the dimensional parameters m, V_0, σ introduced in the model, which is $\sqrt{mV_0}\,\sigma$. So one necessarily has

$$A = \alpha\sqrt{mV_0}\,\sigma\ ,$$

where α is a pure number. On the other hand, it is well known that the molecular parameters actually met in nature do indeed contain $\hbar$, and in particular one has, for example for the noble gases,

$$\sqrt{mV_0}\,\sigma \simeq 2Z\hbar\ ,$$

where Z is the atomic number. In conclusion, Planck's constant had been introduced somehow by hands in the model through the molecular parameters. This is the way in which Galgani and Scotti came to venture the suggestion (see [10, 11]) that, if one can prove that the FPU paradox persists in the thermodynamic limit, then the apparently stationary FPU states may provide a sort of classical analog to the quantum degeneration described by Planck's law. This idea continued to be pursued up to the present days (see [30, 31]).

By the way, it may be noted that the good fit of the FPU spectrum to Planck's law (suggested by the analogy with quantum mechanics) amounts to be perhaps the first clear exhibition of the fact that, for large wavenumbers k, the energies $\overline{E}_k$ decay exponentially fast with k, i.e., with the frequency ω_k. This fact is indeed a quite general one, the nature of which was clearly understood analytically ten years later with the paper of Fucito et al., who, through the intermediary of the paper of Frisch and Morf, transported to the FPU problem general ideas of turbulence theory.

This ends the first phase of the history of the FPU problem in our personal way of reconstructing it.

4.3 A Voice in the Desert: The Paper of Fucito et al. (1982) and the Proposal of a Metastability Scenario. The Work of Parisi and the Analogy with Glasses. Relations with Turbulence Theory

In the first phase of the history we traced back in the previous section, due to the need of concentrating our attention on the papers that are most relevant for our reconstruction, we already had to neglect a considerable amount of papers, among which stay for example several ones of the late J. Ford, the memory of whom is particularly dear to the oldest of the present authors, who exchanged with him tenths of letters on the subject. In the same way, we are going to neglect in the present section many other papers, including several ones worked out by the present authors.

The next relevant step was made with the paper [13] of Fucito et al., where both a new point of view and a new technique were introduced.

The new point of view concerns metastability and was certainly borrowed from the frame of the theory of glasses and of disordered systems, in which distinguished contributions had been given in Roma by Parisi. The idea is that the FPU state is just an apparently (rather than a true) stationary one. This is somehow at variance with the attitude of Izrailev and Chirikov, who were apparently thinking in terms of truly stationary states; indeed they were explicitly making reference to KAM theorem, which is expressed in terms of invariant surfaces (that is, surfaces on which the orbits lie for all times). Instead, in the paper of Fucito et al. reference is made to the quick formation of a state which remains essentially undisturbed for extremely long times, until it eventually precipitates through a "catastrophic mechanism" to the true "final" equilibrium state.

The new technique is simply that of relating the decay of the tail of the spectrum to the singularities of the analytic continuation of a field interpolating the positions of the FPU model. This idea was borrowed from a very interesting paper of Frisch and Morf (see [32]), the aim of which was to understand certain features of turbulence theory (see [33]) as manifestations of quite general relations between the high-frequency tail of the Fourier transform (in complex time) of a temporal signal and the singularities of the analytical continuation of the signal itself. By the way, it can be noted that the existence of a deep analogy between the problem of a dynamical justification of the Boltzmann–Gibbs equipartition principle, and the general problem of turbulence, had been clearly pointed out by von Neumann (see [34]), in the year 1949.[3]

In fact, Fucito et al. were not actually studying the FPU model itself, but rather a variant of it, namely, the so-called φ^4-model (to be presently recalled),

[3] Such an analogy between turbulence and ordinary statistical mechanics permeates the whole paper of von Neumann. See, for example p. 445, where it is said: "The $k^{-5/3}$ law calls for an interpretation akin to (although not identical with) the ultraviolet catastrophe of black-body radiation theory", and reference to "non-ergodic conservation laws" is made. See also p. 447 and finally p. 468, where it is said: "From the point of view of statistical physics, turbulence is the first clear-cut instance calling for a new form of statistical mechanics. ... The existing theories. ... suffice to show that those laws will differ essentially from those of classical (Maxwell–Boltzmann–Gibbsian) statistical mechanics. Thus it is certain that the law of equipartition of energy between all degrees of freedom, which is valid in the latter, is replaced by something altogether different in the former."

which was a very familiar one in field theory (and had also been studied a little before in the spirit of the FPU problem in [35]). In fact, it turns out that the techniques used by Fucito et al. to investigate the φ^4-model cannot be immediately transported to the FPU model itself (a subsequent attempt will be mentioned later), but the transport of the global scenario proved instead to be possible (actually in terms of the work of Zabusky and Kruskal), as shown later by Bambusi and Ponno.

Perhaps, as a preliminary introduction to the description of the paper of Fucito et al., it may be useful to illustrate the main phenomena understood by them, through the Figs. 4.6 and 4.7, which refer the FPU model (with $N=127$ and specific energies $\epsilon=10^{-4}$ and $\epsilon=5\times10^{-3}$ respectively). In each figure, the spectrum (namely, the plot of the time-averaged energies $\overline{E}_k$ versus k/N) is reported at successive times t_j (with $t_{j+1}=10\,t_j$). From Fig. 4.6 one clearly sees that, at any observation time, the spectrum consists of both a packet of low-frequency modes (having a tail which decreases exponentially fast with k/N) which essentially contains the whole available energy, and of a complementary packet of high-frequency modes displaying an essential equipartition of energy at a much smaller energy. One also observes that the slope of the main low-frequency packet decreases as time increases, until it appears to have come to a stop, remaining essentially constant (this is the phenomenon of the apparent stationarity) during a rather long-time interval (covering at least four orders of magnitude in the case of the figure). This occurs for $\epsilon=10^{-4}$. But if one considers a larger specific energy ($\epsilon=5\times10^{-3}$ in the case of Fig. 4.7), then the same phenomenology is speeded up, and within the same final observation time ($t=10^8$), a further phenomenon is exhibited. This is the final attainment of global equipartition, which occurs through a quite different mechanism. Indeed, one might have imagined that the final global equipartition be attained through a successive decreasing of the slope of the tail. Instead, the approach occurs in the following way. The complementary packet of high-frequency modes continues to essentially display partial equipartition at an energy smaller than that of the main low-frequency packet, and what occurs is that the level of the energy of the complementary packet rises as time increases,[4] until global equipartition is attained. Note that, in both figures, the scale of the ordinates is not the same at the various times. In conclusion, one observes the existence of two different mechanisms: first, the quick formation of a packet of low-frequency modes with an exponential tail having an apparently stabilized slope (formation of the metastable state), and, secondly, the final approach to global equipartition through the rising of the equipartition level of the complementary packet.

After this introduction, let us finally come to an illustration of the paper of Fucito et al. They consider the one-dimensional nonlinear Klein–Gordon equation

$$\varphi_{tt}=\varphi_{xx}-m^2\varphi-g\varphi^3\ ,$$

[4] Moreover, the complementary packet extends its size towards the left.

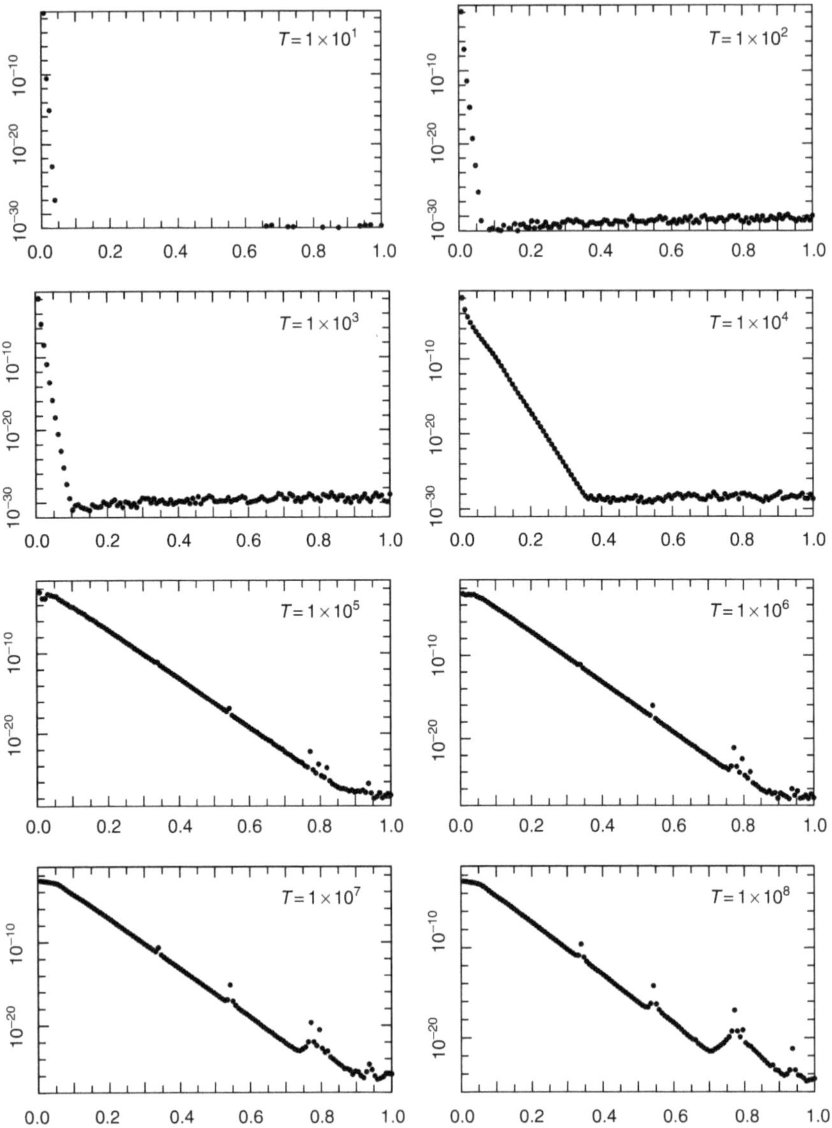

Fig. 4.6. The spectrum at several times $(10, 10^2, \ldots, 10^8)$. First phase: formation of the metastable state (note the change in the vertical scale of the figures). Here, FPU model with initial data of FPU type, $N = 127$, $\epsilon = 1 \times 10^{-4}$)

where the real, one-component, field $\varphi(x,t)$ is defined in the interval $-L/2 \leq x \leq L/2$ with periodic boundary conditions, and m and g are positive parameters. From this partial differential equation, a discretization leading to an analog of the FPU model is immediately obtained. The name φ^4-model is

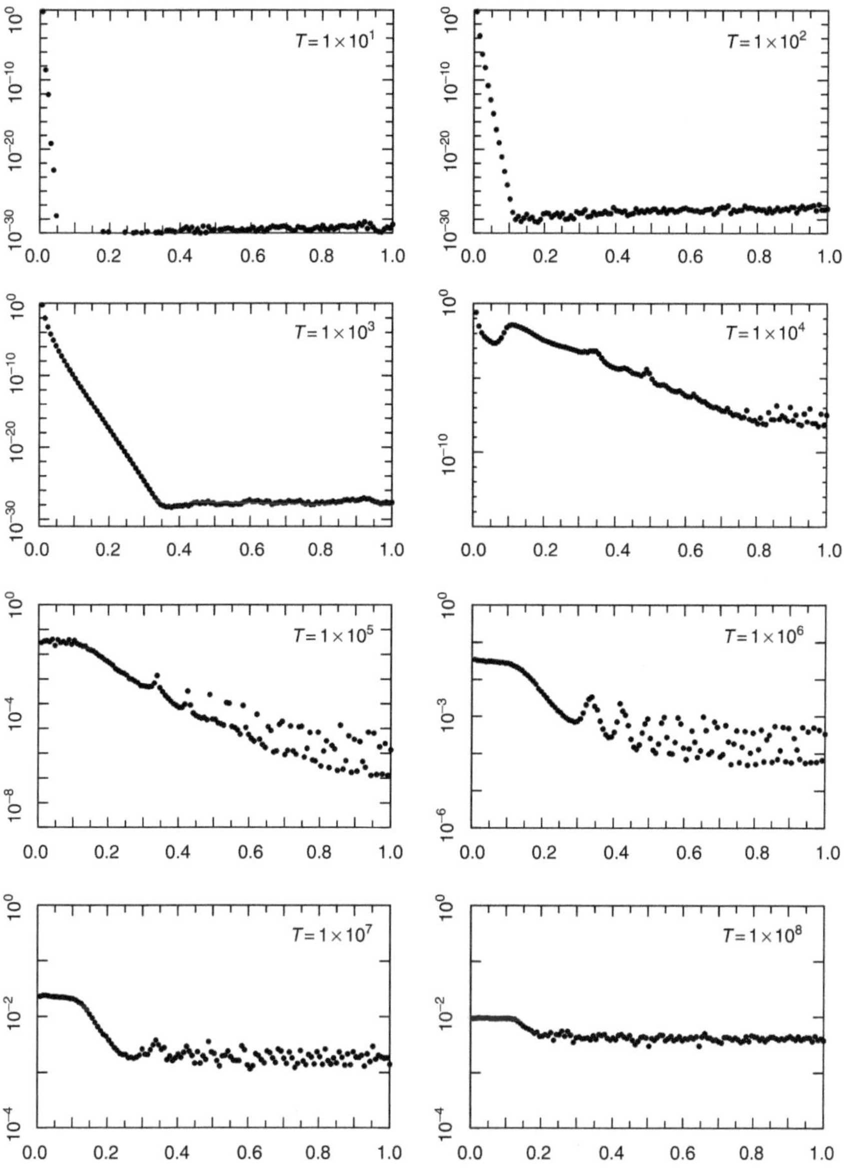

Fig. 4.7. The spectrum at several times $(10, 10^2, \ldots, 10^8)$. Illustration of the final phase following the first one, with the attainment of global equipartition (note the change in the vertical scale of the figures). Same as Fig. 4.6, but now with $\epsilon = 5 \times 10^{-3}$. The time-scale of observation is the same in both cases

due to the fact that the only nonlinearity in the model comes from a term φ^4 in the potential energy.

The quantity they are interested in is the analog of the spectrum previously discussed for the FPU model, i.e., the distribution of energy (in time-average) among the modes, as a function of time. To this end, they introduce the space Fourier transform of the field φ by

$$\hat{\varphi}(k,t) = (2\pi)^{-1/2} \int_{-L/2}^{L/2} \mathrm{d}x \; \mathrm{e}^{-ikx} \varphi(x,t)$$

and define the *power spectrum*[5] $W(k,t)$ by

$$W(k,t) = |\hat{\varphi}(k,t)|^2 \; .$$

Notice that the continuum analog of the spectrum previously defined for the FPU model would rather be twice the quantity $k^2 W(k,t)$.

Anyway, they are interested in investigating the form of the spectrum $W = W(k)$ as a function of time t for large values of k. To this end they make reference to well-known analyticity properties of Fourier transforms, and notice: "We expect the field φ to reach asymptotically a thermal equilibrium distribution, given by a Boltzmann factor $\mathrm{e}^{-\beta H}$ for some value of the inverse temperature β determined by the initial conditions. In this case, at values of the wavenumber k so large that the mass and nonlinear terms of H are negligible, one would have

$$W(k,t) \simeq \text{const.} \times k^{-2} \; . \tag{4.3}$$

This behavior of W corresponds to functions $\varphi(x,t)$ which are not differentiable with respect to x. Now it is known that, since $\varphi(x,0)$ is analytical as a function of x, the solution $\varphi(x,t)$ will remain analytical at any finite time t. Equation (4.3) can only be valid for infinite time. This means that, as time goes on, singularities of $\varphi(x,t)$ appear in the complex x plane which creep towards the real axis and accumulate onto it at infinite times. We show below that these singularities are simple poles."

Indeed, it is well known that one may relate such singularities to the large k behavior of W by means of the theorem of residues. Therefore one "obtains the following asymptotic behavior of $W(k,t)$ at large k:

$$W(k,t) \simeq \text{const.}\, \mathrm{e}^{-2ky_{\mathrm{S}}(t)} \; ,$$

where $y_{\mathrm{S}}(t)$ is the imaginary part of the location of the pole which lies nearest to the real axis."

In conclusion, "the strategy is then to evaluate the most likely value of $y_{\mathrm{S}}(t)$ by extending the approach of Frisch and Morf to a deterministic partial differential equation."

[5] They actually call it just *the spectrum*.

So, one remains with the problem of evaluating the most likely value of $y_S(t)$. To this end, the idea was to exploit a particular feature of the φ^4-model in connection with initial data of FPU type, namely with small k and thus large wavelengths. Indeed, this means that initially, and actually up to times until which energy did not yet flow to high k modes, the profile of the field does not present large curvatures, and thus the term φ_{xx} in the equation of motion can be neglected, with the consequence that the equation of motion reduces to an ordinary one depending parametrically on the space coordinate x. So, up to not too large times, for any x one has an unknown $\varphi = \varphi(t)$ obeying the ordinary differential equation

$$\ddot{\varphi} = -m^2\varphi - g\varphi^3 \ . \tag{4.4}$$

An analytical study of such an equation is easily performed, and this leads to the result that, for initial data of the form

$$\varphi(x,0) = A\cos(k_0 x) \ , \quad \varphi_t(x,0) = 0$$

with k_0 small, the imaginary part of the nearest pole starts descending from infinity towards the real axis, approaching a point with imaginary part $(1/k_0)$ $\ln(ma/g^{1/2})$. This is illustrated in a very beautiful way in a subsequent paper by Bassetti et al. (see [36]), where the relevant poles for the φ^4-model were computed numerically by the technique of the Padé approximants.

So, there exists a first temporal phase of the dynamics, in which the spectrum presents an exponential decay towards the high wave-numbers k, with a slope decreasing as

$$|y_S(t)| = -\ln(tAg^{1/2})/k_0 \ .$$

This corresponds, for the low frequency modes, to an increase of energy as a power of t (formation of the packet). This stage, by the way, is the analog of the one in which, in the terminology of Zabusky and Kruskal, the third-derivative term u_{xxx} can be neglected, i.e.: "Initially, the first two terms (of the KdV equation) dominate and the classical overtaking phenomenon occurs; that is, u steepens in regions where it has a negative slope."

One might thus expect that equipartition will eventually occur, with the slope tending to zero (i.e., with the pole approaching the real axis). But this is not the case. As mentioned previously, the poles do not collapse onto the real axis, because the imaginary part tends to a finite positive value and so the slope stops decreasing (see [36]). At this point, according to Fucito et al. the contribution of the Laplacian starts becoming relevant, and this fact can be looked upon as the addition of a noise to the r.h.s. of (4.4). This further stage of the process is described at p. 710 of the paper by Fucito et al. in terms of a probabilistic analysis performed on the harmonic chain ($g = 0$) in the limit of infinite length ($L \to \infty$), in which use is made of the known fact that the one-point probability distribution function of a classical harmonic field in dimension one is Gaussian. This leads to an extremely slow decrease of the slope.

The authors then turn to a qualitative discussion of the final stage of the process of approach to equipartition. The analysis is made in terms of the variance σ^2 of the Gaussian probability distribution function previously mentioned. They say: "The main effect of the nonlinear terms in this regime will be to change the value of σ^2. If σ^2 were time independent," the previous analysis "would be essentially correct. Let us distinguish between the role of short and long wavelength modes. At the times we are interested in, most of the energy is contained in the long wavelength modes, which may be assumed to be in a kind of thermal equilibrium among themselves. Their contribution to σ^2 may then be considered as essentially constant in time. The short wavelength modes will however also contribute to σ^2. As long as $W(k,t)$ is small in the large k region, their contribution is negligible. As time goes on, however, $W(k,t)$ will start increasing, what will increase the value of σ^2 and fasten therefore the transfer of energy to short wavelength modes. This triggers a catastrophic process which our analytical tools are unable to handle. We cannot therefore draw conclusions about the behavior at very long times before thermal equilibrium is reached."

In conclusion, here for the first time one finds explicitly expressed the conjecture that, for all values of the perturbation, at sufficiently long times one will attain the standard equilibrium state, and the FPU paradox is interpreted as corresponding to a preliminary stage of the process in which the energy, initially given to extremely low-frequency modes, quickly flows to a larger packet of low-frequency modes (with an exponential decay towards the high frequencies) and remains frozen there up to an extremely long time. This is what we informally call *the metastability scenario*. For what concerns the law describing the dynamical evolution towards equipartition, in the subsequent work [39] by Parisi numerical indications were given that the corresponding time scale could be a stretched exponential in terms of the inverse of the specific energy, rather than a simple exponential.

As previously pointed out, the paper of Fucito et al., with its interpretation of the FPU paradox as a metastability phenomenon, did not produce a great impact, and even was essentially forgotten for a long time. For example, if one looks at the 21 papers published quite recently on the FPU problem in a special issue that a journal devoted to it on the occasion of the 50 years from the original work, one will find out that not one of them even mentions that paper, apart from the papers [37, 38], where it is amply discussed (see also [26]).

How could this have happened? In our opinion, the main reason is that at those times the key point under discussion was the choice between the two alternatives previously mentioned about the energy threshold, namely, whether the specific energy threshold $\epsilon_c(N)$ vanishes or not in the limit $N \to \infty$; on the other hand, no mention of a threshold at all was made in the paper of Fucito et al. According to them, equipartition should be attained at all energies. This statement, that no threshold should exist, was particularly emphasized by Parisi in [39]. So, actually, there was some misunderstanding about the sense to be attributed to the word "threshold". Indeed, Parisi was stressing that energy equipartition should be attained (after a sufficiently long time) at any specific energy, and such a conjecture is obviously opposed to the conception of a threshold, if the latter is meant as the specific energy below which equipartition is never attained. On the other hand, there is no opposition, if the threshold is understood in a softer way, namely, in the sense that for smaller energies one meets with a state which is only apparently stationary, and will later evolve, on a much longer time scale, to equipartition, i.e., to the final "true" equilibrium state.[6] However, the relation with the threshold in the sense of Bocchieri et al. was not discussed in an explicit way. In our opinion, this is the fact that generated some confusion. The situation was finally clarified in the paper of Berchialla et al. (described in a subsequent section), where a clear exhibition was given of the fact that two well distinct time-scales of relaxation exist, but only below a certain critical specific energy, which could thus be interpreted as the threshold previously discussed by Izrailev–Chirikov and by Bocchieri et al.

We mention now the very few papers in which the work of Fucito et al. was discussed.

The previously mentioned paper [36] somehow constitutes an appendix to the paper of Fucito et al., because it reported numerical computations of the relevant poles in the φ^4-model, showing a very good agreement with the theoretical predictions and some further details. Something analogous can be said of the paper [40], still devoted to numerical investigations on the φ^4 model. Here, following Frisch and Morf, the analysis concerns the statistical aspects of the field φ, which is shown to present typical non-Gaussian features. By the way, it may be worth mentioning that analogous indications of non-Gaussian behaviors were also reported much later for the process of energy exchanges of the internal degrees of freedom of diatomic molecules produced by atomic collisions (see [41]).

The suggestion that a description analogous to that of Fucito et al. for the φ^4-model could be given also for the FPU model was first advanced and

[6] In the words of Fucito et al.: "One of our main results is that the system reaches equilibrium with a logarithmic dependence on t, so that the nonequilibrium spectrum may persist for extremely long times, and may be mistaken for a stationary state if the observation time is not sufficiently long". By the way, one also finds here the words: "It is amusing to remark that the quasi-equilibrium distribution is similar to Wien's law for black body radiation with a slowly varying Planck's constant, a statement which is clearly inspired by the possible physical interpretation proposed in the work [10].

discussed in [42],[7] (see also [43])[8] in which numerical computations were performed on the FPU model itself (actually on the so called β-model, i.e., the one with $\alpha = 0$). The previously mentioned "slope" of the straight line describing, for large k, the exponential decay of $\overline{E}_k$ versus k in semi-log scale was investigated and was shown to stop decreasing, but the further evolution was not investigated. The accent was rather put on the fact that a quick approach to equilibrium occurs only for high enough energy, and the authors even added the comment (p. 3550): "The numerical results that we have described in the present section yield to the interesting conclusion that a threshold value exists, below which the equipartition of energy is never reached."

The same problem was rediscussed two years later (see [44]), still for the FPU β-model. The accent was still put on the existence of an energy threshold, trying to make a decision between the conjecture of Izrailev–Chirikov and that of Bocchieri et al. (that was there called the Galgani conjecture), and the authors said: "For the N dependence our results seem to be unquestionable and in contrast with the existing theoretical predictions" (of Izrailev and Chirikov). The point of view of Fucito et al. was mentioned in the conclusions, where they added the comment: "But as far as the time dependence is concerned we cannot conclude that the threshold does not vanish as t approaches ∞ ... The situation can be likened to the very slow relaxation behavior in disordered systems, where the evolution towards 'equilibrium' takes place through metastable states, approached at different time scales." Analogous conclusions were reached for the FPU α-model in [45].

4.4 Other Pathways

The metastability perspective, initiated with the work of Fucito et al., was finally recovered 20 years later, with the paper of Berchialla et al. that will be illustrated in the next section. Many more works were written down in the meantime by several authors (for example Kantz et al. entered the game), with an attention to several interesting problems. We cannot follow them here in detail, but it seems to us that, apparently, no reference to the metastability

[7] From the technical point of view, difficulties were met in trying to describe the motion of the poles through a direct transport of the method used by Fucito et al., which was a very special one devised for the φ^4-model. In fact the authors proposed somehow a partial differential equation which, as we now understand, can give a good agreement only for extremely short times, the ones corresponding to the first stage described by Zabusky and Kruskal, because it does not even prevent the formation of a discontinuity (in the terminology of Fucito et al., it does not prevent the falling of the poles on the real axis), and so does not lead to a blocking of the decay of the slope in the spectrum (stage II of KZ).

[8] Here, the idea is suggested that in the α–β model the β-term dominates over the α one, also at very low energies.

perspective can be found there. In the present section, we limit ourselves with a short survey of some other problems that were dealt with.

4.4.1 The Idea of Long Relaxation Times, Boltzmann and Jeans, Nekhoroshev and Landau-Teller

In the meantime, people had started becoming familiar with the fact that the relaxation times to equilibrium can actually be extremely long (see [46]). This in fact had been much discussed by Boltzmann himself and by Jeans, who had conceived of explaining by such a mechanism the observed lack of equipartition in nature (see the quotations in [47]). The same fact was later understood in terms of perturbation theory through the work of Nekhoroshev (see [48]), and through a reconsideration of the work of Landau and Teller of the years 30s on the exchanges of energy of the internal degrees of freedom in atomic collisions (see [49]). Problems of this kind actually became very popular and were much investigated, and would deserve a long discussion. Here we only remark that, while on the one hand the existence of long relaxation times was well understood, on the other hand there was no completely clear awareness of the fact that in a very short time some kind of equilibrium (or apparent equilibrium, or metaequilibrium) is attained (see however the works [50, 51]). Such a quick approach to a metaequilibrium state corresponds to what we now call the quick formation of a packet (presenting a partial thermalization), which is the one accounted for by the first two stages of Zabusky and Kruskal, and of Fucito et al.

4.4.2 The Works around Pettini

The existence of long relaxation times for the FPU problem (and also for the φ^4-model) in the spirit of Nekhoroshev's theorem was first discussed and exhibited by Pettini and Landolfi (see [52]). Indeed, already in the abstract of their paper, they make the following quite clear statement: "Below a critical value $\epsilon_c \ldots$ of the energy density ϵ, the relaxation time τ_R is found to follow a 'Nekhoroshev-like' law, i.e., $\tau_R = \tau_0 \exp(\epsilon_0/\epsilon)^\gamma$", and also add: "A remarkable difference with respect to Nekhoroshev's theorem (where the exponent γ scales as $1/N^2$) is the N independence of numerical experiments results. An important consequence of this fact is the existence of nonequilibrium states of arbitrary lifetimes also at large N values. On the other hand, at high-energy densities ($\epsilon > \epsilon_c$), τ_R is almost independent of ϵ". Such a scenario of Pettini and Landolfi seems to perfectly agree, actually anticipating it, with the one described in the next section along the lines of the work of Berchialla et al. However, from some subsequent works (see, for example the review paper [53]) one may have the impression that the authors rather started adhering to the scenario of Izrailev and Chirikov. We hope to come back to this point on another occasion.

In some subsequent papers (see [54]) a very ingenious method, based on certain considerations on the geometry of phase space, was devised which allowed Pettini and his collaborators to provide a semianalytical estimate of the maximal Lyapunov Characteristic Exponent as a function of the specific energy ϵ_c in the thermodynamic limit. This result, although not yet completely cleaned up from an analytic point of view, constitutes in our opinion one of the most relevant contributions to the subject. The curve of the maximal LCE versus the specific energy ϵ had been numerically investigated by Casartelli et al. in the paper[9] [56] for the FPU model with Lennard-Jones potential. Later, in their paper [52], Pettini and Landolfi found the interesting result that such a curve presents a well-marked knee at a certain value of ϵ. In fact, an analogous remark had been made three years before in a paper of Butera and Caravati (see [55]) for a plane model of rotators (the so-called O(2) planar Heisenberg model), in which the position of the knee had been associated to the presesnce of a certain phase transition (of Kosterlitz and Thouless). For previous works on the model of rotators see [57] and [58].

4.4.3 Metastability and Specific Heats

A very interesting discussion had also been started concerning estimates for the fluctuations of energy in subsystems of the FPU model. The aim was to understand whether the FPU model may be of interest in connection with the problem of the specific heats. In order to obtain some estimates through numerical studies on isolated systems, without having to make recourse to an interaction with a heat reservoir, the attention was addressed to the energy fluctuations of a subsystem of the FPU system of interest: the aim was to compare the fluctuations computed as time-averages with those expected at equilibrium, since the relation of the latter ones with the specific heat is well known. Two apparently opposite results had been obtained in the papers [59, 60] (by Livi et al. and by Perronace and Tenenbaum, respectively). The difference could be explained as due to the fact that two completely different kinds of subsystems had been considered in such papers: a spatial piece of the FPU string in [59], where the time-averages were found to agree with the equilibrium expectations, and a packet of modes of nearby frequencies in [60], where an analog of the FPU paradox was observed, because the time-averages were apparently found to tend to zero as temperature decreases.

So, there naturally arose the idea of eliminating all the problem of the good choice of the subsystem, by estimating the specific heat directly through the energy actually exchanged between the whole FPU system and a heat reservoir (see [15] by Carati and Galgani). Obviously this in turn opens the new problem which might be considered to be a good model for the energy exchanges with the reservoir, a problem we shall not discuss here. We just

[9] This, by the way, is the paper where the now familiar technique of computing the maximal LCE was first introduced.

limit ourselves to mention that in such a way some analogies between the FPU system and the glasses were pointed out, and this fact was instrumental to rediscover and recover the metastability perspective introduced by Fucito et al. Moreover, another interesting fact was observed. Namely, something analogous to the formation of a low-frequency packet (which is a standard result for long-wavelength initial data) occurs even if one starts up with initial data extracted from a Boltzmann–Gibbs distribution at a certain temperature (and so essentially with equipartition of energy among the modes). Indeed it was found (see [16]) that, if the FPU system (with initial data of the just mentioned type) is put in contact with a heat reservoir having a slightly different temperature, then only a small packet of low-frequency modes does manifest a quick reaction to the reservoir, attaining equipartition at the temperature of the latter, whereas the high-frequency modes do not manifest any reaction at all. Presumably, they too will much later attain global equipartition, in analogy with what occurs for an isolated system with long-wavelength initial data.

This fact naturally leads to expect (see [31, 60]) that metastability phenomena may show up in actual measurements of the specific heats (for example of crystals) at low temperatures, more or less in the spirit of the rationale of the time-dependent specific heats, as discussed for example by Birge and Nagel (see [61]). On this very interesting problem we plan to come back elsewhere.

4.4.4 Towards the Natural Packet through Resonance: the FPU Model with Alternating Masses

It will be shown later that, in order to understand the quick formation of an apparently stationary state, a key point is played by some relevant resonances. Such a role, already pointed out for the FPU problem in the pioneering work of Ford of the year 1961 (see [62]), became particularly evident when a modification of the original FPU model was studied (see [63]). We refer to the so-called FPU model with alternating masses which is very familiar in solid state physics, namely, the one in which the successive material points of the FPU chain have masses $m, M, m, M, \ldots$ with $m < M$. The main qualitative consequence of such a modification is that the "dispersion relation", namely, the function $\omega = \omega(k)$, presents now two branches: the "acoustical" one (emanating from near the origin) and the "optical" one, characterized by larger frequencies. The separation between the two branches becomes larger and larger (with the optical one tending to become a horizontal curve, i.e., with all frequencies equal) as the ratio M/m of the two masses is increased. So one meets here (for M/m large) with two clearly distinct subsystems, each of which can be essentially considered as completely resonant, being characterized by essentially just one frequency. Resonant systems had been previously studied in the frame of Nekhoroshev theorem (see [64]), and it had been well understood that chaotic motions in general occur within each single resonant subsystem, whereas the exchange of energy between the two subsystems is in general extremely slow. Furthermore, the strong dependence of the results on the number of elements constituting a subsystem was almost completely eliminated. Notice that the dependence of the estimates on N in the general case is instead quite heavy, and this fact was often interpreted as indicating that "chaos should prevail" in the thermodynamic limit.

4.5 The Resurgence of the Metastability Perspective, and Its Compatibility with the Existence of a Specific Energy Threshold: The Natural Packet and the Two Relaxation Times

A very clear numerical illustration of the phenomenon of metastability, exhibiting on the one hand the existence, at low energies, of two well separated time-scales (i.e., the quick formation of a "natural packet" which persists up to very long times, when the final relaxation to equipartition occurs), and on the other hand the existence of a stochasticity threshold in the sense of Bocchieri et al., was given in [14] by Berchialla et al. The main underlying idea was to measure the width of the packet that is quickly formed by the dynamics itself when the energy is initially given to the mode of lowest frequency. While in the modified FPU model with alternating masses one was meeting with two "fixed" packets, here the packets are naturally formed by the dynamics itself (think of the first figure of the original FPU work), and their width is expected to depend on the initial energy. By the way, here too one meets with a resonance phenomenon, because the low frequencies are given in a first approximation by $\omega(k) \simeq k\pi/(N+1)$, which is just the familiar resonance relation of the continuous linear string. The idea of taking into account such a typical resonance of the low frequency modes, already indicated by Ford, was later reconsidered by Shepelyansky and by Ponno (see [7, 8]).

The first relevant result of the paper of Berchialla et al. is illustrated in Fig. 4.8 (which we familiarly refer to as "the shower"). In abscissas one has the time and in ordinates the energy. Here the results refer to a FPU system with $N = 15$, with the energy given initially to the lowest frequency mode. Having fixed the initial energy (and thus a line parallel to the axis of the abscissas), the various different symbols give an estimate of the time at which the various other modes start sharing energy with the first mode. The correspondence between the symbols and the mode numbers is not explicitly indicated in the figure, but in general it turns out that the times at which the various modes "enter the packet" are increasing with the mode number k. So one clearly sees that, for a sufficiently low energy, in a rather short time a packet of modes is formed which share the energy among themselves, then follows a rather large interval of time in which "nothing happens", until eventually the subsequent modes start entering the packet, and such an energy cascade is not interrupted until all modes did enter the packet. This is the time at which equipartition is attained.

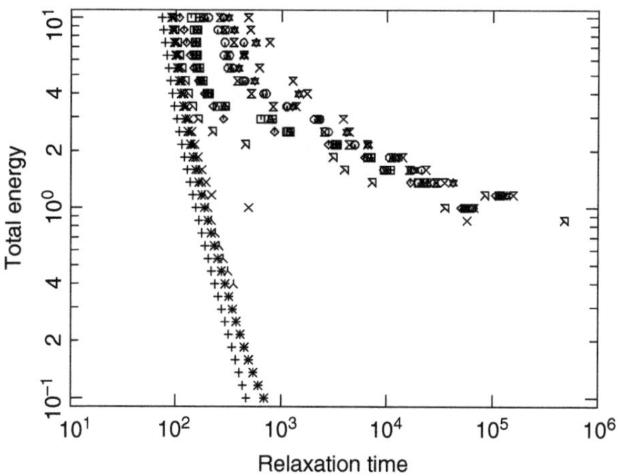

Fig. 4.8. The "shower". Here $N = 15$, and the energy was given initially just to the first mode. Having fixed an initial energy (and thus moving on a horizontal line), the various symbols give the times at which the other modes start sharing a (suitably defined) significant amount of the available energy. Such a time is found to be an increasing function of the mode number k, so that in the figure the mode number should be thought as increasing in going from left to right. The existence of two time-scales below a certain critical energy is clealy exhibited. Above the critical energy, instead, only one time-scale exists, which leads directly to equipartition. Taken from [14]

The relevant point is that such a separation of two time scales occurs only below a certain energy (namely, the one where the two inferior branches of the shower join); this just corresponds to the critical energy E_c of Bocchieri et al., because for higher energies the packet which is quickly formed covers all the available frequencies, i.e., there occurs a quick attainment of equipartition. Notice that the time needed for the quick formation of the low-frequency packet just below the critical energy is smaller than the time required for getting equipartition at larger energies.

Two more phenomena were also exhibited. The first one is that the width of the packet is a function of the specific energy, and is independent of N. The subtle point here is that such an independence with respect to N is exhibited if the width of the packet is plotted versus a quantity which is itself independent of N, and such a quantity is the frequency ω^* of the maximal mode included in the packet (or equivalently the corresponding value k^*/N). This is shown in Fig. 4.9, where the frequency ω^* defining the width of the packet (estimated in a suitable way) is plotted versus the specific energy ϵ. One very well sees that the data correspond to a curve $\omega^*(\epsilon) = c\,\epsilon^{1/4}$ (with a certain constant c), which by the way is just the law obtained later analytically. Notice that the data refer to N ranging from 8 to 1023. Notice also that the value of ϵ for which one has $\omega^* = 2$ (the maximal available frequency) provides a definition

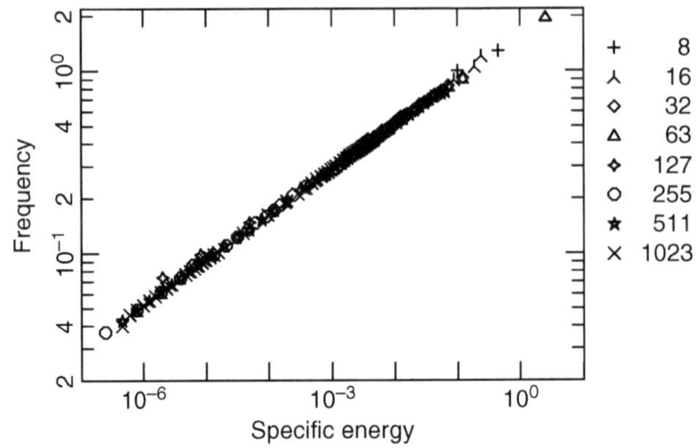

Fig. 4.9. Width of the "packet" (in frequency) versus specific energy, for N ranging from 8 to 1023. From [26] (adapted from [14])

for the critical specific energy in the sense of Bocchieri *et al.*, and that this quantity too is independent of N (i.e., pertains to the thermodynamic limit).

The second phenomenon concerns the way in which the time of formation of the packet depends on the initial conditions. One meets here with a problem that had been raised by Livi et al. (see [65]), who had pointed out that some relevant relaxation times were proportional to N if the energy was initially given to just one mode, whereas the behavior was quite different for other kinds of initial conditions. This fact is confirmed by Fig. 4.10, where the time of formation of the packet is plotted versus N for several kinds of initial conditions. Here one sees that such a time is proportional to N if the energy is given initially to the first mode. However, one also sees that the time is essentially independent of N if the energy is given initially to a packet of modes proportional to N, i.e., to a small packet extending to a maximal frequency Ω.

It should be mentioned that very interesting numerical informations on the formation of the packet had also been obtained by Biello et al. (see [66]), who were able to give quantitative estimates both of its width ω^* (namely, $\omega^* \simeq \epsilon^{1/4}$) and of its time of formation t_{f} (namely, $t_{\mathrm{f}} \simeq \epsilon^{-3/4}$).

We now briefly mention the results of three subsequent papers that are strictly related to the work of Berchialla et al., namely, the papers [67], [68] and [69]. In [67], the attention is addressed to the second time-scale τ_{eq}, namely the final time-scale to equipartition, which is shown to be of stretched exponential type, precisely, of the form $\tau_{\mathrm{eq}} \simeq \exp(\epsilon^{-1/4})$, at variance with the power law $\tau_{\mathrm{eq}} \simeq \epsilon^{-3}$ that had been suggested in [70]. Moreover, this result appears to be independent of N (for N large enough). This is clearly exhibited in Fig. 4.11, which reports the time of relaxation to the final global equipartition as a function of N, for two values of the specific energy ϵ. An analogous result was later obtained in [68], where the law $\tau_{\mathrm{eq}} \simeq \exp(\epsilon^{-1/5})$ was found for initial data with excitations of the high-frequency modes. By the way, in the latter paper

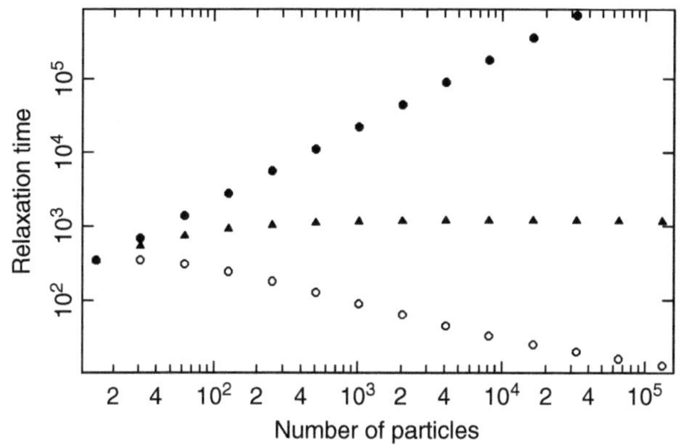

Fig. 4.10. Time of formation (relaxation time) of the packet versus N (with $N = 15, 31, 63, 127, \ldots, 32\,767$) for three types of initial conditions. *Dots*: all the energy initially on the lowest frequency mode. *Triangles* and *circles*: energy initially distributed in two different ways among the first $(N+1)/16$ modes (with zero energy to the higher frequencies). *Triangles*: energy linearly decreasing from the first mode to the last excited one. *Circles*: energy equally distributed among the initially excited modes. In all cases the specific energy is $\epsilon = 0.01$. From [14]

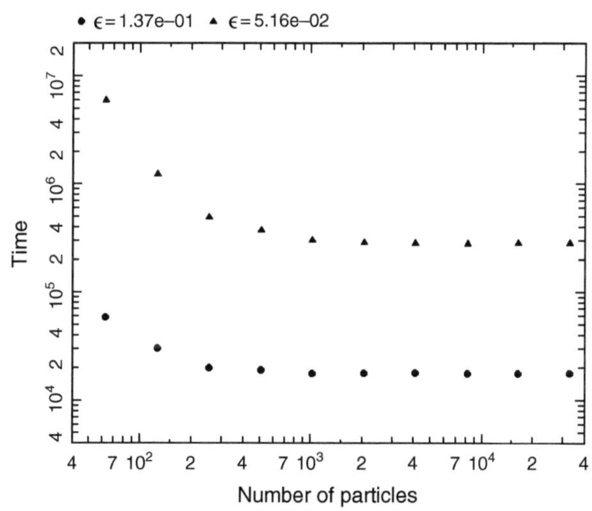

Fig. 4.11. Supporting the validity of the two-times description in the thermodynaimc limit: time of relaxation to the final global equipartition versus number N of particles, for two values of the specific energy ϵ. The relaxation times appear to tend to a definite limit (depending on ϵ) as N increases. *Circles*: $\epsilon = 0.137$, *triangles*: $\epsilon = 0.051$

an astute way was devised for exhibiting the analog of the shower when one deals with initial data of any type (and not just with excitations of the low-frequency modes). Finally, in the paper The figure showd that the relaxation time tends to a constant (depending on ϵ) as N increases, thus supporting te conjecture that the exponential law remains valid in the thermodynamic limit [69]. a vivid illustration (through projections of surfaces of section) is given of the way in which the final global equilibrium is attained. In fact, the system appears to be successively trapped into well different metastable regions, instead of finally merging, from some "ordered" region, to some "chaotic" one, as had been sometimes suggested.

4.6 New Analytical Contributions

We finally come to a brief illustration of some analytical developments that were obtained quite recently.

(a) *Solitons recovered*: The first relevant point is that soliton theory itself started to be reconsidered as a useful tool for analytical studies on the FPU problem. Indeed, in [71] it was shown that the form of the Fourier spectrum of the packet of the metastable state of the FPU system is explained in terms of KdV solitons. In particular, quantitative estimates both of its width (as $\epsilon^{1/4}$) and of its time of formation (as $\epsilon^{-3/4}$) were given. Soliton theory within the FPU problem had in fact been previously reconsidered in [72].
(b) *Shepelyansky and Ponno*: As first pointed out by Ford in his pioneering work (see [62]) of the year 1961, in order to explain the short-time dynamics of the modes (in particular, the quick formation of the packet, as we now say) one has to take into account the fact that the low-frequency modes are almost completely resonant. This idea was reconsidered by Shepelyansky (see [7]), who tried to deduce from it that the specific energy threshold tends to zero for $N \to \infty$, for initial long-wavelength excitations.[10] In [8], it was shown instead that the resonant normal form actually explains the formation of the metastable packet, and moreover that both its width and its time of formation are functions of the specific energy, exactly in the forms previously obtained (as $\epsilon^{1/4}$, and as $\epsilon^{-3/4}$, respectively) in [71] through soliton theory and in [66].

[10] In the Introduction of the paper, the result of Izrailev and Chirikov is mentioned: "According to Izrailev and Chirikov, in the case of low-mode excitation (nonlinear sound waves) the critical energy increases with the number of oscillators in the chain (or the energy per oscillator is constant)". It is then discussed how such authors had neglexted to take into account certain resonances in their semianalytical estimates, with the conclusion: "Such resonances not being considered by Izrailev and Chirikov give a sharp decrease of the chaos border in energy which goes to zero with the increase of the number of particles in the lattice. In this sense the long-wave chaos can exist for arbitrarily small nonlinearity".

(c) *Bambusi and Ponno, and the KdV equation as the resonant normal form for the FPU α-model*: In [17, 38, 73, 74] the attention was given to the resonant normal form of the FPU model for long-wavelength initial data. In [7] and [8], such a normal form had been expressed in terms of the mode coordinates, whereas in the new papers it was pointed out that, if such a normal form is read in terms of the particle coordinates in the continuum interpolation and in the thermodynamic limit, then the normal form is nothing but the KdV equation itself (actually, a pair of such equations, in agreement with the time reversal symmetry of the FPU system). In such a way, the privileged role of the KdV equation for the FPU system with long-wave initial data was recovered, with moreover an understanding of the time-scale of its validity. This happened after a previous understanding, by Bambusi et al., that the nonlinear Schroedinger (NLS) equation plays an analogous role of normal form in the FPU problem with short-wavelength initial data (see [75]).

(d) *Perturbation theory in the thermodynamic limit*: The analytical results of Bambusi and Ponno in [17] could be obtained only for an extremely special class of initial conditions, in which exactly one low-frequency mode was excited, at an energy E proportional to N, and so at a given specific energy $\epsilon = E/N$. Only in this sense does the result hold in the thermodynamic limit. It may be conjectured that such a limitation is only a technical one, to be hopefully removed in the future.

This fact rises the general problem of whether it is possible to extend the methods of classical perturbation theory of nearly integrable Hamiltonian systems to the thermodynamic limit ($N \to \infty$ with a nonvanishing specific energy ϵ). The presently available techniques do not allow it, as they apply only to finite N (or to any N, but with a bounded energy E, i.e., with a vanishing specific energy $\epsilon = E/N$ in the limit $N \to \infty$; see [19]). It was proved quite recently by one of the present authors (see [18]) that a rather simple modification of the known techniques actually allows one to do so. This is obtained at the cost of weakening the results, by renouncing to control all the orbits in phase space (a control which usually is obtained by making use, in the estimates, of the sup norm), and looking instead for results holding only in the mean. This is analogous to the way in which the von Neumann ergodic theorem can be considered as a weaker version of the Birkhoff ergodic theorem, although it still keeps all the relevant physical significance of the result (as particularly pointed out in [25]).

4.7 Conclusions

In the present review, we have illustrated the relevance of a metastability scenario for the interpretation of a large part of the results on the one-dimensional FPU model, in the thermodynamic limit. Such a scenario involves two well separated time-scales for the approach to equilibrium, below a critical specific energy.

It was also mentioned that too little information is presently available for the case of dimension two and especially for the physically significant case of dimension three. Two "simple " possibilities can be conceived. The first one is that the metastability scenario will be proved to be incorrect in the "physical case" of dimension three, in the sense that at any finite specific energy (or temperature) the time-averages of the relevant quantities present a quick relaxation to their equilibrium values. In such a case the "FPU paradox" will turn out to have been removed completely in the thermodynamic limit. This would provide a proof of the conjecture advanced long ago by Izrailev and Chirikov, at least in the way many people understand it, namely as claiming that no "FPU physical phenomenon" essentially exists.

The second "simple" possibility is that the metastability scenario as described above (with two well separated time-scales) will be proved to be correct. In such a case, in a sense the FPU paradox will still turn out to have been removed, because at any temperature the equilibrium state is finally attained. But some paradox will still remain. Indeed it will turn out that, below a certain critical specific energy (i.e., below a certain critical temperature), the FPU model predicts the existence of some metastable state which, for quite long times, may be practically indistinguishable from a true equilibrium state, although providing a statistics quite different from the standard equilibrium one (in this connection see [76, 77, 78]). So one would remain with the problem of ascertaining whether such a physical prediction is in agreement with the observations or not. We are particularly thinking of possible metastability phenomena in the measurements of specific heats at low temperatures, in the spirit of the rationale of the time-dependent specific heats (see [61]).

Naturally, other more complicated scenarios can be conceived. For example, there could exist a "cascade" of growing-time scales of different orders of magnitude as $N \to \infty$, and this, in a larger scale, could look like a continuous growth (we thank a referee for kindly pointing this out to us). To what physical phenomena would such a situation possibly correspond, is not clear to us.

References

1. E. Fermi, J. Pasta and S. Ulam, Studies of nonlinear problems, in E. Fermi: Note e Memorie (Collected Papers), Vol. II, No. 266. 977–988 (Accademia Nazionale dei Lincei, Roma, and The University of Chicago Press, Chicago 1965).
2. D.K. Campbell, P. Rosenau and G. Zaslavsky, Introduction: the Fermi-Pasta-Ulam problem—the first fifty years, Chaos **15**, 015101 (2005), and the papers following it.
3. T.P. Weissert, The genesis of simulation in dynamics; pursuing the Fermi-Pasta-Ulam problem. Springer, New York, 1997.
4. N.J. Zabusky and M.D. Kruskal, Interaction of solitons in a collisionless plasma and the recurrence of initial states, Phys. Rev. Lett. **15**, 240–243 (1965).

5. F.M. Izrailev and B.V. Chirikov, Statistical properties of a nonlinear string, Sov. Phys. Dokl. **11**, 30–34 (1966).
6. F.M. Izrailev, A.I. Khisamutdinov and B.V. Chirikov, Numerical experiments with a chain of coupled anharmonic oscillators, Report 252, Institute of Nuclear Physics, Novosibirsk, URSS, 1968 (English translation: LA 4440 TR, Los Alamos. 1970).
7. D.I. Shepelyansky, Low-energy chaos in the Fermi–Pasta–Ulam problem, Nonlinearity **10**, 1331–1338 (1997).
8. A. Ponno, The FPU problem in the thermodynamic limit: scaling laws of the energy cascade, in P. Collet et al. eds, Chaotic dynamics and transport in classical and quantum systems. Kluwer, Dordrecht, 2005, pp. 431–440.
9. P. Bocchieri, A. Scotti, B. Bearzi and A. Loinger, Anharmonic chain with Lennard-Jones interaction, Phys. Rev. A **2**, 2013–2019 (1970).
10. L. Galgani and A. Scotti, Planck-like distribution in classical nonlinear mechanics, Phys. Rev. Lett. **28**, 1173–1176 (1972).
11. C. Cercignani, L. Galgani and A. Scotti, Zero-point energy in classical nonlinear mechanics, Phys. Lett. A **38**, 403 (1972); C. Cercignani, On a nonquantum derivation of Planck's distribution law, Found. Phys. Lett. **11**, 189–199 (1998).
12. L. Galgani and A. Scotti, Recent progress in classical nonlinear dynamics, Rivista Nuovo Cim. **2**, 189–209 (1972).
13. E. Fucito, F. Marchesoni, E. Marinari, G. Parisi, L. Peliti, S. Ruffo and A. Vulpiani, Approach to equilibrium in a chain of nonlinear oscillators, J. Phys., **43**, 707–713 (1982).
14. L. Berchialla, L. Galgani and A. Giorgilli, Localization of energy in FPU chains, Discr. Cont. Dyn. Syst. A **11**, 855–866 (2004).
15. A. Carati and L. Galgani, On the specific heat of the Fermi–Pasta–Ulam systems and their glassy behavior, J. Stat. Phys. **94** 859–869 (1999).
16. A. Carati, P. Cipriani and L. Galgani, On the definition of temperature in FPU systems, J. Stat. Phys. **115**, 1119–1130 (2004).
17. D. Bambusi and A. Ponno, On metastability in FPU, Comm. Math. Phys., 264(2), 539–561 (2006).
18. A. Carati, An averaging theorem for Hamiltonian dynamical systems in the thermodynamic limit, preprint.
19. D. Bambusi and A. Giorgilli, Exponential stability of states close to resonance in infinite-dimensional Hamiltonian systems, J. Stat. Phys. **71**, 569–606 (1993).
20. G. Benettin, Time scales for energy euipartition in a two-dimensional FPU model, Chaos **15**, 015108 (2004).
21. G. Marcelli and A. Tenenbaum, Quantumlike short-time behavior of a classical crystal, Phys. Rev. E **68**, 041112 (2003).
22. H. Poincaré, Méthodes nouvelles de la mécanique céleste, Vol. I, Chap. 5.
23. E. Fermi, Beweiss das ein mechanisches Normalsystem im allgemeinen quasiergodisch ist, Phys. Zeits. **24**, 261–265 (1923); Über die Existenz quasiergodischer Systeme, Phys. Zeits. **25**, 166–167 (1924), in Note e Memorie (Collected Papers), Vol. I, No. 11: 79–87 Accademia Nazionale dei Lincei, Roma, and The University of Chicago Press, Chicago, 1965.
24. G. Benettin, G. Ferrari, L. Galgani and A. Giorgilli, An extension of the Poincaré-Fermi theorem on the nonexistence of invariant manifolds in nearly integrable Hamiltonian systems, Nuovo Cim. B **72**, 137–148 (1982); G. Benettin, L. Galgani and A. Giorgilli, Poincaré's non-existence theorem and classical

perturbation theory in Nearly-Integrable Hamiltonian Systems, in Dynamics and Stochastic Processes, R. Livi and A. Politi, eds, Advances in Nonlinear World Scientific, Singapore, 1985.
25. J. von Neumann, Physical approach to the ergodic hypotesis, N.A.S. Proc. **18**, 263–266 (1932), also in J. von Neumann, Collected works. H.A. Taub, ed., Pergamon Press, Oxford, 1961. Vol. II, No. 13, pp. 274–277.
26. A. Carati, L. Galgani and A. Giorgilli, Dynamical systems and thermodynamics, in Encyclopedia of Mathematical Physics, Elsevier, Oxford (2006).
27. D. Poggi, S. Ruffo and H. Kantz, Shock waves and time scales to reach equipartition in the Fermi-Pasta-Ulam model, Phys. Rev. E **52**, 307–315 (1995).
28. P. Lorenzoni and S. Paleari, Metastability and dispersive shock waves in the FPU system, Physica D 221 (2006), 110–117.
29. A.N. Kolmogorov, Preservation of conditionally periodic movements with small change in the Hamilton function, Dokl. Akad. Nauk **98**, 527 (1954); English translation in G. Casati and G. Ford, eds, Lecture Notes in Physics No. 93, Springer Verlag, Berlin, 1979. See also G. Benettin, L. Galgani, A. Giorgilli and J.-M. Strelcyn, A proof of Kolmogorov's theorem on invariant tori using canonical transformations defined by the Lie method, Nuovo Cim. B **79**, 201–223 (1984).
30. A. Carati and L. Galgani, Analog of Planck's formula and effective temperature in classical statistical mechanics far from equilibrium, Phys. Rev. E **61**, 4791–4794 (2000); A. Carati and L. Galgani, Einstein's nonconventional conception of the photon, and the modern theory of dynamical systems, in J. Bricmont et al., eds, Chance in Physics, Lecture Notes in Physics No. 574, Springer, Berlin, 2001; A. Carati and Galgani, Planck's formula and glassy behavior in classical nonequilibrium statistical mechanics, Physica A **280**, 105–114 (2000); L. Galgani, Relaxation times and the foundations of classical statistical mechanics in the light of modern perturbation theory, in G. Gallavotti and P.F. Zweifel, eds, non-linear evolution and chaotic phenomena, NATO ASI Series R71B: Vol. 176, Plenum Press, New York, 1988.
31. A. Carati and L. Galgani, The theory of dynamical systems and the relations between classical and quantum mechanics, Found. Phys. **31**, 69–87 (2001).
32. U. Frisch nad R. Morf, Intermittency in nonlinear dynamics and singularities at comlex times, Phys. Rev. A **23**, 2673–2705 (1981).
33. G.K. Batchelor, The theory of homogeneous turbulence, Cambridge U.P., Cambridge, 1960.
34. J. von Neumann, Recent theories of turbulence, in J. von Neumann, Collected works. H.A. Taub, ed., Pergamon Press, Oxford, 1961. Vol. VI, No. 32, pp. 437–472.
35. P. Butera, L. Galgani, A. Giorgilli, A. Tagliani and H. Sabata, Stochasticity thresholds in a lattice field theory, Nuovo Cim. B **59**, 81–86 (1980).
36. B. Bassetti, P. Butera, M. Raciti and M. Sparpaglione, Complex poles, spatial intermittencies, and energy transfer in a classical nonlinear string, Phys. Rev. A **30**, 1033–1039 (1984).
37. A. Carati, L. Galgani and A. Giorgilli, The Fermi-Pasta-Ulam problem as a challenge for the foundations of physics, Chaos **15**, 015105 (2004).
38. A. Ponno and D. Bambusi, Korteweg-de Vries equation and energy sharing in Fermi-Pasta-Ulam, Chaos **15**, 015107 (2004).
39. G. Parisi, On the approach to equilibrium of a Hamiltonian chain of anharmonic oscillators, Europhys. Lett. **40**, 357 (1997).

40. F. Fucito, F.Marchesoni, M. Sparpaglione and A. Vulpiani, Intermittent behaviour in non-linear-Hamiltonian systems far from equilibrium, J. Phys. A **16**, 117–124 (1983).
41. A. Carati, L. Galgani and B. Pozzi, Levy flights in the Landau-Teller model of molecular collisions, Phys. Rev. Lett. **90**, 010601 (2003).
42. R. Livi, M. Pettini, S. Ruffo, M. Sparpaglione and A. Vulpiani, Relaxation to different stationary states in the FPU model, Phys. Rev. A **28**, 3544–3552 (1983).
43. R. Livi, S. Ruffo, M. Pettini and A. Vulpiani, Short-time asymptotics in classical nonlinear wave equations, Nuovo Cim. B **89**, 120–130 (1985).
44. R. Livi, M. Pettini, S. Ruffo, M. Sparpaglione and A. Vulpiani, Equipartition threshold in nonlinear large Hamiltonian systems: the Fermi–Pasta–Ulam model, Phys. Rev. A **31**, 1039–1045 (1985).
45. R. Livi, M. Pettini, S. Ruffo, and A. Vulpiani, Further results on the equipartition threshold in large nonlinear Hamiltonian systems, Phys. Rev. A **31**, 2740–2742 (1985).
46. G. Benettin, L. Galgani and A. Giorgilli, Boltzmann's ultraviolet cutoff and Nekhoroshev's theorem on Arnold diffusion, Nature **311**, 444–445 (1984).
47. A. Carati, L. Galgani and B. Pozzi, The problem of the rate of thermalization and the relations between classical and quantum mechanics, in M. Fabrizio et al., eds, Mathematical models and methods for smart materials, Series of Advances in Mathematics no. 62, World Scientific, Singapore, 2002.
48. N.N. Nekhoroshev, Russ. Math. Surv. **32**, 1 (1977); in O.A. Oleinik, ed. Topics in modern mathematics: Petrovskii Sem. No. 5, Consultant Bureau, New York, 1985. See also G. Benettin, L. Galgani and A. Giorgilli, A proof of Nekhoroshev's theorem for the stability times in nearly integrable Hamiltonian systems, Celestial Mech. **37**, 1–25 (1985); A. Giorgilli, On the problem of stability for near to integrable Hamiltonian systems, Proceedings of the International Congress of Mathematicians Berlin 1998, Vol. III, Documenta Mathematica, extra volume ICM 1998, 143–152 (1998); G. Benettin, The elements of Hamiltonian perturbation theory, in D. Benest, C. Froeschle' and E. Lega eds, Hamiltonian systems and Fourier analysis, Cambridge Scientific Publisher, Cambridge (UK), 2005.
49. L.D. Landau and E. Teller, On the theory of sound dispersion, Phys. Z. Sowjet. **10**, 34–41 (1936), also in Collected Papers of L.D. Landau, ter Haar, ed., Pergamon Press, Oxford 1965, pp. 147–153; G. Benettin, A. Carati and P. Sempio, On the Landau-Teller approximation for the energy exchanges with fast degrees of freedom, J. Stat. Phys. **73**, 175 (1993); G. Benettin, A. Carati and G. Gallavotti, A rigorous implementation of the Landau-Teller approximation for adiabatic invariants, Nonlinearity **10**, 479–505 (1997); G. Benettin, P. Hjorth and P. Sempio, Exponentially long equilibrium times in a one-dimensional collisional model of a classical gas, J. Stat. Phys. **94**, 871–892 (1999); A. Carati, L. Galgani and B. Pozzi, The problem of the rate of thermalization, and the relations between classical and quantum mechanics, in M. Fabrizio, B Lazzari, A. Morro, eds, Mathematical models and methods for smart materials, World Scientific, Singapore, 2002.
50. G. Benettin, L. Galgani and A. Giorgilli, Exponential law for the equipartition times among translational and vibrational degrees of freedom, Phys. Lett. A **120**, 23–27 (1987).

51. O. Baldan and G. Benettin, Classical 'freezing' of fast rotations. A numerical test of the Boltzmann–Jeans conjecture, J. Stat. Phys. **62**, 201 (1991).
52. M. Pettini and M. Landolfi, Relaxation properties and ergodicity breaking in nonlinear Hamiltonian dynamics, Phys. Rev. A **41**, 768–783 (1990).
53. M. Pettini, L. Casetti, M. Cerruti-Sola, R. Franzosi and E.G.D. Cohen, Weak and strong chaos in Fermi-Pasta-Ulam models and beyond, Chaos **15**, 015106 (2004).
54. L. Casetti, R. Livi and M. Pettini, Gaussian model for chaotic instability of Hamiltonian flows, Phys. Rev. Lett. **74**, 375–378 (1995); L. Casetti, C. Clementi and Pettini, Riemannian theory of Hamiltonian chaos and Lyapunov exponents, Phys. Rev. E **54**, 5969–5984 (1996); L. Casetti, M. Pettini and E.G.D. Cohen, Geometric approach to Hamiltonian dynamics and statistical mechanics, Phys. Rep. **337**, 237–341 (2000).
55. P. Butera and G. Caravati, Phase transitions and Lyapunov characteristic exponents, Phys. Rev. A **36**, 962–964 (1987).
56. M. Casartelli, E. Diana, L. Galgani and A. Scotti, Numerical computations on a stochastic parameter related to the Kolmogorov entropy, Phys. Rev. A **13**, 1921–1925 (1976).
57. G. Benettin, L. Galgani and A. Giorgilli, Classical Perturbation Theory for systems of weakly coupled rotators, Nuovo Cim. **89** B, 89–102 (1985).
58. G. Benettin, L. Galgani and A. Giorgilli, Numerical investigations on a chain of weakly coupled rotators in the light of classical perturbation theory, Nuovo Cim. **89** B, 103–119 (1985).
59. R. Livi, M. Pettini, S. Ruffo and A. Vulpiani, Chaotic behavior in nonlinear Hamiltonian systems and equilibrium statistical mechanics, J. Stat. Phys. **48**, 539–559 (1987).
60. A. Perronace and A. Tenenbaum, Classical specific heat of an atomic lattice at low temperature, revisited, Phys. Rev. E **57**, 100–107 (1998); Erratum, Phys. Rev. E **57**, 6215 (1998).
61. N.O. Birge and S.R. Nagel, Specific-heat spectroscopy of the glass transition, Phys. Rev. Lett. **54**, 2674 (1985); N.O. Birge, Specific-heat spectroscopy of glycerol and propylene near the glass transition, Phys. Rev. **34**, 1631–1642 (1986).
62. J. Ford, Equipartition of energy for nonlinear systems, J. Math. Phys. **2**, 387–393 (1961).
63. L. Galgani, A. Giorgilli, A. Martinoli and S. Vanzini, On the problem of energy equipartition for large systems of the Fermi–Pasta–Ulam type: analytical and numerical estimates, Physica D **59**, 334–348 (1992).
64. G. Benettin, L. Galgani and A. Giorgilli, Realization of holonomic constraints and freezing of high frequency degrees of freedom in the light of classical perturbation theory, II, Comm. Math. Phys. **121**, 557–601 (1989); G. Benettin, J. Fröhlich and A. Giorgilli, A Nekhoroshev-type theorem for Hamiltonian systems with infinitely many degrees of freedom, Comm. Math. Phys, **119**, 95–108 (1988).
65. H. Kantz, Vanishing stability thresholds in the thermodynamic limit of nonintegrable conservative systems, Physica D **39**, 322 (1989); H. Kantz, R. Livi and S. Ruffo, Equipartition thresholds in chains of anharmonic oscillators, J. Stat. Phys. **76**, 627 (1994).
66. J.A. Biello, P.R. Kramers, and Y. Lvov, Stages of energy transfer in the FPU model, Discr. Cont. Dyn. Syst. B (Suppl.) 113–122 (2003).

67. L. Berchialla, A. Giorgilli and S. Paleari, Exponentially long times to equipartition in the thermodynamic limit, Phys. Lett. A **321**, 167–172 (2004).
68. S. Paleari and T. Penati, Equipartition times in a FPU system, Discr. Cont. Dynam. Syst. (Suppl. Volume) 1–10 (2005).
69. A. Giorgilli, S. Paleari and T. Penati, Local chaotic behavior of the FPU system, Discr. Cont. Dynam. Syst. B **5**, 1–14 (2005).
70. J. De Luca, A. Lichtenberg and S. Ruffo, Finite times to equipartition in the thermodynamic limit, Phys. Rev. E **60**, 3781–3786 (1999).
71. A. Ponno, Soliton theory and the Fermi-Pasta-Ulam problem in the thermodynamic limit, Europh. Lett. **64**, 606–612 (2003).
72. A. Ponno, L. Galgani and F. Guerra, Analytical estimate of stochasticity thresholds in Fermi-Pasta-Ulam and phi-4 models, Phys. Rev. E **61**, 7081–7086 (2000).
73. A. Ponno and D. Bambusi, Energy cascade in FPU models, in G. Gaeta, et al. eds, Symmetry and perturbation theory 2004, World Scientific, Singapore, 2005, pp. 263–270.
74. D. Bambusi and A. Ponno, Resonance, metastability and blow-up in FPU, Chap. 6 in this volume.
75. D. Bambusi, A. Carati and A. Ponno, The nonlinear Schroedinger equation as a resonant normal form, Discr. Cont. Dyn. Sys. B **2**, 109–128 (2002).
76. E. Olivieri and M.E. Vares, Large deviations and metastability, Cambridge U.P., Cambridge, 2005.
77. H. Larralde and F. Leyvraz, Metastability for Markov processes with detailed balance, Phys. Rev. Lett. **94**, 160201 (2005).
78. A. Carati, Thermodynamics and time-averages, Physica A **348**, 110–120 (2005).

5

Resonance, Metastability and Blow up in FPU

Dario Bambusi[1] and Antonio Ponno[2]

[1] Dipartimento di Matematica Via Saldini 50, 20133 Milano, Italy
bambusi@mat.unimi.it
[2] Università di Padova, Dipartimento di, Matematica
ponno@math.unipd.it

Abstract. We consider the FPU model with nonlinearity starting with terms of order $n \geq 3$. We compute the resonant normal form in the region where only one low-frequency mode is excited and deduce rigorous results on the correspondence between the dynamics of the normal form and that of the complete system. As n varies, we give a criterion in order to deduce whether the FPU phenomenon (formation of a metastable packet of modes) is present or not. The criterion is that, if the normal form equation has smooth solutions then the FPU phenomenon is present, while it is absent if the solutions of the normal form equations have blow up in a finite time. In particular the phenomenon should be present for $n \leq 5$ and absent for $n \geq 7$.

5.1 Introduction

In the present contribution some analytic results on the Fermi–Pasta–Ulam (FPU) problem are presented. Our purpose is to use the methods of rigorous perturbation theory for infinite dimensional Hamiltonian systems to explain some of the features of the FPU phenomenology. In particular we have the following results:

(1) For low-energy and long-wavelength initial data, canonical perturbation theory allows us to put the Hamiltonian of the system in resonant normal form up to a small remainder. The equations of motion of the normal form consist of two partial differential equations (PDEs) which describe well the dynamics of the system within a certain time-scale.
(2) In the case of the so-called FPU α-model, the normal form equations are two uncoupled Kortweg–de Vries (KdV) equations. Thus all the remarkable features of the KdV equation turn out to pertain, *in an approximate way*, also to the FPU model. In particular the α-model will behave as an integrable system over the time-scale of validity of the normal form. We emphasize that, as predicted in [1], such a time-scale is of order $\epsilon^{-3/4}$, ϵ being the specific energy (i.e. the energy per degree of freedom) of the

D. Bambusi and A. Ponno: *Resonance, Metastability and Blow up in FPU*, Lect. Notes Phys. **728**, 191–205 (2008)
DOI 10.1007/978-3-540-72995-2_5

system. Such a time-scale has to be considered as short in connection with the problem of thermalization. Indeed, thermalization, at least in dimension one, seems to take place on a much longer time-scale, namely a "Nekhoroshev-like" stretched exponential of $1/\epsilon$. So the first conclusion, at least from a rigorous point of view, is that KdV is relevant for the dynamics of the FPU (as heuristically predicted since the pioneering works of Zabusky and Kruskal [2]) on short, power law time-scales, as opposed to the longer, exponential ones over which one can observe relaxation to equilibrium.

(3) The KdV equation allows us to explain (rigorously), in the case of the α-chain, the phenomenon of formation of the packet [3, 4, 5, 6, 7, 8], namely the fact (first observed by Fermi, Pasta and Ulam themselves) that, if one low-frequency mode is initially excited, then the energy quickly flows to a small packet of modes whose energy, in time average, decreases exponentially with the mode index. The packet turns out to be stable over the time-scale covered by the normal form, namely $\epsilon^{-3/4}$.

(4) The situation of having a resonant normal form which is integrable seems to be quite exceptional: indeed, if one introduces action-angle variables one sees that the normalization procedure consists in eliminating just one among infinitely many angles; thus the system in normal form is in principle expected to have only one integral of motion independent of the energy. Instead, the KdV has a complete set (infinitely many) of integrals of motion in involution! We are thus naturally led to study the β-model and systems with perturbations of higher order, in order to check the occurrence and the relevance of integrability. The normal form of the β-model turns out to consist of two uncoupled modified KdV (mKdV) equations which are integrable too. The same qualitative behavior of the α-model is thus expected also in the β-model (a rigorous proof is still missing due to the lack of a deeper knowledge of the mKdV equation).

(5) The situation changes with higher order FPUs where the normal form equations we get are higher order generalized KdVs (gKdV) that are no longer integrable. One is thus naturally led to ask whether the FPU phenomenon persists in such cases, and whether the integrability of the normal form plays a fundamental role or not. We think that the answer is no, and that there should be a weaker property entraining the formation of a metastable packet of modes. We guess that such a property actually is related to the smoothness of the solutions of the Cauchy problem of the normal form PDEs. In particular if the solutions of the PDEs under investigation have blow up in a finite time we guess that metastability should be lost, while we think that the phenomenon of formation of a metastable packet of modes should be present in all models that do not display blow up. We also give a heuristic argument for that.

For the case of one-dimensional FPU chains it turns out that the metastable packet is expected to exist in the case of nonlinearity of degree less than or equal to 5, while it is expected not to exist when the nonlinearity has degree higher than or equal to 7. The case of nonlinearity of order 6 is critical and therefore the existence of the packet strongly depends, in principle, on the initial conditions.

5.2 Normal Form

Consider the Hamiltonian system

$$H(q,p) = \sum_{j=-N}^{N-1} \frac{p_j^2}{2} + U(q_{j+1} - q_j) \ , \tag{5.1}$$

$$U(x) = \frac{x^2}{2} + \frac{x^n}{n} \ , \quad n \geq 3 \tag{5.2}$$

$$q_{j+2N} = q_j \ , \quad p_{j+2N} = p_j \ , \tag{5.3}$$

describing a periodic chain of $2N$ particles interacting through nonlinear springs. The canonical variables are $q = (q_{-N}, \dots, q_{N-1}), p = (p_{-N}, \dots, p_{N-1})$. Hamiltonians of the form (5.1) were first introduced and studied by Fermi, Pasta and Ulam (FPU) in the case $n = 3, 4$; such model Hamiltonians are commonly referred to as FPU models (in the case $n = 3$ and $n = 4$ one refers to the α-model and β-model, respectively). Due to the periodic boundary conditions (5.3), the total linear momentum of the system is preserved. So one can restrict oneself to the case $\sum_j p_j = \sum_j q_j = 0$.

The equations of motion associated with (5.1) are given by

$$\ddot{q}_j = q_{j+1} + q_{j-1} - 2q_j + (q_{j+1} - q_j)^{n-1} - (q_j - q_{j-1})^{n-1} \tag{5.4}$$

We are interested in initial data in which only one Fourier mode is excited, say the one with wave number $k_0 \ll N$.

Remark 5.1. In this case the dynamics is equivalent to the dynamics of a shorter FPU chain; in particular, if k_0 divides N it turns out that the dynamics is equivalent to that of a chain of length $2N/k_0$, in which only the first Fourier mode is excited. More precisely, the solution lies on an invariant submanifold on which the Fourier modes with index which is not an integer multiple of k_0 (modulo N) are exactly zero (see [9]).

We will use as a small parameter

$$\mu := \frac{k_0}{N} \ . \tag{5.5}$$

To begin with, we construct a canonical transformation setting the system in normal form in the region of long-wavelength states. We rewrite the FPU system in terms of the new variables r_j defined by

$$r_j := q_j - q_{j-1} \quad , \quad \sum_j r_j = 0 \; ; \tag{5.6}$$

the change of variables $q \to r$ is well defined and invertible. Introducing also the discrete Laplacian Δ_1 defined by

$$(\Delta_1 r)_j := r_{j+1} + r_{j-1} - 2r_j \; , \tag{5.7}$$

the FPU equations take the form

$$\ddot{r} = \Delta_1 (r + r^{n-1}) \; . \tag{5.8}$$

We introduce now an interpolating function $r = r(x,t)$ for the sequence r_j, namely a (smooth) function with the property that the sequence

$$r_j(t) \equiv r(j,t) \tag{5.9}$$

fulfills the FPU equations (5.8). Moreover we will assume that the function $r(x,t)$ is $2N/k_0 \equiv 2/\mu$ periodic and has zero average. Thus we postulate that the function r fulfills (5.8) with an obvious extension of the definition of Δ_1 to smooth functions.

We remark that up to now we did essentially nothing: the continuous system we got is equivalent to the original FPU system.

Since we are interested in states with long wavelength and small amplitude we will look for solutions of the form

$$r(x,t) = \mu^{\frac{2}{n-2}} u(\mu x, \mu t) \; ; \tag{5.10}$$

it turns out that such a re-scaling gives a correct balance between dispersive and nonlinear terms in the equations of motion (see below). It has to be remarked at this stage that by the choice (5.10) one links the small parameter μ defined by (5.5) to the specific energy of the system (energy per degree of freedom). Indeed, from

$$\epsilon \equiv \frac{E}{2N} \sim \frac{1}{2N} \sum_n r_n^2 = \mu^{\frac{4}{n-2}} \frac{1}{2N} \sum_n u^2(\mu n) \quad ,$$

it follows that

$$\mu \sim \epsilon^{(n-2)/4} \quad . \tag{5.11}$$

We remark that if u is a smooth function (e.g. analytic) then it turns out that the Fourier coefficients of r and therefore of the original variables of the FPU model will decay fast (e.g. exponentially) with the index k. Correspondingly, one expects that it should be possible to approximate the frequencies of the FPU with their low-mode expansion. Indeed, when k/N is small, one has

$$\omega(k) \equiv 2 \sin \left(\frac{k\pi}{2N} \right) = \frac{k\pi}{N} - \frac{k^3 \pi^3}{24 N^3} + \mathrm{O}((k/N)^5) \quad . \tag{5.12}$$

By substituting (5.10) into (5.8) one gets the equation of motion for u, namely

$$u_{\tau\tau} = \mu^{-2}\Delta_\mu \left(u + \mu^2 u^{n-1}\right) \quad . \tag{5.13}$$

Here Δ_μ is the difference operator

$$(\Delta_\mu u)(y) := u(y+\mu) + u(y-\mu) - 2v(y) \ . \tag{5.14}$$

Notice that (5.13) is written in terms of the rescaled variables $y = \mu x$ and $\tau = \mu t$ introduced in (5.10); moreover, since the function $r(x)$ is $2/\mu$-periodic, then the function $u(y)$ is 2-periodic. Equation (5.13) is Hamiltonian with Hamiltonian function

$$K(u,v) = \int_{-1}^{1} \left(\frac{-v\Delta_\mu v}{2\mu^2} + \frac{u^2}{2} + \frac{\mu^2 u^n}{n} \right) dy \ , \tag{5.15}$$

where v is the variable canonically conjugated to u. From now on we will study the system (5.15).

Inserting the formal expansion of the operator Δ_μ (5.14), given by

$$\frac{\Delta_\mu}{\mu^2} = \partial_y^2 + \frac{\mu^2 \partial_y^4}{12} + \mathrm{O}(\mu^4) \ , \tag{5.16}$$

into the Hamiltonian (5.15), one gets

$$K = H_0 + \mu^2 P + \mathcal{R}_1 \ , \tag{5.17}$$

where

$$H_0(u,v) := \int_{-1}^{1} \left[\frac{v(-\partial_y^2 v) + u^2}{2} \right] dy = \sum_{k\neq 0} \frac{\hat{u}_k^2 + (k\pi)^2 \hat{v}_k^2}{2} \ , \tag{5.18}$$

$$P(u,v) := \int_{-1}^{1} \left[-\frac{v\partial_y^4 v}{24} + \frac{u^n}{n} \right] dy \ , \tag{5.19}$$

and $\mathcal{R}_1 \sim \mathrm{O}(\mu^4)$ is the remainder of the expansion. In (5.18) $\hat{u}_k$ and $\hat{v}_k$ are the Fourier coefficients of u and v, respectively. More precisely, the Fourier coefficients of a (zero-average) function $u(y)$ are defined here as

$$u(y) = \frac{1}{\sqrt{2}} \sum_{k>0} \left[\hat{u}_k \ \cos(\pi k y) + \hat{u}_{-k} \ \sin(\pi k y) \right] \tag{5.20}$$

and clearly $\hat{v}_k$ is the momentum conjugated to $\hat{u}_k$.

Remark 5.2. The equations of motion of $H_0 + \mu^2 P$, in second-order form, are

$$u_{\tau\tau} = \left[u + \mu^2 \frac{1}{12} u_{yy} + \mu^2 u^{n-1} \right]_{yy} \ , \tag{5.21}$$

which is a generalized Boussinesq (gB) equation, sometimes considered as a starting point in approaching the FPU problem. In particular the (5.21) has been shown to be integrable in the case $n = 3$ (even though the actual integration is non-trivial); in the other cases the equation is non-integrable.

The strategy for studying the dynamics of (5.17) is well known from the theory of finite dimensional systems: substitute P with its average $\langle P \rangle$ with respect to the flow of the Hamiltonian vector field X_{H_0}, and study the dynamics of this system, then use some rigorous method in order to prove that the solutions of the averaged system, namely of the system

$$H_0 + \mu^2 \langle P \rangle \ , \tag{5.22}$$

are close to the solutions of the original one at least over some long time-scale.

We start now by giving a precise definition of $\langle P \rangle$. So we consider the equations of motion of H_0, i.e.

$$\begin{cases} u_\tau = -\partial_y^2 v \ , \\ v_\tau = -u \end{cases} \iff \begin{cases} \frac{\mathrm{d}\hat{u}_k}{\mathrm{d}\tau} = \hat{v}_k \\ \frac{\mathrm{d}\hat{v}_k}{\mathrm{d}\tau} = -(\pi k)^2 \hat{u}_k \end{cases} . \tag{5.23}$$

Denote by $(u, v) = \Psi^\tau(u_0, v_0)$ the solution of the Cauchy problem with initial datum (u_0, v_0), which in terms of Fourier coefficients is given by

$$\hat{u}_k(\tau) = \widehat{(u_0)}_k \cos(\pi k \tau) + \frac{\widehat{(v_0)}_k}{k\pi} \sin(\pi k \tau) \ , \tag{5.24}$$

$$\hat{v}_k(\tau) = \widehat{(v_0)}_k \cos(\pi k \tau) - k\pi \widehat{(u_0)}_k \sin(\pi k \tau) \ . \tag{5.25}$$

In the following we will show that there is a third representation in which the computations are particularly simple. In order to keep in mind that all we will do is completely independent of the coordinate system, we will use the abstract notation z for a phase space point, namely for the pair (u, v). Correspondingly, the flow of H_0 will be denoted by $\Psi^\tau(z)$

Definition 5.1. *The average of P is defined by*

$$\langle P \rangle(z) := \frac{1}{2} \int_0^2 P(\Psi^\tau(z)) \mathrm{d}\tau \tag{5.26}$$

Remark 5.3. In the jargon of Birkhoff normal forms one usually says that $\langle P \rangle$ is the resonant part of P, and indeed, when expressed in the complex variables usually employed in order to solve the homological equation it contains all (and only) the resonant monomials present in P.

As anticipated above the explicit computation of $\langle P \rangle$ (which is very complicated in Fourier space) is simpler in different variables that we presently introduce. Consider the non-canonical change of variables

$$\xi := \frac{u + v_y}{\sqrt{2}} \ , \quad \eta := \frac{u - v_y}{\sqrt{2}} \ . \tag{5.27}$$

Correspondingly the equations of motion take the form

$$\frac{\mathrm{d}z}{\mathrm{d}\tau} = J\,\nabla H(z) \iff \left(\xi_\tau = -\partial_y \frac{\delta H}{\delta \xi} \,,\; \eta_\tau = \partial_y \frac{\delta H}{\delta \eta} \right) \,, \tag{5.28}$$

where ∇H denotes the L^2 gradient of $H(\xi,\eta)$ and the Poisson tensor is now defined by

$$J = \begin{pmatrix} -1 & 0 \\ 0 & 1 \end{pmatrix} \partial_y \,. \tag{5.29}$$

In the variables (ξ,η) the various parts of the Hamiltonian take the form

$$H_0(\xi,\eta) = \int_{-1}^{1} \frac{\xi^2+\eta^2}{2} \mathrm{d}y \tag{5.30}$$

$$P(\xi,\eta) = \int_{-1}^{1} \left[-\frac{[\partial_y(\xi-\eta)]^2}{48} + \frac{(\xi+\eta)^n}{n 2^{n/2}} \right] \mathrm{d}y \,, \tag{5.31}$$

and in particular the equations of motion of H_0 and the flow Ψ^τ assume the simple form

$$[\xi_\tau = -\xi_y \,,\quad \eta_\tau = \eta_y] \iff [\xi(y,\tau) = \xi_0(y-\tau) \,,\; \eta(y,\tau) = \eta_0(y+\tau)] \,. \tag{5.32}$$

It is now easy to obtain the following

Proposition 5.1. *In the variables* ξ,η *the average of the perturbation is given by*

$$\langle P\rangle(\xi,\eta) = -\int_{-1}^{1} \left[\frac{(\xi_y)^2 + (\eta_y)^2}{48} \right] + \frac{1}{2^{n/2} n} \sum_{l=0}^{n} C_l^n [\xi^l][\eta^{n-l}] \,. \tag{5.33}$$

In the above expression, $C_l^n \equiv n!/(l!(n-l)!)$ *while* $[f^j] \equiv \int_0^L f^j \,\mathrm{d}x/2$*; we will refer to the latter as to the moment of* f *of order* j*, or simply the* j*-th moment of* f*.*

For the proof see [10, 11].

The equations of motion associated to $H_0 + \mu^2\langle P\rangle$ are

$$\begin{cases} \xi_\tau = -\xi_y - \mu^2(1/24)\xi_{yyy} - \mu^2(1/2^{n/2}) \sum_{l=1}^{n-1} C_l^{n-1} [\eta^{n-l-1}](\xi^l)_y \\ \eta_\tau = \eta_y + \mu^2(1/24)\eta_{yyy} + \mu^2(1/2^{n/2}) \sum_{l=1}^{n-1} C_l^{n-1} [\xi^{n-l-1}](\eta^l)_x \end{cases} . \tag{5.34}$$

These are generalized Kortweg–de Vries (gKdV) equations.

One can easily check that for $n=3$ and $n=4$ they yield, respectively, the KdV and the modified KdV equation (see below), which are both integrable. Moreover it is now clear that they appear here as the resonant normal forms of the corresponding FPU models.[1]

[1] For an interesting related result see [12]

Notice that, as a consequence of averaging, the two second-order momenta of ξ and η are constants of motion for system (5.34). Moments of order greater than two will be time-dependent, and as a consequence the above equations are actually coupled and of integro-differential type for $n \geq 6$ (as we will show below the case $n = 5$ is a special one).

5.3 Metastability or Blow up

In this section we discuss the dynamics of the normal-form equations (5.34) from a qualitative point of view.

5.3.1 $n = 3$: α-Model

To begin with, let us consider the case $n = 3$, and report the results obtained in [9]. The equations of motion are

$$\xi_\tau = -\xi_y - \mu^2 \frac{1}{24}\, \xi_{yyy} - \mu^2 \frac{1}{2\sqrt{2}}\, (\xi^2)_y \ , \tag{5.35}$$

$$\eta_\tau = \eta_y + \mu^2 \frac{1}{24}\, \eta_{yyy} + \mu^2 \frac{1}{2\sqrt{2}}\, (\eta^2)_y \ , \tag{5.36}$$

i.e. two uncoupled KdV equations in translating frames (to the right for the first equation and to the left for the second one). Their dynamics is very well understood and, in particular, it is known that any solution of the Cauchy problem is almost periodic and the solution has the same regularity as the initial datum. One can easily conclude from the theory of [13] that if one initially gives energy only to the first Fourier mode then, up to infinite times, the solution remains analytic, and therefore the energy remains exponentially localized in the first Fourier modes. Thus corresponding to initial data on a low-frequency Fourier mode one will see that, on a time-scale of the order of μ^{-3}, the energy will flow to a small packet of modes which oscillate with an amplitude exponentially decreasing with the mode index, and this situation will last forever. Thus, in KdV there is a formation of a small packet *à la* FPU which is stable for infinite times (the time averages of the mode-energies relax).

In order to get conclusions on the dynamics of the FPU one needs a precise relation between the KdV and the FPU. From the fact that the KdV is a first-order normal form of the system one expects that the solutions of KdV describe well the FPU only up to times of order $\mu^{-3} \sim \epsilon^{-3/4}$. This was proved in [10] by a rigorous theory that will be summarized in Sect. 5.4 (see Theorem 5.3 below for a precise statement).

Afterwords one can expect that higher order corrections come into play and modify the dynamics. However, one has to remark that the KdV is an integrable system which is nondegenerate (in KAM sense) and therefore one expects higher order corrections not to modify qualitatively the dynamics. In the spirit of Nekhoroshev's theory, the dynamics of the KdV should be unchanged up to times exponentially growing with a power of ϵ^{-1}. The small packet of modes which was quickly formed should persist for exponentially long times; this was actually verified numerically in [7]. Thus we can call it a *metastable state*.

It is worth mentioning that this phenomenon is related to two properties, namely the fact that the linearized FPU is resonant (more precisely the dispersion relation is approximatively linear for low frequencies), and the very surprising fact that the first order normal form is integrable. Indeed, if the linearized system were nonresonant, the perturbation would not be able to efficiently spread energy among the modes, thus preventing the formation of the packet of modes.

5.3.2 $n = 4$: β-Model

In the case $n = 4$ the situation is similar to the case $n = 3$. The equations of motion have the form

$$\xi_\tau = -\left(1 + \frac{3\mu^2}{4}[\eta^2]\right)\xi_y - \frac{\mu^2}{24}\xi_{yyy} - \frac{\mu^2}{4}(\xi^3)_y \ , \tag{5.37}$$

$$\eta_\tau = \left(1 + \frac{3\mu^2}{4}[\xi^2]\right)\eta_y + \frac{\mu^2}{24}\eta_{yyy} + \frac{\mu^2}{4}(\eta^3)_y \ , \tag{5.38}$$

which, taking into account the conservation of the momenta of order two, reduce to two uncoupled mKdV equations in translating frames, again two integrable equations.

However the mKdV is less studied than the KdV, thus at present a precise theorem is not available. However we think that this is only a technical problem, and that, developing the techniques of [13], one should be able to show that the β-model has the same features of the α-model; indeed numerical computations display a behavior very similar to that of the α-FPU [7].

5.3.3 $n = 5$: γ-Model

It is interesting to consider explicitly also the case $n = 5$. In such a case the equations of motion (5.34) read

$$\xi_\tau = -\left(1 + \frac{\mu^2[\eta^3]}{\sqrt{2}}\right)\xi_y - \frac{\mu^2}{24}\xi_{yyy} - \frac{3\mu^2}{2\sqrt{2}}[\eta^2](\xi^2)_y - \frac{\mu^2}{4\sqrt{2}}(\xi^4)_y \ , \tag{5.39}$$

$$\eta_\tau = \left(1 + \frac{\mu^2[\xi^3]}{\sqrt{2}}\right)\eta_y + \frac{\mu^2}{24}\eta_{yyy} + \frac{3\mu^2}{2\sqrt{2}}[\xi^2](\eta^2)_y + \frac{\mu^2}{4\sqrt{2}}(\eta^4)_y \ . \tag{5.40}$$

Notice that in this case the (time dependent) momenta of third order appear on the r.h.s. of the two equations. By the change of variables

$$\xi(y,\tau) = \xi'(y - a(\tau), \tau) \quad , \quad \eta(y,\tau) = \eta'(y + b(\tau), \tau) \ ,$$

where

$$\frac{\mathrm{d}a}{\mathrm{d}\tau} \equiv 1 + \frac{\mu^2[\eta^3](\tau)}{\sqrt{2}} \quad , \quad \frac{\mathrm{d}b}{\mathrm{d}\tau} \equiv 1 + \frac{\mu^2[\xi^3](\tau)}{\sqrt{2}} \ ,$$

(5.39)–(5.40) become two uncoupled KdV equations of fourth order.

5.3.4 $n \geq 6$: Higher Models

In the case $n \geq 6$ the normalized equations consist of a pair of coupled integrodifferential gKdV equations. Their behavior can be studied heuristically by using, for example scaling arguments (details will be published elsewhere), and one can show that, if one starts with analytic (in space) initial data having a pole at a certain distance from the real axis, then the corresponding solution is expected to remain analytic over $\mathbb{R}$ if $n \leq 5$, while the pole is expected to reach the real axis if $n \geq 7$, the case $n = 6$ being "critical." Since the whole behavior seems to depend only on the homogeneity degree of the nonlinearity and on the fact that the momenta of order two are conserved, it is useful to use a model problem where a lot is known, namely the standard gKdV

$$\xi_t = \xi_{yyy} + (\xi^{n-1})_y \ . \tag{5.41}$$

Here (at least on unbounded domains) it is known [14] that corresponding to smooth initial data there are two possibilities: (1) the solution remains smooth for all times or (2) the H^1 (energy) norm of the solution becomes infinite at a finite time. The situation depends on the value of n. If $n \leq 5$ the solution is smooth for all times, if $n \geq 7$ every solution is expected to blows up in finite time, and finally, if $n = 6$ then the behavior of the solution depends on the initial datum, and there are rigorous results on the existence of blow up solutions [15].

As a consequence one might conjecture the existence of a corresponding situation for the FPU: if $n \leq 5$ a metastable packet of modes is expected to exists, if $n \geq 7$ the FPU phenomenon is expected to disappear, and finally, the case $n = 6$ is critical in the sense that the packet will exist only for some initial data.

It is important to remark that in the case $n = 5$ the normal form (5.39)–(5.40) are not integrable. However, according to our conjecture one would expect to observe a FPU phenomenology similar to the cases $n = 3, 4$, where the normal form equations are integrable. This fact forces to look for a mechanism weaker than integrability to explain the FPU paradox. The stability of the soliton solutions of (5.41) for $n \leq 5$ and their blowing-up for $n \geq 7$ suggest that a sort of Soliton Turbulence, analogous to the one proposed by Zakharov in the context of the NLS equations, might be the relevant phenomenon in FPU.

From a physical point of view, stable, space-localized solitary wave structures, trap the energy on a spatial scale which is lower bounded. As a consequence, in Fourier space the energy cannot flow above some critical wave-number. On the contrary, in the supercritical cases $n \geq 7$, to the collapse in real space of the solitary wave structures there corresponds energy injection at any Fourier wavenumber, which we interpret as the mechanism triggering off the approach to equipartition in a short time.

In conclusion we conjecture the existence of a critical degree of nonlinearity, $n_c = 6$, in system (5.4), below which the FPU phenomenology (metastability of partial equipartition states) should be possible, and above which it could disappear. The relevant quantities characterizing the metastable state in the subcritical cases ($n \leq 5$), are the time-scale T_1 of formation of the metastable state (consisting of a packet of modes sharing the energy) and almost equipartited, and the size $\Delta k/N$ of such a packet in Fourier space. From the scaling previously introduced in terms of the small parameter μ and from its link with the specific energy given in (5.11), one easily gets

$$T_1 \sim \mu^{-3} \sim \epsilon^{-\frac{3(n-2)}{4}} \quad ,$$

and

$$\Delta k/N \sim \mu \sim \epsilon^{\frac{n-2}{4}} \quad ,$$

which generalize to the case of arbitrary n those introduced in [1].

5.4 Rigorous Results

We give here the precise relation between the FPU and its normal form (5.34) and deduce some rigorous results on the formation of the packet.

First of all we have to introduce a norm in order to measure the error between the correct equations and the equations in normal form. This is conveniently done in terms of Fourier coefficients. Moreover, since we are interested in exponentially localized packets we will use a norm that controls the Fourier coefficients with an accuracy growing exponentially with the index.

Definition 5.2. *Having fixed two positive constants s, σ consider the Hilbert space $\ell^2_{\sigma,s}$ of the complex sequences $v \equiv \{v_k\}_{k\in\mathbb{Z}-\{0\}}$ such that*

$$\|v\|^2_{\sigma,s} := \sum_k |v_k|^2 |k|^{2s} e^{2\sigma|k|} < \infty \quad . \tag{5.42}$$

We will identify a 2 periodic function v with its Fourier coefficients defined by (5.24). We will say that $v \in \ell^2_{\sigma,s}$ if its Fourier coefficients have this property. Correspondingly we define the phase spaces $\mathcal{P}_s$ by

$$\mathcal{P}_s := \ell^2_{\sigma,s+1} \times \ell^2_{\sigma,s} \ni (v,u) \ , \tag{5.43}$$

and denote

$$\|(v,u)\|^2_s := \|v\|^2_{\sigma,s+1} + \|u\|^2_{\sigma,s} \ ; \tag{5.44}$$

σ is fixed. The ball of radius R centered at the origin of $\mathcal{P}_s$ will be denoted by $B_s(R)$.

Then the flow Ψ^τ of the system H_0 is unitary in all the spaces $\mathcal{P}_s$.

Theorem 5.1. *For any $r \geq 5$ there exists a constant $\mu_* \equiv \mu_{*r}$, such that, if*

$$\mu < \mu_*$$

then there exists an analytic canonical transformation $\mathcal{T} : B_r(1) \to B_r(2)$ which averages K see (5.17), *namely such that*

$$K \circ \mathcal{T} = H_0 + \mu^2 \langle P \rangle + \mathcal{R} \ ; \tag{5.45}$$

the vector field $X_{\mathcal{R}}$ of the remainder is analytic in a complex ball of radius 1 and fulfills the estimate

$$\sup_{\|z\|_r \leq 1} \|X_{\mathcal{R}}(z)\|_0 \leq C_r \mu^{4 - \frac{12}{6+r}} \ . \tag{5.46}$$

Moreover for any $1 \leq r_1 \leq r$ the transformation $\mathcal{T}$ maps $B_{r_1}(1)$ into $\mathcal{P}_{r_1}$ and fulfills

$$\sup_{\|z\|_{r_1} \leq 1} \|z - \mathcal{T}(z)\|_{r_1} \leq C \mu^{2 - \frac{6}{6+r}} \ . \tag{5.47}$$

The idea of the proof is to approximate the system by a system with finitely many degrees of freedom, say M, to construct the canonical transformation that averages the finite dimensional system, and finally to choose M in a such a way that the order of magnitude of the error due to the averaging procedure is equal to the order of magnitude of the error due to the cutoff procedure. For the details see [10].

Theorem 5.1 shows that the equations of motions of the FPU are a perturbations of the normal form equations described in the previous section. However they are a very singular perturbation since the remainder has a vector field which is small only when considered as an operator extracting r derivatives, with $r \geq 5$. As pointed out in [16] the use of such a normal form for dynamical previsions is far from trivial. In the present case one can use a technique developed by Schneider and Wayne in [17] and generalized to a form suitable for our purpose [10, 16].

Since when dealing with the FPU system we are interested in Fourier modes, we explicit state their definition. Thus, for a given FPU state (q_j, p_j), we have

$$p_j = \frac{1}{\sqrt{2N}} \sum_{k=-N}^{N-1} \hat{p}_k e^{i\frac{jk\pi}{N}} \tag{5.48}$$

and similarly for q_j. We denote by

$$E_k := \frac{|\hat{p}_k|^2 + \omega_k^2 |\hat{q}_k|^2}{2} \ , \quad k = -N...., N-1 \tag{5.49}$$

the energy of the kth mode.

Let $\xi^a(y,\tau), \eta^a(y,\tau)$ be a solution of the normal form equations (5.34). Correspondingly, we define an approximate solutions $z^a \equiv (r^a, s^a)$ of the FPU by

$$r^a(x,t) := \frac{\xi^a(\mu x,\mu t) + \eta^a(\mu x,\mu t)}{\sqrt{2}} \tag{5.50}$$

$$s^a_x(x,t) := \frac{\xi^a(\mu x,\mu t) - \eta^a(\mu x,\mu t)}{\sqrt{2}} \tag{5.51}$$

and an initial datum for the FPU by $r_{0,j} := r^a(j,0)$, $s_{0,j} := s^a(j,0)$. We will denote by $(r_j(t), s_j(t))$ the corresponding true solution of the FPU.

Then we have the following:

Theorem 5.2. *Fix a positive T_f; assume that for all times $|t| < T_f/\mu^3$ the approximate solution is such that $(\xi^a, \eta^a) \in \mathcal{P}_{78}$ and has norm smaller than 1 and fix some $\sigma > 0$. Then there exists μ_* depending on T_f and on $\left\|(\xi^a(t), \eta^a(t))\right\|_{78}$ only, such that, if $\mu < \mu_*$ then for all times t fulfilling*

$$|t| \le \frac{T_f}{\mu^3} \tag{5.52}$$

one has

$$\sup_j \left(|r_j(t) - r^a(j,t)| + |s_j(t) - s^a(j,t)|\right) \le C\mu \quad ; \tag{5.53}$$

moreover

$$\left|\frac{E_k(t)}{N} - \mu^4\left[\frac{|\hat{\xi}^a_K(t)|^2 + |\hat{\eta}^a_K(t)|^2}{2} + \frac{|\hat{\xi}^a_{-K}(t)|^2 + |\hat{\eta}^a_{-K}(t)|^2}{2}\right]\right| \le C\mu^5 \tag{5.54}$$

for all k such that $\frac{k}{N} = \mu K$ with $|K| \le \frac{|\ln \mu|}{2\sigma}$, and

$$\frac{|E_k(t)|}{N} \le \mu^5 \tag{5.55}$$

for all k such that $\frac{k}{N} = \mu K$ with $|K| > \frac{|\ln \mu|}{2\sigma}$, whereas $E_k(t) = 0$ otherwise.

We recall that the Fourier coefficients $\hat{\xi}^a_K$, and $\hat{\eta}^a_K$ are defined by (5.20).

We remark that in view of the discussion of the previous section (see also the references quoted therein) we can ensure rigorously that correspondingly to smooth initial data the solution of the normal form equation is smooth enough to apply the above theorem in the case $n = 3$ (see, e.g. [13]), we expect that in the case $n = 4$ the same result should be true and that, due to integrability of the mKdV it should be possible to obtain a proof by the same methods as for the case $n = 3$. We expect that the smoothness property of the solutions should be true also in the case $n = 5$, and also when $n = 6$ provided the energy is small enough, however here no ideas for the proof are available. On the contrary we expect that this assumption is violated in the case $n \geq 7$.

In particular in the case $n = 3$, using the theory of the KdV equation developed in [13], one can prove a precise result for the formation of the packet.

It is convenient to state the result in terms of "specific quantities," thus we will label the modes with the index

$$\kappa := \frac{k}{N} \ ;$$

correspondingly we denote by

$$\mathcal{E}_\kappa := \frac{E_k}{N} \tag{5.56}$$

the specific energy in the mode with index κ.

Theorem 5.3. *Fix a constant C_0 and a positive (large) time T_f; then there exist positive constants μ_*, C_1, C_2, dependent only on C_0 and on T_f, such that the following holds. Consider an initial datum with*

$$\mathcal{E}_{\kappa_0}(0) = C_0 \mu^4 \ , \quad \mathcal{E}_\kappa(0) \equiv \mathcal{E}_\kappa(t)\big|_{t=0} = 0 \ , \quad \forall \kappa \neq \kappa_0 \ , \tag{5.57}$$

and assume $\mu < \mu_*$. Then, there exists $\sigma > 0$ such that, along the corresponding solution, one has

(i)

$$\mathcal{E}_\kappa(t) \leq \mu^4 C_1 \mathrm{e}^{-\sigma\kappa/\mu} + C_2 \mu^5 \ , \quad \text{for } |t| \leq \frac{T_f}{\mu^3} \tag{5.58}$$

for all $\kappa > 0$.

(ii) There exists a sequence of almost periodic functions $\{F_n\}_{n\in\mathbb{N}}$ such that, defining the specific energy distribution

$$\mathcal{F}_{n\kappa_0} = \mu^4 F_n \ , \quad \mathcal{F}_\kappa = 0 \text{ if } \kappa \neq n\kappa_0 \tag{5.59}$$

one has

$$|\mathcal{E}_\kappa(t) - \mathcal{F}_\kappa(t)| \leq C_2 \mu^5 \ , \quad |t| \leq \frac{T_f}{\mu^3} \ . \tag{5.60}$$

Remark 5.4. Since $F_n(t)$ are almost periodic functions of time their time average defined by

$$\bar{F}_n := \lim_{T\to\infty} \frac{1}{T} \int_0^T F_n(t) \mathrm{d}t \tag{5.61}$$

exists (see, e.g. [18]). It follows that up to the error the time average of $\mathcal{E}_\kappa(t)$ relaxes to the limit distribution obtained by rescaling $\bar{F}_n$ as in (5.59).

References

1. A. Ponno, Soliton theory and the Fermi–Pasta–Ulam problem in the thermodynamic limit. Europhys. Lett., **64**(5), 606–612 (2003).
2. N.J. Zabusky and M.D. Kruskal, Interaction of solitons in a collisionless plasma and the recurrence of initial states. Phys. Rev. Lett., 15, 240–243 (1965).
3. E. Fermi, J.R. Pasta and S.M. Ulam, Studies of nonlinear problems. In *Collected works of E. Fermi, Vol. 2.* Chicago University Press, Chicago, 1965.
4. L. Galgani and A. Scotti, Planck-like distribution in classical nonlinear mechanics. Phys. Rev. Lett. **28**, 1173–1176, (1972).
5. J.A. Biello, P.R. Kramer, Y. Lvov, Stages of energy transfer in the FPU model, DCDS 113–122 (2003); supplement volume for the Proceedings of the AIMS-4th International Conference on Dynamical Systems and Differential Equations, Wilmington, NC, USA, 2002.
6. L. Berchialla, L. Galgani and A. Giorgilli, Localization of energy in FPU chains. Discrete Contin. Dyn. Syst., **11**(4), 855–866 (2004).
7. L. Berchialla, A. Giorgilli and S. Paleari, Exponentially long times to equipartition in the thermodynamic limit. Phys. Lett. A, **321**, 167–172 (2004).
8. A. Giorgilli, S. Paleari and T. Penati, Local chaotic behaviour in the Fermi–Pasta–Ulam system. Discr. Contin. Dyn. Syst. Ser. B, **5**(4), 991–1004 (electronic) (2005).
9. B. Rink, Symmetric invariant manifolds in the Fermi–Pasta–Ulam lattice. Phys. D, 175(1–2), 31–42 (2003).
10. D. Bambusi and A. Ponno, On metastability in FPU. Comm. Math. Phys., **264**(2), 539–561 (2006).
11. A. Ponno and D. Bambusi, Energy cascade in Fermi–Pasta–Ulam model. Symmetry and Perturbation Theory 2004, pp. 263–270. World Scientific, 2005.
12. B. Rink, Symmetry and resonance in periodic FPU chains. Comm. Math. Phys., **218**(3), 665–685 (2001).
13. T. Kappeler and J. Pöschel, *KdV & KAM*, Vol. 45 of *Ergebnisse der Mathematik und ihrer Grenzgebiete. 3. Folge. A Series of Modern Surveys in Mathematics [Results in Mathematics and Related Areas. 3rd Series. A Series of Modern Surveys in Mathematics].* Springer-Verlag, Berlin, 2003.
14. Y. Martel and F. Merle, Stability of blow-up profile and lower bounds for blow-up rate for the critical generalized-KdV equation. Ann. of Math. (2), **155**(1), 235–280 (2002).
15. Frank Merle, Existence of blow-up solutions in the energy space for the critical generalized KdV equation, J. Amer. Math. Soc., **14**(3), 555–578 (2001).
16. D. Bambusi, Galerkin averaging method and Poincaré normal form for some quasilinear PDEs. Ann. Scuola Norm. Sup. Pisa Cl. Sci. (5), **4**(4), 669–702 (2005).
17. G. Schneider and C.E. Wayne, Counter-propagating waves on fluid surfaces and the continuum limit of the Fermi–Pasta–Ulam model. *International Conference on Differential Equations, Vol. 1, 2* Berlin, 1999, pp. 390–404. World Sci. Publishing, River Edge, NJ, 2000.
18. A.M. Fink, *Almost periodic differential equations.* Lecture Notes in Mathematics, Vol. 377. Springer-Verlag, Berlin, 1974.

6

Center Manifold Theory in the Context of Infinite One-Dimensional Lattices

Guillaume James and Yannick Sire

Laboratoire Mathématiques pour l'Industrie et la Physique
UMR CNRS 5640, INSA de Toulouse, 135 avenue de Rangueil, 31077 Toulouse Cedex 4, France.
james@insa-toulouse.fr, sire@insa-toulouse.fr

Abstract. Center manifold theory has been used in recent works to analyze small amplitude waves of different types in nonlinear (Hamiltonian) oscillator chains. This led to several existence results concerning traveling waves described by scalar advance-delay differential equations, pulsating traveling waves determined by systems of advance-delay differential equations, and time-periodic oscillations (including breathers) obtained as orbits of iterated maps in spaces of periodic functions. The Hamiltonian structure of the governing equations is not taken into account in the analysis, which heavily relies on the reversibility of the system. The present work aims at giving a pedagogical review on these topics. On the one hand, we give an overview of existing center manifold theorems for reversible infinite-dimensional differential equations and maps. We illustrate the theory on two different problems, namely the existence of breathers in Fermi–Pasta–Ulam lattices and the existence of traveling breathers (superposed on a small oscillatory tail) in semi-discrete Klein–Gordon equations.

6.1 Introduction

The seminal work of Fermi, Pasta and Ulam in 1955 [12] had a broad impact on the development of the theory of nonlinear lattices. The Fermi–Pasta–Ulam (FPU) model consists of a chain of masses nonlinearly coupled to their nearest neighbors. For a general nearest-neighbors interaction potential V, the governing equations read

$$\frac{\mathrm{d}^2 x_n}{\mathrm{d}t^2} = V'(x_{n+1} - x_n) - V'(x_n - x_{n-1}), \qquad n \in \mathbb{Z} \tag{6.1}$$

where we note x_n the displacement of the nth mass with respect to an equilibrium position (in this version of the model the chain is of infinite extent). The anharmonic interaction potential V satisfies $V'(0) = 0$, $V''(0) = 1$. System (6.1) is Hamiltonian, with total energy

G. James and Y. Sire: *Center Manifold Theory in the Context of Infinite One-Dimensional Lattices*, Lect. Notes Phys. **728**, 207–238 (2008)
DOI 10.1007/978-3-540-72995-2_6

$$H = \sum_{n=-\infty}^{+\infty} \frac{1}{2}\left(\frac{\mathrm{d}x_n}{\mathrm{d}t}\right)^2 + V(x_{n+1} - x_n). \tag{6.2}$$

In their numerical work (for finite chains with polynomial interaction potentials) Fermi et al surprisingly observed a near-recurrence of some initial conditions (corresponding to a low-frequency linear mode) instead of the expected thermalization of the system. The FPU-recurrences have been later understood with the discovery of the properties of solitons solutions of the Kortweg–de Vries (KdV) equation [2, 3, 4]. In parallel to the mathematical justification of the KdV-limit in the FPU lattice [5, 6, 7], the mathematical theory of exact solitary waves in this system has been the object of important developments in the last 10 years [8, 9, 10, 11, 12, 13, 14, 15, 16].

In addition, Iooss [16] provides for the FPU system a precise description of the set of small amplitude traveling waves with near-sonic speed (both above and below the sound velocity). Traveling waves with velocity c are solutions of a scalar advance-delay differential equation, which is viewed as a (reversible) infinite-dimensional differential equation in the moving frame coordinate $\xi = n - ct$. Small amplitude solutions are determined by a "reduced" differential equation on an invariant local center manifold. If $V^{(3)}(0) \neq 0$, the principal part of the reduced equation corresponds to a stationary KdV equation in the moving frame coordinate. This approach is not limited to solitary wave solutions, since in a given parameter regime *all* the small amplitude traveling waves of the FPU system lie on a finite-dimensional center manifold.

More generally, the center manifold reduction method has been used to analyze small amplitude traveling waves in different types of nonlinear lattices, e.g. for generalized discrete nonlinear Schrödinger equations in certain parameter regimes [17]. In this context, infinite-dimensional center manifold theory has been initially introduced by Iooss and Kirchgässner [18] to study small amplitude traveling waves in the Klein–Gordon (KG) lattice. The KG lattice consists of a chain of nonlinear oscillators in an anharmonic potential V, linearly coupled to their nearest neighbors. The equations of motion read

$$\frac{\mathrm{d}^2 x_n}{\mathrm{d}t^2} + V'(x_n) = \gamma(x_{n+1} - 2x_n + x_{n-1}), \ n \in \mathbb{Z}, \tag{6.3}$$

where $\gamma > 0$ determines the coupling strength, $V'(0) = 0$, $V''(0) = 1$. This system is Hamiltonian, with total energy

$$H = \sum_{n=-\infty}^{+\infty} \frac{1}{2}\left(\frac{\mathrm{d}x_n}{\mathrm{d}t}\right)^2 + V(x_n) + \frac{\gamma}{2}(x_{n+1} - x_n)^2.$$

Subsequently, the analysis of traveling waves by Iooss and Kirchgässner has been extended to pulsating traveling waves by the present authors [19, 20, 21]. These solutions are searched to satisfy

$$x_n(t) = x_{n-p}(t - T), \tag{6.4}$$

where $p \geq 1$ is a fixed integer and the propagation time $T \in \mathbb{R}^*$ is taken as a parameter (traveling waves correspond to fixing $p = 1$ and $c = 1/T$ in the above definition). Systems (6.3)–(6.4) is equivalent to a (reversible) p-dimensional system of coupled advance-delay differential equations for $(x_1, \ldots, x_p)$. The center manifold theorem reduces systems (6.3)–(6.4) locally to a finite-dimensional ordinary differential equation on a center manifold, whose dimension depends on p and parameters γ, T.

In certain parameter regimes, this simplification allows one to show the existence of small amplitude traveling waves [18] and pulsating traveling waves [20] (for an even potential V) in the form of *nanopterons*, a denomination introduced by Boyd [22]. Nanopterons consist in a single pulse (here a modulated plane wave whose envelope is spatially localized), superposed on a nondecaying oscillatory tail which is small with respect to the central pulse amplitude. For the KG system, fully localized solutions (i.e. solutions without an oscillatory tail) may exist for exceptional values of parameters (γ, T, parameters in the potential V) but this situation in *non-generic.* In the small amplitude limit, nanopterons possess an exponentially small oscillatory tail with respect to the central pulse amplitude, and consequently they are very close to spatially localized solutions. However, nanopterons are solutions of infinite energy due to their nondecaying oscillatory tail. For the KG system and in the small amplitude regime, the central pulse of nanopterons can be approximated at leading order using the nonlinear Schrödinger (NLS) equation [23]. Figure 6.1 shows a corresponding solution profile followed up numerically in the high amplitude regime.

Pulsating traveling waves of this type have been numerically computed by the present authors in [24], and [25] treats the case of traveling waves. The existence of traveling kinks showing similar oscillatory tails is reported in several references (see, e.g. [26]). Pulsating solitary waves in the FPU lattice are analyzed by Iooss and one of us (G.J.) in [27], using the same center manifold reduction method. In addition to (subsonic) *nanopterons* close to the

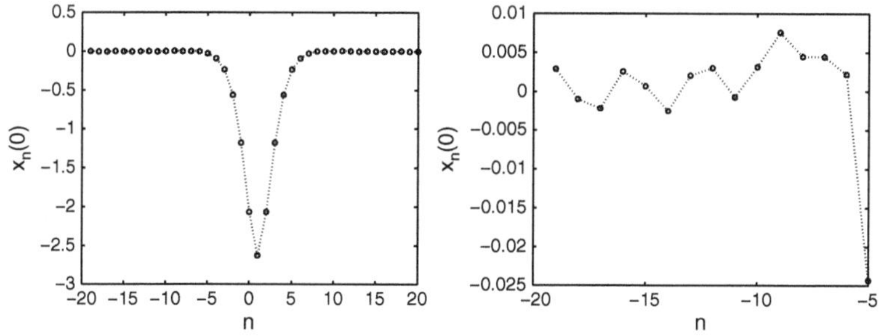

Fig. 6.1. Solution of (6.3)–(6.4) for $p = 2$, $T = 8.1$, $\gamma = 0.9$, $V(x) = 1 - \cos(x)$. This pulsating wave satisfies $x_n(t) = -x_{n-1}(t - T/2)$. *Left figure* : solution at $t = 0$. *Right figure* : zoom on the oscillating tail (not visible at the scale of *left figure*)

NLS limit, one proves the existence of supersonic solitary waves superposed on an (arbitrarily small) extended tail moving at subsonic speed.

The above mentionned applications of center manifold theory concerned reversible systems of advance-delay differential equations. In addition, the method has been extended by one of us (G.J.) to infinite-dimensional quasilinear maps in Hilbert spaces [28]. The center manifold theorem proved in this context can be used to study time-periodic oscillations in infinite oscillator chains. For example, the FPU system (6.1) can be viewed (after rescaling time) as an infinite-dimensional reversible map $(x_n, x_{n-1}) \mapsto (x_{n+1}, x_n)$ in a space of 2π-periodic functions of t (one uses the fact that V' is locally invertible). The oscillation frequency enters the map as a bifurcation parameter as time has been rescaled. Small amplitude orbits of the map lie on a finite-dimensional center manifold, thereby being fully determined by the finite-dimensional "reduced" mapping on the center manifold. The center manifold theorem for quasilinear maps has been subsequently applied to other nonlinear lattices, including diatomic FPU chains [32] and spin lattices [33].

The center manifold dimension is determined by the number of resonant phonons in a given parameter regime. A phonon denotes a normal mode $x_n(t) = a^{\exp[i(kn-\Omega t)]} + \text{c.c.}$ solution of the linearized equations of motion ($\Omega(k)$ being fixed by a dispersion relation). As one looks for solutions with frequency ω close to some value ω_0, a phonon is said to be resonant (a shortcut for nearly-resonant) if $\Omega/\omega_0 \in \mathbb{Z}$.

The reduction scheme described above allows one to prove for the FPU model (with hardening interaction potentials) the existence of small amplitude "breathers" bifurcating above the top of the phonon frequency band [29] (other types of proofs based on variational methods are also available [30, 31]). A breather consists of a time-periodic solution of (6.1) (excluding equilibria) which is spatially localized, i.e. satisfies

$$\lim_{n \to \pm\infty} \| x_n(.) - c_\pm \|_{L^\infty(\mathbb{R})} = 0 \tag{6.5}$$

for some constants $c_\pm \in \mathbb{R}$. Near the top of the phonon band, the reduced mapping determines (roughly speaking) the first-order Fourier coefficient of relative displacements, and breather solutions correspond to orbits of the reduced mapping homoclinic to 0.

Let us mention a connection between breather solutions and spatially localized pulsating traveling waves satisfying (6.4). Pulsating traveling waves can be seen as time-periodic oscillations in a frame moving at some constant velocity, and consequently spatially localized ones appear as breather solutions in this moving frame. Such solutions are commonly referred to as *traveling breathers*. Traveling breathers can be generated in certain systems by perturbing a breather solution in the direction of a "pinning mode" [25]. The expression "traveling breather" is also often used to denote nanopteron solutions, due to the fact that their spatially extended tail can be made exponentially small, which makes them very close to fully localized traveling breather solutions.

The aim of the present work is double. On the one hand, we present a review of [29] (Sects. 6.2.2 and 6.4.1) and [20] (Sects. 6.3.1 and 6.4.2), based on center manifold theory, which examine the existence of breathers and nanopterons in FPU and KG lattices respectively. Numerical computations from references [24, 34] are also presented to illustrate the mathematical results. On the other hand we give an overview of the general center manifold theorems which have been applied in this context, considering both infinite-dimensional differential equations [35, 36] and quasilinear maps [28]. We only examine the aspects of the theory which are relevant to the above class of nonlinear lattices. In particular, related tools in semigroup theory as well as the center manifold theory for dissipative systems are not reviewed here.

Sect. 6.2 treats the center manifold reduction for maps. After recalling the finite-dimensional case (Sect. 6.2.1), we start by the description of an infinite-dimensional situation (time-periodic oscillations in the FPU lattice, Sect. 6.2.2). This illustration is aimed at making the general theory for quasilinear maps (Sect. 6.2.3) more accessible. Sect. 6.3 is concerned with the center manifold reduction for infinite-dimensional differential equations. We start by an example on advance-delay differential equations (pulsating traveling waves in the KG lattice, Sect. 6.3.1), and state the general center manifold theorem in Sect. 6.3.2. This version of the center manifold theorem in infinite dimensions concerns semilinear equations and is due to Vanderbauwhede and Iooss [36]. The quasilinear case has been treated by Mielke [35]. Section 6.4 focuses on applications of the theory, for the existence of breathers in FPU lattices (Sect. 6.4.1) and the existence of nanopterons in KG lattices (Sect. 6.4.2). The analysis consists in studying the dynamics of the related maps and flows on the center manifolds.

6.2 Center Manifold Reduction for Maps

6.2.1 Finite-Dimensional Case

We consider a nonlinear mapping in $\mathbb{R}^N$

$$u_{n+1} = L u_n + N(u_n, \mu) \tag{6.6}$$

where $L \in M_N(\mathbb{R})$, $\mu \in \mathbb{R}^p$ is a parameter, and N is a smooth nonlinear map (C^k in a neighborhood of 0 with $k \geq 2$) satisfying $N(0,0) = 0$, $D_u N(0,0) = 0$. An orbit of (6.6) is a sequence $\{u_n\}_{n \in \mathbb{Z}}$ in $\mathbb{R}^N$. For $\mu = 0$ the mapping (6.6) admits $u_n = 0$ as a fixed point.

In what follows we denote by a *generalized eigenvector* of L (corresponding to an eigenvalue λ) a nonzero element of the kernel of $(L - \lambda I)^n$ for some integer $n \geq 0$ (a basis of generalized eigenvectors can be used to write the matrix L in Jordan form). One can split $\mathbb{R}^N$ into invariant subspaces under L spanned by such generalized eigenvectors, each subspace corresponding to a single eigenvalue λ. More globally we shall note X_c as the subspace of

$\mathbb{R}^N$ spanned by the (generalized) eigenvectors of L whose eigenvalues have modulus 1, and note X_{h} the complementary subspace spanned by the other (generalized) eigenvectors. The subspaces X_{c} and X_{h} are denoted respectively as center and hyperbolic subspace. We assume $X_{\mathrm{c}}, X_{\mathrm{h}} \neq \{0\}$. The center manifold theorem states the following [37, 38, 39].

Theorem 6.1. *There exits a neighborhood $\Omega \times \Lambda$ of 0 in $\mathbb{R}^N \times \mathbb{R}^p$ and a map $\psi \in C^k(X_{\mathrm{c}} \times \Lambda, X_{\mathrm{h}})$ (with $\psi(0,0) = 0$, $D_u\psi(0,0) = 0$) such that for all $\mu \in \Lambda$ the manifold*

$$\mathcal{M}_\mu = \left\{ u \in \mathbb{R}^N \,|\, u = x + \psi(x,\mu), x \in X_{\mathrm{c}} \right\} \tag{6.7}$$

has the following properties.

(i) Local invariance: $\mathcal{M}_\mu$ is locally invariant under $L + N(.,\mu)$, i.e. if $u \in \mathcal{M}_\mu \bigcap \Omega$ then $Lu + N(u,\mu) \in \mathcal{M}_\mu$.

(ii) Reduction: If $(u_n)_{n\in\mathbb{Z}}$ is a solution of (6.6) such that $u_n \in \Omega$ for all $n \in \mathbb{Z}$ then $u_n \in \mathcal{M}_\mu$ for all $n \in \mathbb{Z}$.

Properties (i) and (ii) are clearly true for $N = 0$ since the mapping is linear (then $\mathcal{M}_\mu = X_{\mathrm{c}}$), and the theorem states that they remain true locally in the nonlinear case. This result guarantees (for $\mu \approx 0$) that small amplitude solutions of (6.6) belong to an invariant manifold whose dimension (equal to $\dim X_{\mathrm{c}}$) is lower than N.

We now want to generalize this result for infinite-dimensional maps. The case when $L + N(.,\mu)$ is C^k in a Banach space X has been treated in [38, 39]. In a more general context, more recent results on local invariant manifolds for C^k maps in Banach spaces can be found in references [40, 41].

Here we shall consider situations when the operator L is unbounded (theory developed in [28]). We shall restrict the discussion to maps in Hilbert spaces, but the same theory could be carried out in Banach spaces (with some additional technicalities involved in the cut-off of nonlinear terms [36]). The following section is aimed at motivating the theory.

6.2.2 Example in Infinite Dimensions

For a large class of infinite one-dimensional lattices, time-periodic oscillations can be viewed as solutions of an ill-posed recurrence relation on a loop space (i.e. a space of periodic functions), which can be locally analyzed using a center manifold reduction [28]. As an example, consider the FPU lattice (6.1). We look for time-periodic solutions of (6.1) having a given frequency ω. Since V' is locally invertible, one can reformulate (6.1) using the force variable $y_n(t) = V'(x_n - x_{n-1})(t/\omega)$ (y_n is 2π-periodic in t). This yields the equations

$$\omega^2 \frac{\mathrm{d}^2}{\mathrm{d}t^2} W(y_n) = y_{n+1} - 2y_n + y_{n-1}, \quad n \in \mathbb{Z}, \tag{6.8}$$

where $W = (V')^{-1}$. Note that the frequency ω appears in (6.8) as a bifurcation parameter. For formulating (6.8) as a mapping in a loop space, we introduce the variable $u_n = (y_{n-1}, y_n) \in D$ where D is a space of 2π-periodic functions of t which has to be precised. Problem (6.8) takes the form of a mapping

$$u_{n+1} = F_\omega(u_n), \tag{6.9}$$

where $F_\omega(y_{n-1}, y_n) = (\, y_n \,,\, \omega^2(\mathrm{d}^2/\mathrm{d}t^2)W(y_n) + 2y_n - y_{n-1}\,)$. Here we analyze problem (6.8) using a discrete *spatial dynamics* approach. This concept originates from the continuous case of partial differential equations. It has been introduced in [42] for studying elliptic partial differential equations in cylindrical domains, formulated as ill-posed evolution problems in the unbounded space coordinate.

We now define appropriate function spaces on which the operator F_ω is acting. We look for $(u_n)_{n\in\mathbb{Z}}$ as a sequence in the following space D

$$D = \left\{ u \in H^2 \times H^2, u \text{ even in } t, \int_0^{2\pi} u \;\mathrm{d}t = 0 \right\},$$

where H^n denotes the classical Sobolev space $H^n(\mathbb{R}/2\pi\mathbb{Z})$ (the conditions of eveness in t and zero time-average simplify the problem but are not essential). The recurrence relation (6.9) holds in

$$X = \left\{ u \in H^2 \times H^0, u \text{ even in } t, \int_0^{2\pi} u \;\mathrm{d}t = 0 \right\}$$

and the operator $F_\omega : D \to X$ is smooth in a neighborhood of zero.

By a solution of (6.9) we mean a sequence $(u_n)_{n\in\mathbb{Z}}$ in D satisfying (6.9) for all $n \in \mathbb{Z}$. A particular solution is the fixed point $u = 0$ corresponding to the lattice at rest. An important feature of (6.9) is that the derivative of F_ω at $u = 0$ (denoted by $L_\omega = DF_\omega(0)$) is unbounded in X (of domain D) and thus the recurrence relation (6.9) is ill-posed (in fact there are no solutions of (6.9) for most initial conditions $u_0 \in D$).

Solutions of (6.9) homoclinic to $u = 0$ (i.e. satisfying $\lim_{|n|\to+\infty} \| u_n \|_D = 0$) have been extensively studied. They correspond to discrete breather solutions of (6.8), i.e. spatially localized solutions with time-period 2π (we refer the reader to [43] for a general review on discrete breathers). One property making the existence of homoclinic solutions more likely is the invariance $y_n \to y_{-n}$ in (6.8). Indeed, this invariance implies that (6.9) is reversible with respect to the symmetry R defined by $R(a, b) = (b, a)$, i.e. if u_n is a solution of (6.9) then $R u_{-n}$ is also a solution. Consequently, if the unstable manifold $W^{\mathrm{u}}(0)$ intersects the fixed set $\mathrm{Fix}\,(R)$ then their intersection also belongs to the stable manifold $W^{\mathrm{s}}(0)$ and homoclinic orbits exist. Note that problem (6.8) has the invariance $y_n \to y_n(t+\pi)$ and thus F_ω commutes with the symmetry T defined by $(T\,u)(t) = u(t+\pi)$. As a consequence, TR defines another reversibility symmetry.

In the sequel, we shall consider small amplitude solutions of (6.9), therefore the first step is to examine the linearized problem around $u = 0$. The operator L_ω is given by $L_\omega(y_{n-1}, y_n) = (\, y_n\,,\, (\omega^2 \frac{d^2}{dt^2} + 2)y_n - y_{n-1}\,)$. We denote by σ_k, σ_k^{-1} ($k \geq 1$) the eigenvalues of L_ω, given by the dispersion relation

$$\sigma^2 + (\omega^2 k^2 - 2)\sigma + 1 = 0 \tag{6.10}$$

(by convention we denote by σ_k the solution of (6.10) satisfying $|\sigma_k| \geq 1$ and $\operatorname{Im} \sigma_k \leq 0$). The invariance $\sigma \to \sigma^{-1}$ in (6.10) is due to reversibility. The invariant subspace under L_ω associated with the pair of eigenvalues σ_k, σ_k^{-1} is spanned by the vectors $(\cos(kt), 0)$ and $(0, \cos(kt))$. When ω is large, the eigenvalues are real negative and lie strictly off the unit circle (both inside and outside). When ω decreases, σ_k moves towards the unit circle. As ω reaches the first critical value $\omega = 2$ (maximal phonon frequency), the eigenvalues σ_1, σ_1^{-1} collide and yield a double (nonsemi-simple) eigenvalue $\sigma = -1$ (see Fig. 6.2). When ω is further decreased, σ_1, σ_1^{-1} rotate on the unit circle and converge towards $\sigma = 1$ as $\omega \to 0^+$. More generally, the pair of eigenvalues σ_k, σ_k^{-1} collide at $\sigma = -1$ when $\omega = 2/k$.

The above analysis shows that the fixed point $u = 0$ of (6.9) is hyperbolic when $\omega > 2$. In this case, we shall see that the stable and unstable manifolds $W^s(0)$, $W^u(0)$ can intersect, depending on the properties of V. More precisely, for $\omega \approx 2$ the pair of eigenvalues σ_1, σ_1^{-1} is close to -1 and (6.9) may admit small amplitude homoclinic orbits to $u = 0$.

In order to perform a local bifurcation analysis, we restrict our attention to the case when $\omega \approx 2$. We introduce the small parameter $\mu = \omega^2 - 4$ and write (6.9) in the form

$$u_{n+1} = L\, u_n + N(u_n, \mu), \tag{6.11}$$

where $L = L_2$ is the linearized operator with $\omega = 2$ and

$$N((y_{n-1}, y_n), \mu) = \left(0\,,\, \frac{d^2}{dt^2} g(y_n, \mu)\right),$$

with $g(y, \mu) = 4(W(y) - y) + \mu W(y)$. One has $\|\, N(u, \mu)\,\|_X = O(\|u\|_D^2 + \mu\,\|u\|_D) \ll \|\, L\, u\,\|_X$ as $(u, \mu) \approx 0$, hence a good starting point for studying (6.11) is to start from the linear case.

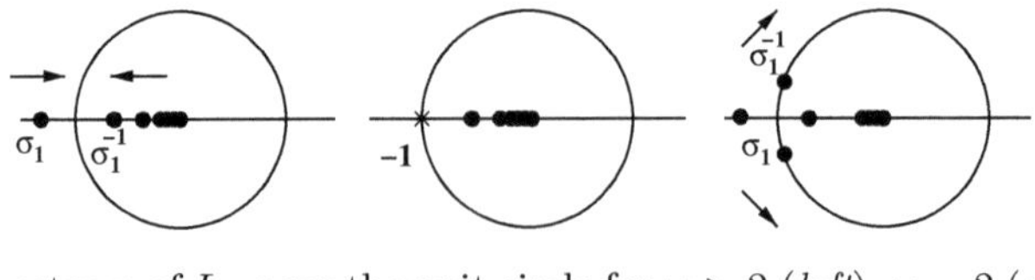

Fig. 6.2. Spectrum of L_ω near the unit circle for $\omega > 2$ (*left*), $\omega = 2$ (*center*), $\omega < 2$ and $\omega \approx 2$ (*right*)

The solutions of

$$u_{n+1} = L\,u_n, \quad u_n \in D, \tag{6.12}$$

can be computed using Fourier series. Due to the existence of a spectral gap separating $\sigma_1 = -1$ from the remaining (hyperbolic) part of the spectrum, solutions of (6.12) can be splitted in the following two classes. The first kind of solutions grows at least like $|\sigma_2|^{|n|}$ as $n \to +\infty$ or $-\infty$. The second kind has the form

$$u_n = (y_{n-1}, y_n), \quad y_n = (-1)^n\,(\alpha + \beta\,n)\,\cos t$$

and diverges at most polynomially as $|n| \to +\infty$. Consequently, solutions u_n of (6.12) which remain bounded in D as $|n| \to +\infty$ necessarily belong to the two-dimensional linear subspace $X_c = \text{Span}\,\{\,(\cos t, 0)\,,\,(0, \cos t)\,\}$ for all $n \in \mathbb{Z}$. The space X_c is denoted as *center space* and is the invariant space under L associated with the double eigenvalue $\sigma_1 = -1$. The spectral projection π_c on X_c reads $\pi_c\,u = (1/\pi)\int_0^{2\pi} u(t)\cos t\,dt\,\cos t$.

In the nonlinear case (6.11), one can locally prove a similar result where the center space is replaced by a two-dimensional invariant center manifold [29]. For $\mu \approx 0$, there exists a smooth local manifold $\mathcal{M}_\mu \subset D$ (which can be written as a graph over X_c) invariant under F_ω and the symmetries R, T. One has $\mathcal{M}_\mu = \{u \in D / u = u^c + \psi(u^c, \mu), u^c \in X_c \cap \Omega\}$, where $\psi : X_c \times \mathbb{R} \to (I - \pi_c)D$ is a smooth map satisfying $\psi(u^c, \mu) = \text{O}(\|u^c\|^2 + \|u^c\|\,|\mu|)$ and Ω is a small neighborhood of 0 in D. More precisely, we have $\psi(\,(a,b)\,\cos t\,,\,\mu\,) = (\,\varphi(b,a,\mu)\,,\,\varphi(a,b,\mu)\,)\;(a, b\,\in \mathbb{R})$ where

$$\varphi(a,b,\mu) = -\frac{1}{16}V^{(3)}(0)\cos{(2t)}\left(ab + \frac{1}{2}a^2 - \frac{7}{2}b^2\right) + \text{h.o.t.} \tag{6.13}$$

For $\mu \approx 0$, the center manifold $\mathcal{M}_\mu$ contains all solutions u_n of (6.11) staying in Ω for all $n \in \mathbb{Z}$ (small amplitude solutions). Their central component $u_n^c = \pi_c\,u_n$ is given by the two-dimensional mapping

$$u_{n+1}^c = f(u_n^c, \mu), \tag{6.14}$$

where $f(u^c, \mu) = L\,u^c + \pi_c\,N(u^c + \psi(u^c, \mu), \mu)$. The map $f : X_c \times \mathbb{R} \to X_c$ is smooth in the neighborhood of 0 and inherits the symmetries of (6.9) ($f(\cdot, \mu)$ is reversible under R and commutes with T). Setting $u_n^c = (a_n, b_n)\cos t$, the mapping (6.14) reads

$$\begin{aligned} a_{n+1} &= b_n, \\ b_{n+1} &= -a_n - 2b_n - b_n\mu + c_1 b_n^3 + c_2 a_n b_n^2 + \tfrac{1}{2}c_2 a_n^2 b_n + \text{h.o.t.}, \end{aligned} \tag{6.15}$$

where $c_1 = \frac{1}{2}V^{(4)}(0) - \frac{17}{16}(V^{(3)}(0))^2$, $c_2 = -\frac{1}{8}(V^{(3)}(0))^2$.

Consequently, the problem of finding small amplitude solutions of (6.9) for $\omega \approx 2$ reduces to the study of the two-dimensional reversible mapping (6.15). Note that we are not limited to localized solutions of (6.9) (i.e. those satisfying

$\lim_{|n|\to+\infty} \| u_n \|_D = 0$) since the reduced mapping (6.15) describes the set of *all* small amplitude solutions when $\mu \approx 0$. For $\mu \approx 0$, each small amplitude solution of (6.15) corresponds to a solution y_n of (6.8) given by

$$y_n(t) = b_n \cos t + \varphi(b_{n-1}, b_n, \mu) \tag{6.16}$$

with $\omega^2 = 4 + \mu$ in (6.8).

6.2.3 Infinite-Dimensional Maps with Unbounded Linear Part and Spectral Separation

The above example suggests the following mathematical framework for studying time-periodic oscillations in infinite one-dimensional lattices.

We consider a Hilbert space X and a closed linear operator $L : X \to X$ of domain D (L is a priori unbounded). We equip D with the scalar product $\langle u, v \rangle_D = \langle Lu, Lv \rangle_X + \langle u, v \rangle_X$, hence D is a Hilbert space continuously embedded in X. We denote by $\mathcal{U} \times \mathcal{V}$ a neighborhood of 0 in $D \times \mathbb{R}^p$ and consider a nonlinear map $N \in C^k(\mathcal{U} \times \mathcal{V}, X)$ ($k \geq 2$) satisfying $N(0,0) = 0$, $D_u N(0,0) = 0$. We look for sequences $(u_n)_{n\in\mathbb{Z}}$ in $\mathcal{U}$ satisfying

$$\forall n \in \mathbb{Z}, \quad u_{n+1} = L\, u_n + N(u_n, \mu) \quad \text{in } X, \tag{6.17}$$

where $\mu \in \mathcal{V}$ is a parameter. In particular, $u = 0$ is a fixed point of (6.17) when $\mu = 0$. Note that the initial value problem for (6.17) is in general ill-posed.

We assume that L has the property of *spectral separation*, i.e. L satisfies Assumption H1 below.

Assumption H1. *The operator L has nonempty hyperbolic ($|z| \neq 1$) and central ($|z| = 1$) spectral parts. Moreover, there exists an annulus $\mathcal{A} = \{ z \in \mathbb{C}, r \leq |z| \leq R \}$ ($r < 1 < R$) such that the only part of the spectrum of L in $\mathcal{A}$ lies on the unit circle.*

The situation corresponding to Assumption H1 is sketched in Fig. 6.3. Under Assumption H1, the hyperbolic part of the spectrum is isolated from its central part. This assumption is essential in the proof of the center manifold reduction theorem.

We do not require the center space X_c (invariant subspace under L corresponding to the central part of the spectrum) to be finite-dimensional. However, the center manifold reduction theorem is more efficient in this case since the local study of (6.17) is amenable to that of a finite-dimensional mapping. The subspace X_c is finite-dimensional when the spectrum of L on the unit circle consists in a finite number of eigenvalues with finite multiplicities.

The spectral projection π_c on the center space can be defined in the following way (see, e.g. [44])

$$\pi_c = \frac{1}{2i\pi} \int_{\mathcal{C}(R)} (zI - L)^{-1}\, dz - \frac{1}{2i\pi} \int_{\mathcal{C}(r)} (zI - L)^{-1}\, dz,$$

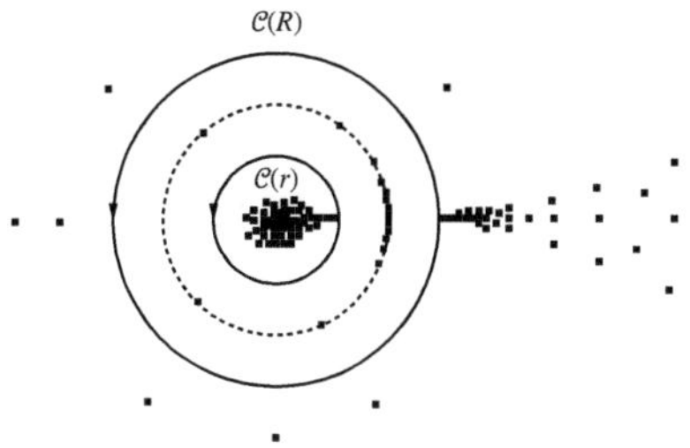

Fig. 6.3. Spectrum of L (*dots*), unit circle (*dashed*) and oriented circles $\mathcal{C}(r)$ and $\mathcal{C}(R)$

where $\mathcal{C}(r)$ denotes the circle of center $z = 0$ and radius r (see Fig. 6.3). One has $\pi_c \in \mathcal{L}(X, D)$, $X_c = \pi_c X \subset D$ and $\pi_c L = L \pi_c$. In the sequel we note $\pi_h = I - \pi_c$ and $D_h = \pi_h D$.

The condition of spectral separation is satisfied in a large class of infinite one-dimensional lattices where one looks for time-periodic solutions, e.g. in multicomponent Klein–Gordon lattices [28] or diatomic FPU chains [32]. In these examples, the center space is finite-dimensional and π_c takes a simple form in term of Fourier coefficients. These properties should be preserved in higher-dimensional lattices with appropriate boundary conditions (e.g. periodic) in the transverse directions.

We now state the center manifold reduction theorem for quasilinear maps with spectral separation [28].

Theorem 6.2. *Assume that L has the property of spectral separation (Assumption H1). Then there exist a neighborhood $\Omega \times \Lambda$ of 0 in $D \times \mathbb{R}^p$ and a map $\psi \in C^k(X_c \times \Lambda, D_h)$ (with $\psi(0,0) = 0$, $D_{u^c}\psi(0,0) = 0$) such that for all $\mu \in \Lambda$ the manifold*

$$\mathcal{M}_\mu = \{\, u \in D \,/\, u = u^c + \psi(u^c, \mu), u^c \in X_c \,\}$$

has the following properties.
(i) $\mathcal{M}_\mu$ is locally invariant under $L + N(., \mu)$, i.e. if $u \in \mathcal{M}_\mu \cap \Omega$ then $Lu + N(u, \mu) \in \mathcal{M}_\mu$.
(ii) If $(u_n)_{n\in\mathbb{Z}}$ is a solution of (6.17) and $u_n \in \Omega$ for all $n \in \mathbb{Z}$, then $u_n \in \mathcal{M}_\mu$ for all $n \in \mathbb{Z}$ and $u_n^c = \pi_c u_n$ satisfies the recurrence relation in X_c

$$\forall n \in \mathbb{Z}, \quad u_{n+1}^c = f(u_n^c, \mu), \tag{6.18}$$

where $f \in C^k((X_c \cap \Omega) \times \Lambda, X_c)$ is defined by $f(., \mu) = \pi_c\,(L + N(., \mu)) \circ (I + \psi(., \mu))$ and $f(., \mu)$ is locally invertible.
(iii) Conversely, if $(u_n^c)_{n\in\mathbb{Z}}$ is a solution of (6.18) such that $u_n^c \in \Omega$ for all $n \in \mathbb{Z}$, then $u_n = u_n^c + \psi(u_n^c, \mu)$ satisfies (6.17).

This result reduces the local study of (6.17) to that of the recurrence relation (6.18) in the smaller space X_c, which is particularly interesting when X_c is finite-dimensional.

The center manifold reduction technique has the advantage of being a constructive method, i.e. the Taylor expansion of the reduction function ψ at $(u^{\mathrm{c}}, \mu) = 0$ (and thus the expansion of f) can be computed at any order. One computes the Taylor expansion of ψ by expanding each side of equation

$$\psi(L\, u^{\mathrm{c}} + \pi_{\mathrm{c}}\, N(u^{\mathrm{c}} + \psi(u^{\mathrm{c}}, \mu), \mu), \mu) = L\, \psi(u^{\mathrm{c}}, \mu) + \pi_{\mathrm{h}}\, N(u^{\mathrm{c}} + \psi(u^{\mathrm{c}}, \mu), \mu) \quad (6.19)$$

with respect to (u^{c}, μ) and identifying terms of equal order ((6.19) expresses the fact that $\mathcal{M}_\mu$ is invariant under $L + N(.,\mu)$). This procedure yields a hierarchy of linear problems which can be solved by induction, starting from the lowest order (see (6.13) and (6.15) for the first terms in the FPU system). Since the method provides explicit expansions of ψ and f, one can in general precisely describe the shape and symmetries of small amplitude bifurcating solutions.

In addition, one can show that the invariances of (6.17) are preserved throughout the reduction procedure. More precisely, if $L + N(.,\mu)$ commutes with a linear isometry $T \in \mathcal{L}(X) \cap \mathcal{L}(D)$ then $\mathcal{M}_\mu$ is invariant under T and $f(.,\mu)$ also commutes with T.

In Sect. 6.2.2 we have considered a situation when (6.17) is reversible with respect to a symmetry $R \in \mathcal{L}(D)$ ($R^2 = I$). This means that if u_n is a solution of (6.17), then Ru_{-n} is also a solution. This situation arises when the map (6.17) satisfies

$$((L + N(.,\mu)) \circ R)^2 u = u \quad (6.20)$$

for all $u \in R^{-1}(\mathcal{U})$ such that $LRu + N(Ru, \mu) \in R^{-1}(\mathcal{U})$. Property (6.20) formally implies

$$\text{if } u \in D \text{ and } LR\,u \in D, \text{then } (LR)^2 u = u \quad (6.21)$$

((6.21) is the formal differentiation of (6.20) at $u = 0$). In the particular case when $\mathcal{U} = D = X$ and $L + N(.,\mu)$ is invertible, note that (6.20) can be written

$$R\,(L + N(.,\mu)) \circ R = (L + N(.,\mu))^{-1},$$

hence $L + N(.,\mu)$ is conjugate to its inverse via the symmetry R. If (6.17) is reversible under a symmetry $R \in \mathcal{L}(D)$, then under some technical assumptions given in [28], $\mathcal{M}_\mu$ is invariant under R and $f(.,\mu)$ inherits reversibility in R.

Note that the property of *local attractivity* of center manifolds (when L has no eigenvalue outside the unit disc) is not reviewed here. Indeed we are mainly concerned with reversible mappings. Due to property (6.21), the spectrum of their linear part has the invariance $z \to z^{-1}$, hence their hyperbolic spectral part is located both inside and outside the unit disc.

6.3 Center Manifold Reduction for Infinite-Dimensional Differential Equations

It is well known [37, 45, 46] that the center manifold theorem for maps stated in Sect. 6.2.1 has an analogue for differential equations in $\mathbb{R}^n$

$$\frac{\mathrm{d}x}{\mathrm{d}t} = Lx + N(x, \mu), \tag{6.22}$$

with $L \in M_n(\mathbb{R})$, $N \in C^k(\mathbb{R}^n \times \mathbb{R}^p, \mathbb{R}^n)$ (with $k \geq 2$), $N(0,0) = 0$, $D_x N(0,0) = 0$. For $\mu = 0$, (6.22) admits $x = 0$ as an equilibrium. One defines X_c as the subspace spanned by the (generalized) eigenvectors of L whose eigenvalues lie on the imaginary axis. If $X_\mathrm{c} \neq \{0\}$, there exists a local invariant manifold $\mathcal{M}_\mu$ of (6.22), which contains all small amplitude solutions for $\mu \approx 0$, and can be written locally as a graph over the center space X_c (for $\mu = 0$, $\mathcal{M}_0$ is tangent to X_c at $x = 0$).

In this section, we examine the infinite-dimensional case, again focusing on the context of reversible systems. The general theory which is (partly) summarized in Sect. 6.3.2 originates from [35, 36, 42]. In the following section, we start by an illustration of center manifold theory on a class of reversible advance-delay differential equations. The material is issued from [18, 20, 21].

6.3.1 Example in Infinite Dimensions

We consider the KG system (6.3), and assume the interaction potential V analytic in a neighborhood of $x = 0$, with the following Taylor expansion

$$V(x) = \frac{1}{2}x^2 + \frac{\alpha}{3}x^3 + \frac{\beta}{4}x^4 + \mathrm{O}(|x|^5), \quad x \to 0. \tag{6.23}$$

We search for solutions in the form of pulsating traveling waves satisfying (6.4). Problem (6.3)–(6.4) can be formulated as a (reversible) system of p second-order differential equations with advance and delay:

$$\frac{\mathrm{d}^2}{\mathrm{d}t^2}\begin{bmatrix} x_1 \\ \vdots \\ x_n \\ \vdots \\ x_p \end{bmatrix} + \begin{bmatrix} V'(x_1) \\ \vdots \\ V'(x_n) \\ \vdots \\ V'(x_p) \end{bmatrix} = \gamma \begin{bmatrix} x_2(t) - 2x_1(t) + x_p(t+T) \\ \vdots \\ x_{n+1}(t) - 2x_n(t) + x_{n-1}(t) \\ \vdots \\ x_1(t-T) - 2x_p(t) + x_{p-1}(t) \end{bmatrix}. \tag{6.24}$$

In the following we restrict ourselves to the case $p = 2$ [20]. Our analysis generalizes the one performed by Iooss and Kirchgässner [18], who treated the case of traveling waves ($p = 1$). In this case, one obtains $x_n(t) = x_0(t - nT)$, where the function x_0 is determined by the scalar differential equation with advance and delay:

$$\frac{\mathrm{d}^2 x_0}{\mathrm{d}t^2} + V'(x_0) = \gamma\,(x_0(t+T) - 2x_0 + x_0(t-T)). \tag{6.25}$$

For the more technical case $p > 2$ we refer the reader to [21].

In the case $p = 2$, solutions can be rewritten [20]:

$$\begin{aligned} x_n(t) &= w_1(\tfrac{t}{T} - \tfrac{n-1}{2}) \text{ if } n \text{ is odd,} \\ x_n(t) &= w_2(\tfrac{t}{T} - \tfrac{n-1}{2}) \text{ if } n \text{ is even} \end{aligned} \tag{6.26}$$

when w_1 *and* w_2 are solutions of the following differential advance-delay system in the variable $\xi = \frac{t}{T} - \frac{n-1}{2}$

$$\frac{\mathrm{d}^2}{\mathrm{d}\xi^2}\begin{bmatrix} w_1 \\ w_2 \end{bmatrix} + T^2 \begin{bmatrix} V'(w_1) \\ V'(w_2) \end{bmatrix} = \gamma T^2 \begin{bmatrix} w_2(\xi - \frac{1}{2}) - 2w_1(\xi) + w_2(\xi + \frac{1}{2}) \\ w_1(\xi + \frac{1}{2}) - 2w_2(\xi) + w_1(\xi - \frac{1}{2}) \end{bmatrix}. \tag{6.27}$$

We note that the solutions of (6.27) such that $w_1 = w_2$ correspond to traveling wave solutions of (6.3), since $x_n(t) = x_{n-1}(t - \frac{T}{2})$. Besides, if V is even then (6.27) admits symmetric solutions such that $w_1 = -w_2$. These solutions correspond to solutions of (6.3) of the form

$$x_n(t) = (-1)^{n+1} w_1 \left(\frac{t}{T} - \frac{(n-1)}{2}\right). \tag{6.28}$$

These solutions are described by the scalar equation

$$\frac{\mathrm{d}^2 w_1}{\mathrm{d}\xi^2} + T^2 V'(w_1) = -\gamma T^2 \left(w_1\left(\xi + \frac{1}{2}\right) + 2w_1(\xi) + w_1\left(\xi - \frac{1}{2}\right)\right). \tag{6.29}$$

We can reformulate the problem (6.27) as an evolution problem with respect to the ξ variable in suitable function spaces. For this purpose, we introduce as in [18] the variable $U = (w_1, w_2, w_1', w_2', X_1(\xi, v), X_2(\xi, v))^{\mathrm{T}}$, where $v \in [-1/2, 1/2]$ and $X_1(\xi, v) = w_1(\xi + v)$, $X_2(\xi, v) = w_2(\xi + v)$. Equation (6.27) can be rewritten

$$\frac{\mathrm{d}U}{\mathrm{d}\xi} = L_{\gamma,T} U + T^2 M(U), \tag{6.30}$$

where

$$L_{\gamma,T} = \begin{pmatrix} 0 & 0 & 1 & 0 & 0 & 0 \\ 0 & 0 & 0 & 1 & 0 & 0 \\ \alpha_1 & 0 & 0 & 0 & 0 & \alpha_2(\delta_{1/2} + \delta_{-1/2}) \\ 0 & \alpha_1 & 0 & 0 & \alpha_2(\delta_{-1/2} + \delta_{1/2}) & 0 \\ 0 & 0 & 0 & 0 & \partial_v & 0 \\ 0 & 0 & 0 & 0 & 0 & \partial_v \end{pmatrix}, \tag{6.31}$$

$\delta_a X_i(\xi, v) = X_i(\xi, a)$, $\alpha_2 = T^2\gamma$, $\alpha_1 = -T^2(1 + 2\gamma)$ and

$$M(U) = (0, 0, f(w_1), f(w_2), 0, 0)^{\mathrm{T}}, \tag{6.32}$$

$$f(u) = -\alpha\, u^2 - \beta\, u^3 + \text{h.o.t.} \tag{6.33}$$

We now consider (6.30) in appropriate function spaces. For this purpose we introduce the Banach spaces

$$X = \mathbb{R}^4 \times (C^0[-1/2, 1/2])^2, \tag{6.34}$$

$$D = \left\{U \in \mathbb{R}^4 \times (C^1[-1/2, 1/2])^2 / X_1(0) = w_1, X_2(0) = w_2\right\}, \tag{6.35}$$

$$Y = \left\{U \in \mathbb{R}^4 \times (C^1[-1/2, 1/2])^2 / X_1 = w_1 = X_2 = w_2 = 0\right\}. \tag{6.36}$$

The operator $L_{\gamma,T}$ maps D into X continuously and $M : D \to Y$ is analytic in a neighborhood of 0. We look for solutions of (6.30) in the class $C^0(\mathbb{R}, D) \cap C^1(\mathbb{R}, X)$.

We observe that the symmetry R on X defined by

$$R(w_1, w_2, w_1', w_2', X_1(v), X_2(v))^{\mathrm{T}} = (w_1, w_2, -w_1', -w_2', X_1(-v), X_2(-v))^{\mathrm{T}}$$

anticommutes with $L_{\gamma,T} + T^2 M$. Therefore, if $U(\xi)$ is a solution of (6.30) then $RU(-\xi)$ is also a solution. In that case the system (6.30) is said reversible under R. This property is due to the invariance $\xi \to -\xi$ of (6.27).

We also notice the invariance of (6.30) under the permutation σ of the two components (w_i, w_i', X_i), $i = 1, 2$ (invariant solutions under this symmetry correspond to traveling waves).

Now we describe the spectrum of $L_{\gamma,T}$, which consists in isolated eigenvalues of finite multiplicity. The study of the spectrum is rather technical, so we shall restrict ourselves to its qualitative description and refer to [20] for the proofs of the results.

Linearizing (6.27) around $(w_1, w_2) = (0, 0)$ and searching for solutions in the form $(w_1, w_2) = \mathrm{e}^{\mathrm{i}q\xi}(\hat{u}_1, \hat{u}_2)^{\mathrm{T}}$ (with $q \in \mathbb{C}$) yields the following dispersion relation

$$(-q^2 + T^2(1 + 2\gamma))^2 - 4(\gamma T^2)^2 \cos^2(q/2) = 0. \tag{6.37}$$

Equation (6.37) gives the eigenvalues $\pm \mathrm{i}q$ of $L_{\gamma,T}$. The spectrum of $L_{\gamma,T}$ is invariant under reflection with respect to the real and imaginary axis, and unbounded on both sides of the imaginary axis (in addition $L_{\gamma,T}$ is not bisectorial).

The central part of the spectrum is finite-dimensional and corresponds to the solutions $q \in \mathbb{R}$ of (6.37). We sketch it on Fig. 6.4 (see [20] for a detailed study of the spectrum, and in particular its description in the parameters regions uncovered by Fig. 6.4).

The hyperbolic part of the spectrum (not represented in Fig. 6.4) is located outside a band of width $\nu > 0$ around the imaginary axis (ν depends on γ, T). Therefore, the hyperbolic spectral part is isolated from the central part.

Double roots $q \in \mathbb{R}$ of (6.37) satisfy

$$2q(-q^2 + T^2(1 + 2\gamma)) = (\gamma T^2)^2 \sin(q). \tag{6.38}$$

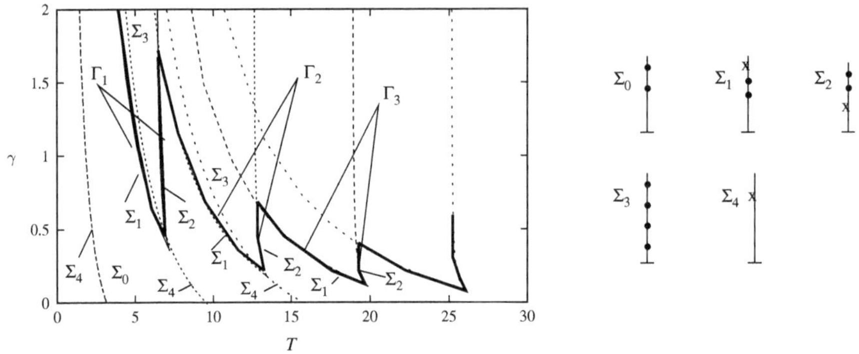

Fig. 6.4. On the right: location of eigenvalues iq de $L_{\gamma,T}$ on the imaginary axis (only the upper part of the axis is represented). Symbols Σ_1, Σ_2, Σ_4 refer to different curves in the plane of parameters T, γ sketched on the left (these curves are represented by bold or dotted lines) and Σ_0, Σ_3 refer to regions delimited by such curves. Eigenvalues iq of $L_{\gamma,T}$ are real solutions of (6.37) (here $q > 0$). A simple eigenvalue is represented by a point (•) and a double one by a cross (×). The latter are semi-simple in the case of curves Σ_4, and nonsemi-simple in the case of curves Σ_1, Σ_2. The curve Δ corresponding to the bifurcations studied in this section is represented by the bold line comprising Σ_1 and Σ_2. In addition to $\Sigma_1, \Sigma_2, \Sigma_4$, other bifurcation curves (corresponding to additional changes in the spectrum on the imaginary axis) are represented by dotted lines

Solutions of (6.37)–(6.38) correspond to double eigenvalues of $L_{\gamma,T}$ (except for isolated values of (T, γ) where they correspond to triple eigenvalues [20]). Double nonsemi-simple eigenvalues show up for parameters T, γ taken on an infinite set of curves $\Gamma_1, \Gamma_2, \Gamma_3, \ldots$ forming "tongues" in the parameter plane. A part of the first curves is represented in Fig. 6.4 by the bold line (the curves can be continued to higher values of γ following the dotted lines). The bold line actually represents the boundary of the region located below the tongues Γ_k. We shall denote this curve by Δ.

The existence of double roots as parameters belong to Γ_k corresponds to the appearance (or disappearance) of a pair of real roots of (6.37) when parameters (T, γ) cross the curve Γ_k. More precisely, under the curve Δ (see region Σ_0 in Fig. 6.4), real solutions of (6.37) consist in two pairs of simple roots $\pm q_1$, $\pm q_2$ (they coincide on the curves denoted Σ_4, leading to a pair of semi-simple double eigenvalues). An additionnal pair of real (double) roots $\pm q_0$ appears when we reach Δ from below (see the curves labelled by Σ_1, Σ_2 in Fig. 6.4). This corresponds to two pairs of simple hyperbolic eigenvalues of $L_{\gamma,T}$ colliding on the imaginary axis and leading to a pair of non semi-simple double eigenvalues $\pm$iq_0 (see the spectra labelled by Σ_1, Σ_2 in Fig. 6.4, right).

In the following, we exclude from Δ the neighborhoods of points where $sq_0 + rq_1 + r'q_2 = 0$ for $s, r, r' \in \mathbb{Z}$ and $0 < |s| + |r| + |r'| \leq 4$ (strong resonances), and we note Δ_0 this new set. We choose then $(\gamma, T) \approx (\gamma_0, T_0)$

with $(T_0, \gamma_0) \in \Delta_0$. The above nondegeneracy condition is required to have the simple normal form structure provided in (6.40), as one considers the flow on a center manifold for $(\gamma, T) \approx (\gamma_0, T_0)$.

Small amplitude solutions of (6.27) correspond to solutions of (6.30) on a center manifold $\mathcal{M}_{\gamma,T}$, invariant by the flow [36], eight-dimensional for $(\gamma, T) \approx (\gamma_0, T_0)$. The dimension of $\mathcal{M}_{\gamma,T}$ is equal to the number of "marginal" modes of (6.27) linearized around $(w_1, w_2) = 0$ with $(\gamma, T) = (\gamma_0, T_0)$ (we call "marginal" mode a mode which does not diverge exponentially when $\xi \to +\infty$ or $-\infty$). Besides, the coordinates $u_c = (A, B, C_1, C_2, \bar{A}, \bar{B}, \bar{C}_1, \bar{C}_2)^{\mathrm{T}} \in \mathbb{C}^8$ of solutions on $\mathcal{M}_{\gamma,T}$ are given by a reversible differential equation. The latter is simplified when written in normal form [45].

In summary, applying the center manifold reduction theorem to the evolution problem (6.30) we have obtained a reduction result for the original system (6.27). The reduction result can be stated as follows (see [20] for the computation of expression (6.39) and the justification of the normal form (6.40)). Fix $(T_0, \gamma_0) \in \Gamma_m \cap \Delta_0$. For $(\gamma, T) \approx (\gamma_0, T_0)$, the small amplitude solutions of (6.27) have the form

$$\begin{pmatrix} w_1(\xi) \\ w_2(\xi) \end{pmatrix} = A(\xi) \begin{pmatrix} (-1)^m \\ 1 \end{pmatrix} + C_1(\xi) \begin{pmatrix} -1 \\ 1 \end{pmatrix} + C_2(\xi) \begin{pmatrix} 1 \\ 1 \end{pmatrix} + \text{c.c.} + \Psi(u_c(\xi), \gamma, T), \tag{6.39}$$

where $\Psi : \mathbb{C}^8 \times \mathbb{R}^2 \to \mathbb{R}^2$ is a C^k function in a neighborhood of $(0, \gamma_0, T_0)$ and $\Psi(u_c, \gamma, T) = \mathrm{O}(\|u_c\|^2 + \|u_c\|(|\gamma - \gamma_0| + |T - T_0|))$ when $(u_c, \gamma, T) \to (0, \gamma_0, T_0)$.

Coordinates (A, B, C_1, C_2) are given by a reversible differential equation written in normal form (normal form of order 3):

$$\begin{aligned}
\frac{\mathrm{d}A}{\mathrm{d}\xi} &= \mathrm{i}q_0 A + B + \mathrm{i}A\mathcal{P}(|A|^2, I, Q, \gamma, T) + \text{h.o.t.}, \\
\frac{\mathrm{d}B}{\mathrm{d}\xi} &= \mathrm{i}q_0 B + [\mathrm{i}B\mathcal{P} + A\mathcal{S}](|A|^2, I, Q, \gamma, T) + \text{h.o.t.}, \\
\frac{\mathrm{d}C_j}{\mathrm{d}\xi} &= \mathrm{i}q_j C_j + \mathrm{i}C_j \mathcal{Q}_j(|A|^2, I, Q, \gamma, T) + \text{h.o.t.}, \\
& \quad j = 1, 2,
\end{aligned} \tag{6.40}$$

where Q is the vector $Q = (|C_1|^2, |C_2|^2)$, $I = \mathrm{i}(A\bar{B} - \bar{A}B)$ and $\mathcal{P}, \mathcal{S}, \mathcal{Q}_j(., \gamma, T)$ are polynomials with real coefficients (which are C^k with respect to $(\gamma, T) \approx (\gamma_0, T_0)$). The principal part of (6.40) is a polynomial of degree 3 in A, B, C_1, C_2 and complex conjugates. Higher order terms are $\mathrm{O}(\|u_c\|^4)$.

Equation (6.40) is reversible under the symmetry

$$\mathcal{R} : (A, B, C_1, C_2) \mapsto (\bar{A}, -\bar{B}, \bar{C}_1, \bar{C}_2),$$

and equivariant under the isometry

$$\sigma : (A, B, C_1, C_2) \mapsto ((-1)^m A, (-1)^m B, -C_1, C_2).$$

Polynomials $\mathcal{P}$ and $\mathcal{S}$ in the normal form (6.40) have the form

$$\mathcal{P} = r_1(\gamma, T) + r_2\,|A|^2 + r_3\,I + \mathrm{O}(|\mu|\|(A,B)\|^2 + \|(C_1,C_2)\|^2), \tag{6.41}$$

$$\mathcal{S} = s_1(\gamma, T) + s_2\,|A|^2 + s_3\,I + \mathrm{O}(|\mu|\|(A,B)\|^2 + \|(C_1,C_2)\|^2), \tag{6.42}$$

where we note $\mu = |T - T_0| + |\gamma - \gamma_0| \approx 0$, $r_i, s_i \in \mathbb{R}$ and $r_1(\gamma_0, T_0) = s_1(\gamma_0, T_0) = 0$. For $(T,\gamma) \approx (T_0, \gamma_0) \in \Gamma_m \bigcap \Delta_0$ and when (T,γ) belongs to the region of the parameters plane located below Δ (see Fig. 6.4), we have $s_1 > 0$ and the linear part of (6.40) possesses two symmetric pairs of hyperbolic eigenvalues close to $\pm iq_0$, and two pairs of purely imaginary simple eigenvalues close to $\pm iq_1, \pm iq_2$.

Besides, the coefficient s_2 of (6.42) (corresponding to a cubic term of the nonlinear system (6.40)) is given by the expression [20]

$$\left(2 - \frac{q_0}{\tan(q_0/2)}\right) s_2 = T_0^2 \left(-6\beta + 8\alpha^2 - \frac{4\alpha^2 T_0^2}{2\gamma_0 T_0^2 \cos(q_0) - T_0^2(1+2\gamma_0) + 4q_0^2}\right), \tag{6.43}$$

where the coefficients α, β appear in the Taylor expansion of V at the origin (see (6.23)).

Consequently, the problem of finding small amplitude solutions of (6.27) for $(\gamma, T) \approx (\gamma_0, T_0)$ reduces to the study of the eight-dimensional reversible differential equation (6.40). We are not limited to localized solutions of (6.27) since (6.40) describes the set of *all* small amplitude solutions when $(\gamma, T) \approx (\gamma_0, T_0)$.

6.3.2 Infinite-Dimensional Differential Equations with Spectral Separation

We consider an infinite-dimensional differential equation in a Banach space X, having the form

$$\frac{\mathrm{d}x}{\mathrm{d}t} = Lx + N(x, \mu), \tag{6.44}$$

where $L : D \subset X \to X$ is a closed linear operator with domain D (L is a priori unbounded). We equip D with the graph norm. The nonlinear term N maps $D \times \mathbb{R}^p$ into a Banach space Y continuously embedded in X, and $N \in C^k(D \times \mathbb{R}^p, Y)$ ($k \geq 2$), $N(0,0) = 0$, $D_x N(0,0) = 0$. Moreover $\mu \in \mathbb{R}^p$ is a parameter. We look for solutions of (6.44) in $C^0(\mathbb{R}, D) \cap C^1(\mathbb{R}, X)$.

In what follows we denote by $\mathcal{L}(X, D)$ the set of bounded linear operators from X into D and note $\mathcal{L}(X)$ as a shortcut for $\mathcal{L}(X, X)$.

One makes the following spectral assumptions on L. One requires on the one hand a spectral separation between the central and hyperbolic parts of the spectrum, and on the other hand the finite-dimensionality of the center space.

Assumption H2: *The spectrum of L on the imaginary axis consists in a finite number of isolated eigenvalues, each with a finite-dimensional generalized eigenspace.*

Assumption H2 implies [44] the existence of a spectral projection π_{c} corresponding to the purely imaginary part of the spectrum, i.e. a continuous projection onto the finite-dimensional center space which commutes with L. It can be written

$$\pi_{\mathrm{c}} = \frac{1}{2\mathrm{i}\pi} \int_{\mathcal{C}} (zI - L)^{-1}\, \mathrm{d}z,$$

where $\mathcal{C}$ denotes a closed path of index 1 surrounding the purely imaginary part of the spectrum, and not including any other elements. One has $\pi_{\mathrm{c}} \in \mathcal{L}(X, D)$, the center space is given by $X_{\mathrm{c}} = \pi_{\mathrm{c}} X \subset D$ and $\pi_{\mathrm{c}} L = L \pi_{\mathrm{c}}$. In the sequel we note $\pi_{\mathrm{h}} = I - \pi_{\mathrm{c}}$ and $X_{\mathrm{h}} = \pi_{\mathrm{h}} X$, $D_{\mathrm{h}} = \pi_{\mathrm{h}} D$, $Y_{\mathrm{h}} = \pi_{\mathrm{h}} Y$, $L_{\mathrm{h}} = L_{|X_{\mathrm{h}}}$.

We also need sufficient regularity for the solutions $x(t) \in X_{\mathrm{h}}$ of the affine differential equation,

$$\frac{\mathrm{d}x}{\mathrm{d}t} = L_{\mathrm{h}}\, x + f(t) \tag{6.45}$$

when the inhomogeneous term $f(t)$ lies in Y_{h}. In what follows we denote by $C^k_{\mathrm{b}}(\mathbb{R}, E)$ the space of functions $\mathbb{R} \to E$ with bounded and continuous derivatives up to order k.

Assumption H3: *For all $f \in C^0_{\mathrm{b}}(\mathbb{R}, Y_{\mathrm{h}})$, (6.45) has a unique solution x in $C^0_{\mathrm{b}}(\mathbb{R}, D_{\mathrm{h}}) \cap C^1_{\mathrm{b}}(\mathbb{R}, X_{\mathrm{h}})$. Moreover, the application $f \mapsto x$ is continuous from $C^0_{\mathrm{b}}(\mathbb{R}, Y_{\mathrm{h}})$ into $C^0_{\mathrm{b}}(\mathbb{R}, D_{\mathrm{h}})$.*

In the case of maps [28], the property of spectral separation H1 is equivalent to a property similar to H3. This is not true however in the case of infinite-dimensional differential equations, where the property of spectral separation is necessary *but not sufficient* for property H3 to be satisfied.

Note that Assumptions H2 and H3 are automatically satisfied if (6.44) is finite-dimensional. Moreover, in many applications the operator L is bi-sectorial and Assumption H3 follows from simpler resolvent estimates [35, 36]. However, reversible advance-delay differential equations typically involve linearized evolution operators which are not bi-sectorial (as in Sect. 6.3.1), which requires different techniques to prove property H3 [18].

We note that property H3 (formulated in a space of bounded functions of t) implies by perturbation [20, 47] a similar property in the spaces of exponentially growing functions used in reference [36] (see hypothesis (ii) p. 127), provided the exponential growth rate is chosen small enough.

The center manifold theorem states the following [36].

Theorem 6.3. *Assume properties H2 and H3 are satisfied. There exists a neighborhood $\mathcal{U} \times \mathcal{V}$ of $(0, 0)$ in $D \times \mathbb{R}^p$ and a map $\psi \in C^k(X_{\mathrm{c}} \times \mathbb{R}^p, D_{\mathrm{h}})$ (with $\psi(0, 0) = 0$, $D\psi(0, 0) = 0$) such that for all $\mu \in \mathcal{V}$ the manifold*

$$\mathcal{M}_\mu = \{\, x \in D \,/\, x = x_c + \psi(x_c, \mu), x_c \in X_c \,\}$$

has the following properties.
(i) $\mathcal{M}_\mu$ is a local integral manifold for (6.44).
(ii) If $x : \mathbb{R} \to D$ is a solution of (6.44) and $x(t) \in \mathcal{U}\ \forall t \in \mathbb{R}$ then $x(t) \in \mathcal{M}_\mu$ for all $t \in \mathbb{R}$ and $x_c = \pi_c x$ is a solution of

$$\frac{dx_c}{dt} = Lx_c + \pi_c N(x_c + \psi(x_c, \mu), \mu). \tag{6.46}$$

(iii) If $x_c : \mathbb{R} \to X_c$ is a solution of (6.46) with $x_c \in \mathcal{U}\ \forall t \in \mathbb{R}$, then $x = x_c + \psi(x_c, \mu)$ is a solution of (6.44).

In addition, we give some comments in the case of a system with symmetries (as in the example of Sect. 6.3.1). If (6.44) is equivariant under a linear isometry T on X (i.e. $L + N(., \mu)$ commutes with T), then $\mathcal{M}_\mu$ is invariant under T and (6.46) inherits the equivariance under T. Moreover, if (6.44) is reversible under an unitary symmetry R on X (i.e. $L + N(., \mu)$ anticommutes with R), then $\mathcal{M}_\mu$ is invariant under R and (6.46) is also reversible under R.

Note that Theorem 6.3 does not describe the local attractivity of the center manifold, which occurs when L has no eigenvalue with strictly positive real part. Indeed the present work is mainly concerned with reversible systems, in which the spectrum of L is symmetric with respect to the imaginary axis.

6.4 Breathers and Traveling Breathers in Nonlinear Oscillator Chains

We have seen in Sect. 6.2.2 that *small amplitude* time-periodic oscillations in the FPU lattice can be described as orbits of a map in a loop space, restricted to a finite-dimensional invariant center manifold. In the same way, small amplitude pulsating traveling waves in the KG lattice can be seen as trajectories of an infinite-dimensional differential equation, lying on a finite-dimensional center manifold (Sect. 6.3.1). In this section, we study the related maps and flow on the corresponding center manifolds, focusing mainly on the existence of homoclinic orbits (corresponding to spatially localized oscillations of the lattice).

The case of FPU lattices is examined in Sect. 6.4.1, where we investigate the existence of breathers and some other types of nonlinear oscillations. Sect. 6.4.2 examines the existence of nanopterons in KG lattices.

6.4.1 Breathers and "Dark" Breathers in Fermi–Pasta–Ulam Lattices

Reduced Mapping on a Center Manifold

We now study a class of small amplitude bifurcating solutions of the mapping (6.15), corresponding to trajectories of (6.11) on a local center manifold.

For this purpose we first write (6.15) in *normal form*, i.e. we perform a polynomial change of variables close to the identity which simplifies (6.15) by keeping only its "essential" terms. The transformation $a_n = P(\alpha_n)$, $b_n = P(\beta_n)$, $P(x) = x - \frac{c_2}{12}x^3$ yields the normal form

$$\alpha_{n+1} = \beta_n, \quad \beta_{n+1} + 2\beta_n + \beta_{n-1} = h(\beta_{n-1}, \beta_n, \mu), \tag{6.47}$$

with

$$h(\beta_{n-1}, \beta_n, \mu) = -\mu\,\beta_n + B\beta_n^3 + \mathrm{O}(\,|\beta_n|\,((|\beta_{n-1}| + |\beta_n|)^2 + |\mu|)^2),$$

$$B = \frac{1}{2}V^{(4)}(0) - (V^{(3)}(0))^2 \tag{6.48}$$

(the principal part of h does not depend on β_{n-1} any more). Note that this transformation preserves the symmetries of (6.15). In particular, the second equation in (6.47) has the invariance $n \to -n$.

As a second step, setting $u_n = (-1)^n\beta_n$ allows us to recover the case of a bifurcation at a double eigenvalue $+1$ (this transformation yields an autonomous system due to the symmetry $-I$ in (6.47)). We obtain

$$u_{n+1} - 2u_n + u_{n-1} = h(u_{n-1}, -u_n, \mu), \tag{6.49}$$

$$h\,(u_{n-1}, -u_n, \mu) = \mu\,u_n - Bu_n^3 + \text{h.o.t.} \tag{6.50}$$

For studying small solutions of (6.49) when $\mu \approx 0$ (including homoclinic or heteroclinic ones), it is practical to consider $U_n = (u_n, v_n)$, $v_n = u_n - u_{n-1}$ and write (6.49) in the form

$$U_{n+1} = G_\mu(U_n), \quad G_\mu(U_n) = \begin{pmatrix} 1 & 1 \\ 0 & 1 \end{pmatrix} U_n + h\,(u_n - v_n, -u_n, \mu)\begin{pmatrix} 1 \\ 1 \end{pmatrix}. \tag{6.51}$$

The invariance $n \to -n$ in (6.49) implies that (6.51) is reversible under the symmetry $R_1\,(u, v) = (u - v, -v)$, the fixed set $\Delta_1 = \mathrm{fix}(R_1)$ being the axis $v = 0$.

The map (6.51) can be investigated using an approximation by a flow (see [48] and references therein). If φ_μ denotes the time-one map of the flow generated by the integrable vector field

$$\frac{du}{dt} = v, \quad \frac{dv}{dt} = \mu\,u - Bu^3, \tag{6.52}$$

there exist a local diffeomorphism h_μ close to the identity such that

$$G_\mu = h_\mu\varphi_\mu h_\mu^{-1} + \mathrm{O}(\,(|u| + |v|)\,((|u| + |v|)^2 + |\mu|)^2). \tag{6.53}$$

When μ and B have the same sign, the vector field has two nonzero symmetric equilibria. It has two symmetric orbits homoclinic to 0 when $\mu > 0$ and $B > 0$

(then the nonzero equilibria are elliptic). Moreover, there exist two symmetric heteroclinic orbits connecting the non-zero equilibria for $\mu < 0$ and $B < 0$.

The persistence of homoclinic solutions for (6.51) when $B > 0$ and $\mu > 0$ is small enough follows from the reversibility of (6.51), and the proof is based on the approximation (6.53). More precisely, the invariance $n \to -n$ in (6.18) implies that (6.51) is reversible under the symmetry $R_1\,(u, v) = (u - v, -v)$ and the involution $R_2 = R_1\, G_\mu$ (i.e. $(G_\mu R_i)^2 = I$). By perturbation (starting from the vector field), one shows that the unstable manifold at $U_n = 0$ has transversal intersections U_1^1 and U_0^2 respectively with the invariant set under R_1 and R_2. Then the corresponding solutions U_n^1 and U_n^2 are homoclinic to 0 since $R_1 U_{-n+2}^1 = U_n^1$ and $R_2 U_{-n}^2 = U_n^2$. Moreover, their first components satisfy $u_{-n+1}^1 = u_n^1$ and $u_{-n}^2 = u_n^2$. The persistence of heteroclinic solutions for $B < 0$ and $\mu < 0$ small enough follows from similar arguments, but we use instead $R_3 = -R_1$ and $R_4 = -R_2$. One finds heteroclinic solutions U_n^3 and U_n^4 connecting the nonzero equilibria, with $R_3 U_{-n+2}^3 = U_n^3$ and $R_4 U_{-n}^4 = U_n^4$. Their first components satisfy $u_{-n+1}^3 = -u_n^3$ and $u_{-n}^4 = -u_n^4$.

We sum up these results in the following lemma [28, 29].

Lemma 6.1. *Assume* $B = (1/2)V^{(4)}(0) - (V^{(3)}(0))^2 \neq 0$. *For* $\mu \approx 0$, *the recurrence relation (6.15) has the following solutions.*
(i) For $\mu > 0$ *and* $B > 0$, *(6.15) has at least two homoclinic solutions* b_n^1, b_n^2 *(and also* $-b_n^1$, $-b_n^2$*) such that* $\lim_{n\to\pm\infty} b_n^i = 0$. *These solutions have the symmetries* $b_{-n+1}^1 = -b_n^1$, $b_{-n}^2 = b_n^2$ *and satisfy* $0 < (-1)^n\, b_n^i \leq C\, \mu^{1/2}\, |\sigma_1|^{-|n|}$, *with* $|\sigma_1| = 1 + \mathrm{O}(\mu^{1/2}) > 1$.
(ii) If μ *and* B *have the same sign, (6.15) has a period 2 solution* $b_n^0 = (-1)^n b^*$, *with* $b^* = \mathrm{O}(|\mu|^{1/2})$.
(iii) For $\mu < 0$ *and* $B < 0$, *(6.15) has at least two heteroclinic solutions* b_n^3, b_n^4 *(and also* $-b_n^3$, $-b_n^4$*) such that* $\lim_{n\to\pm\infty} |b_n^i \mp b_n^0| = 0$. *These solutions have the symmetries* $b_{-n+1}^3 = b_n^3$ *and* $b_{-n}^4 = -b_n^4$. *Moreover,* b_n^3, b_n^4 *are* $\mathrm{O}(|\mu|^{1/2})$ *as* $n \to \pm\infty$ *and* $\mathrm{O}(|\mu|)$ *for bounded values of* n.

Note that for $B < 0$ and $\mu > 0$ ($\mu \approx 0$), the local stable and unstable manifolds of $(a_n, b_n) = 0$ do not intersect and thus (6.15) has no small amplitude homoclinic solution.

Breathers and "Dark" Breathers Corresponding to Homoclinics

According to Theorem 6.2, each solution b_n^i of Lemma 6.1 corresponds to a solution y_n^i of (6.8), given by

$$y_n^i(t) = b_n^i \cos t + \varphi(b_{n-1}^i, b_n^i, \mu) \tag{6.54}$$

with $\omega^2 = 4 + \mu$ in (6.8). This leads to the following result [28, 29].

Theorem 6.4. *Suppose* $B = \frac{1}{2}\, V^{(4)}(0) - (V^{(3)}(0))^2 \neq 0$. *For* $\omega \approx \omega_0 = 2$, *problem (6.8) admits the following solutions with* $y_n \in H^2(\mathbb{R}/2\pi\mathbb{Z})$ *for all* $n \in \mathbb{Z}$ *(with* y_n *even in* t, *and having* 0 *time-average).*

(i) For $\omega > \omega_0$ and $B > 0$, (6.8) has two homoclinic solutions y_n^1, y_n^2 (with the symmetric solutions $T\,y_n^1$, $T\,y_n^2$) such that $\lim_{n\to\pm\infty} \|y_n^i\|_{H^2} = 0$. These solutions satisfy $y^1_{-n+1} = T\,y_n^1$, $y^2_{-n} = y_n^2$ and have the form

$$y_n^i = b_n^i \, \cos t + \mathrm{O}(|\omega - \omega_0|), \tag{6.55}$$

where $0 < (-1)^n b_n^i \leq C\,(\omega-\omega_0)^{1/2}\,|\sigma_1|^{-|n|}$, $|\sigma_1| = 1 + \mathrm{O}((\omega-\omega_0)^{1/2}) > 1$.
(ii) If $\omega - \omega_0$ and B have the same sign, (6.8) has a solution y_n^0 with period 2 in n, having the form $y_n^0 = y(t+n\pi)$, with $y(t) = b^*\,\cos t + \mathrm{O}(|\omega-\omega_0|)$ (with y even in t, and having 0 time-average) and $b^* = \mathrm{O}(|\omega-\omega_0|^{1/2})$.
(iii) If $\omega < \omega_0$ et $B < 0$, (6.8) admits two heteroclinic solutions y_n^3, y_n^4 (together with $T\,y_n^3$, $T\,y_n^4$) such that $\lim_{n\to-\infty} \|y_n^i - Ty_n^0\|_{H^2} = 0$, $\lim_{n\to+\infty} \|y_n^i - y_n^0\|_{H^2} = 0$. These solutions satisfy $y^3_{-n+1} = y_n^3$ and $y^4_{-n} = T\,y_n^4$. Moreover, $\|y_n^3\|_{H^2}$, $\|y_n^4\|_{H^2}$ are $\mathrm{O}(|\omega-\omega_0|^{1/2})$ when $n \to \pm\infty$ and $\mathrm{O}(|\omega-\omega_0|)$ for bounded values of n.

Solutions y_n^1, y_n^2 correspond to breather solutions of (6.1), i.e. time-periodic oscillations (with frequency ω) satisfying (6.5) (Figure 6.5 shows a typical solution profile). Their amplitude tends to 0 when the frequency $\omega \to \omega_0$, and y_n^1, y_n^2 decay exponentially as $n \to \pm\infty$ (the decay is slow for $\omega \approx \omega_0$ since $|\sigma_1| \approx 1$). One can check that the total energy (6.2) of the corresponding solutions of (6.1) tends to 0 as $\omega \to \omega_0$.

Remark 6.1. If $B < 0$, small amplitude breathers (with y_n even in t, and having 0 time-average) bifurcating from $y_n = 0$ as $\omega \to \omega_0^+$ do not exist. This is due to the fact that the map (6.15) does not admit small amplitude homoclinics.

Solution y_n^0 corresponds to a spatially periodic traveling wave (of period 2 with respect to n), whose amplitude tends to 0 when $\omega \to \omega_0$. Its time-average is 0 and a translation of one site on the lattice is equivalent to an half-period phase shift ($y_{n+1}^0(t) = y_n^0(t+\pi)$). Note that the associated displacements x_n^0 (solutions of (6.1) corresponding to y_n^0) are in general different from the binary oscillations solutions of (6.1). These solutions are defined by $x_n^b(t) = (-1)^n f(t)$ with $f'' + V'(2f) - V'(-2f) = 0$ and f close to 0. Indeed, the time-average of the *relative displacements* $z_n^b = x_n^b - x_{n-1}^b$ is 0 whereas in our case the time-average of the *interaction forces* $V'(x_n^0 - x_{n-1}^0)$ is zero. Nevertheless, these two families of solutions coincide when the potential V is even.

Solutions y_n^3, y_n^4 are called "dark" breathers [49], due to the fact that their amplitude is almost vanishing at some point of the lattice and is larger at infinity (see Fig. 6.6 for an illustration), breather solutions having opposite characteristics. The adjective "dark" originates from the context of nonlinear optics where such waves have been also considered. The amplitude of these solutions is $\mathrm{O}(|\omega-\omega_0|^{1/2})$ when $n \to \pm\infty$ (they converge to y_n^0, with an index shift at $-\infty$) and their amplitude is $\mathrm{O}(|\omega-\omega_0|)$ at the center.

It is interesting to compare the sign of the bifurcation coefficient B to the hardness of the potential V in a neighborhood of $x = 0$. Recall that a potential V (such that $V'(0) = 0$, $V''(0) = 1$) is hard in a neighborhood of $x = 0$ if

$$\frac{3}{5}V^{(4)}(0) - (V^{(3)}(0))^2 > 0. \tag{6.56}$$

The potential V is soft if $\frac{3}{5}V^{(4)}(0) - (V^{(3)}(0))^2 < 0$. One can check that if V is soft then $B < 0$. It is interesting to notice that conditions (6.56) and $B > 0$ are not equivalent but rather close nevertheless.

The condition $B > 0$ corresponds to the modulational instability of the nonlinear normal modes of frequency $\omega \approx \omega_0$, yielding their spatial localization (see [50] for a formal study through multi-scale expansions). This condition leads to a similar instability for binary oscillations [51].

A numerical study has been performed in [34], where the solutions given by Theorem 6.4 have been followed up numerically in the high amplitude regime. More precisely, this study concerns polynomial interaction potentials

$$V(u) = \frac{u^2}{2} + \frac{\alpha}{3}u^3 + \frac{\beta}{4}u^4, \tag{6.57}$$

and breather solutions are approximated by spatially periodic solutions having large periods. When $\omega \approx \omega_0$ (we have here $\omega_0 = 2$), the principal part of y_n^2 reads

$$y_n^2(t) \simeq (-1)^n \sqrt{\frac{2\mu}{B}} \frac{\cos t}{\cosh(n\sqrt{\mu})} \tag{6.58}$$

and one can obtain a similar expression for y_n^1 (see [34], Sect. II). Expression (6.58) is obtained by solving the differential equation (6.52), which yields after discretization the principal part of (6.49)–(6.50). Figure 6.5 compares a numerically computed breather profile and approximation (6.58), ω being close to ω_0. The solution is weakly localized (its "support" extends on approximately 30 sites on the lattice). When ω goes far away from ω_0, the solution becomes more localized and its amplitude increases, hence approximation (6.58) becomes less accurate. Nevertheless, one observes (see for instance [34], Fig. 3) that the relative displacements $z_n = x_n - x_{n-1}$ are still precisely approximated by the right side of (6.58), *even for highly localized solutions* concentrated approximately on six sites on the lattice (note that y_n and z_n are equal at leading order in the small amplitude regime, since $V'(z_n) = z_n + O(z_n^2)$).

A similar study has been performed in the case of dark breathers [34]. The approximation derived in this context is

$$y_n^3(t) \simeq (-1)^n \sqrt{\frac{\mu}{B}} \tanh\left(\frac{(-n + 1/2)\sqrt{-\mu}}{\sqrt{2}}\right) \cos(t). \tag{6.59}$$

Figure 6.6 compares a numerically computed dark breather solution and the approximation (6.59), ω being close to ω_0 (in Fig. 6.6, the righ side of (6.59)

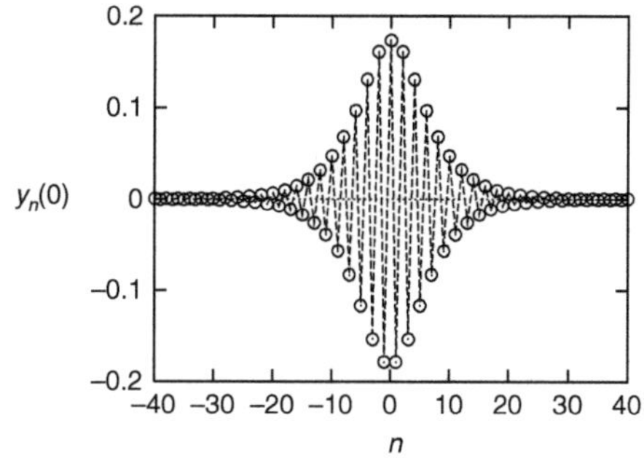

Fig. 6.5. Comparison between the numerically computed solution y_n^2 for the potential (6.57) (*circles*) and the approximation (6.58) (*dotted lines*), at time $t = 0$. In this example, $B = 2.64$, $\alpha = -0.3$ and $\omega = 2.01$ ($\mu \approx 0.04$)

is actually compared to the relative displacements z_n, very close to y_n at this amplitude). Once again, the approximation remains valid for steeper kink envelopes (see [34], Fig. 11).

Finally, note that for $\beta > 0$ and $B < 0$, there exist breather solutions whose energy and amplitude do not tend to 0 when $\omega \to \omega_0$ (see, e.g. [34] for numerically computed profiles). The existence of these solutions is in accordance with the results [30, 31], based on variational techniques (condition $\beta > 0$ implies that $V(u)$ is super-quadratic when $u \to \pm\infty$). The fact that the breather amplitude does not tend to 0 as $\omega \to \omega_0$ is a consequence of the remark following Theorem 6.4. The situation is sketched in Fig. 6.7.

The spectral stability of the above mentioned breather solutions has been studied in several works for potential (6.57), see e.g [34, 52, 53]. The spectral stability of dark breathers has been studied in reference [34].

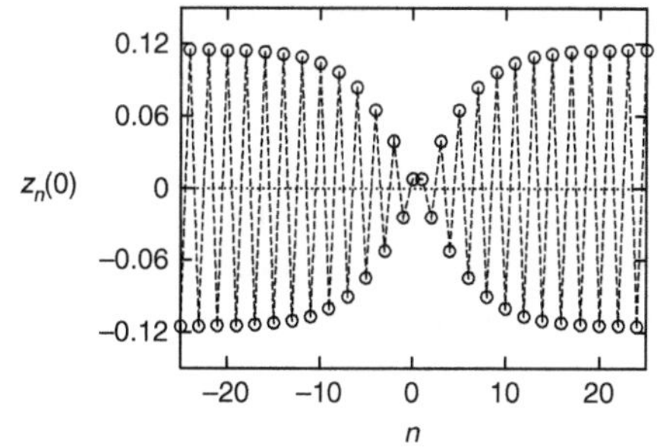

Fig. 6.6. Dark breather solution of (6.1) for potential (6.57) with $\alpha = 0$, $\beta = -1$ and $\omega = 1.99$. We represent relative displacements $z_n = x_n - x_{n-1}$ at time $t = 0$. This profile corresponds to the solution y_n^3 of Theorem 6.4. *Circles* stand for a numerically computed solution and dashed lines for an analytical approximation of z_n (right hand side of (6.59))

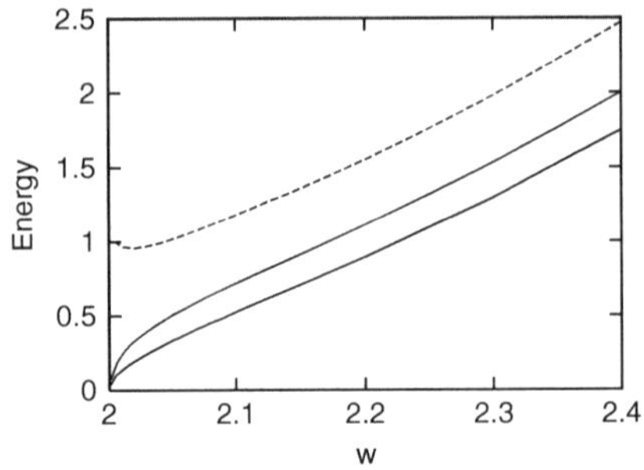

Fig. 6.7. Energy (6.2) of breather solutions of (6.1) as a function of their frequency ω, for potential (6.57) with $\beta = 1$. The *dashed line* corresponds to the case $\alpha = -1$ ($B < 0$). The minimal energy for the numerically computed breather family is reached for $\omega \approx 2.02$ and equal to $H \approx 0.95$. *Continuous lines* correspond to $\alpha = -0.6$ and $\alpha = 0$ ($B > 0$) (from top to bottom). In that case, there exist breather solutions with arbitrary small energy

6.4.2 Traveling Breathers in Klein–Gordon Lattices

Local Analytical Study

As seen in Sect. 6.3.1, system (6.3)–(6.4) taken for $p = 2$ reduces locally (for parameter values (T, γ) near Δ_0) to a eight-dimensional reversible normal form (6.40), which provides the coordinates of small amplitude solutions on a center manifold. In what follows, we shall describe bifurcating homoclinic solutions of this reduced system, following [20].

The *principal part* of the normal form (6.40) (obtained by neglecting the terms of order ≥ 4) is integrable. Indeed, fixing the first integrals $|C_i|^2$ ($i = 1, 2$) of the truncated system allows one to recover the classical 1:1 reversible resonance case [54]. If $s_2(\gamma_0, T_0) < 0$, the truncated normal form admits solutions homoclinic to 0, which bifurcate from 0 when $(T, \gamma) \to (T_0, \gamma_0)$. By assigning nonzero (and small enough) values to the first integrals $|C_i|^2$, one obtains also solutions homoclinic to periodic orbits and 2−tori. Close to Γ_{2k+1}, homoclinic solutions to 0 correspond to approximate solutions of (6.3) in the form of pulsating solitary waves (close to Γ_{2k}, we have bifurcations of solitary waves). Solutions homoclinic to periodic orbits and 2-tori correspond to nanopterons.

We now examine the problem of *persistence* of the homoclinics as higher order terms are taken into account in the normal form, which provides exact solutions of (6.3)–(6.4) when this property is satisfied. The case of solitary waves (with an oscillatory tail) bifurcating near Γ_{2k} has been addressed by Iooss and Kirchgässner. For pulsating solitary waves bifurcating near Γ_{2k+1}, one obtains a persistence result in the case when V is even (the general case is still open). This result is summarized below.

We fix $(T_0, \gamma_0) \in \Delta_0 \bigcap \Gamma_{2k+1}$ such that $s_2(\gamma_0, T_0) < 0$ (for the following persistence results to apply, one has to avoid in addition certain resonant cases corresponding to exceptional values of the parameters [55]). We consider $(\gamma, T) \approx (\gamma_0, T_0)$, (T, γ) being chosen under the curve Δ in the parameter plane (see Fig. 6.4). If V is even, the full normal form (6.40) inherits an invariance under $-I$, hence one has also the invariance under $-\sigma$. Consequently, one can search for solutions on the invariant subspace $\mathrm{Fix}(-\sigma)$ where $C_2 = 0$. Corresponding solutions of (6.27) satisfy $w_2 = -w_1$, which gives solutions of (6.3)–(6.4) having the property

$$x_{n+1}(t) = -x_n\left(t - \frac{T}{2}\right). \tag{6.60}$$

These solutions satisfy the simpler scalar (6.29).

Restricting to the invariant subspace $C_2 = 0$, there remains only one pair of simple purely imaginary eigenvalues (close to $\pm iq_1$) in addition to weakly hyperbolic ones. This situation is denoted as reversible $(iq_0)^2 iq_1$ (near-) resonance and has been treated in [55]. On $\mathrm{Fix}(-\sigma)$, small amplitude reversible solutions under $\mathcal{R}$ or $-\mathcal{R}$ homoclinic to a periodic orbit persist *for the full normal form*, above a critical size of the limiting periodic orbit [55]. This minimal size is $\mathrm{O}[\exp(-c/\mu^{1/2})]$, $c > 0$. There exist four families of reversible solutions under $\mathcal{R}$ of this type, since each component (A, B) and C_1 can be changed into its opposite (the same holds for reversible solutions under $-\mathcal{R}$). Fixing $(\gamma, T) \approx (\gamma_0, T_0)$, these reversible solutions appear in one-parameter families, parametrized by the amplitude of the limiting periodic orbit. On the contrary, reversible solutions homoclinic to 0 do not *generically* persist for the full normal form [55]. However their persistence is a codimension-1 phenomenon, hence such solutions may exist if parameters (T, γ) are chosen on some isolated curves in the parameter plane.

We sum up the above results in the following theorem [20].

Theorem 6.5. *We fix* $(T_0, \gamma_0) \in \Delta_0 \bigcap \Gamma_{2k+1}$ *such that* $s_2(\gamma_0, T_0) < 0$. *We consider parameter values* $(T, \gamma) \approx (T_0, \gamma_0)$ *below the curve* Δ_0 *in the parameter plane (Fig. 6.4). Then the principal part of the normal form (6.40) admits small amplitude reversible solutions (under* $\mathcal{R}$ *or* $\mathcal{R}\sigma$*) homoclinic to* $2-$*tori.*

If V *is even, the full normal form (6.40) admits* $C_2 = 0$ *as an invariant subspace. On this subspace (except for* (T_0, γ_0) *on a subset of* $\Delta_0 \bigcap \Gamma_{2k+1}$ *with zero Lebesgue measure corresponding to resonant cases), the full normal form (6.40) admits small amplitude reversible solutions (under* $\pm\mathcal{R}$*), homoclinic to periodic orbits. These solutions correspond to exact pulsating solitary waves for system (6.3), superposed at infinity on periodic oscillations, and such that* $x_{n+1}(t) = -x_n(t - \frac{T}{2})$. *For a given value of* (γ, T)*, these reversible solutions exist in one-parameter families, parametrized by the amplitude of oscillations at infinity. The lower bound of this amplitude is* $\mathrm{O}[\exp(-c/\mu^{1/2})]$, *where* $\mu = |T - T_0| + |\gamma - \gamma_0|$, $c > 0$.

Solutions of Theorem 6.5 correspond to fixing $p = 2$ in (6.4). The case $p > 2$ is examined in reference [21]. The simplest bifurcation yields a normal form having a similar structure, with p pairs of simple purely imaginary eigenvalues in addition to the bifurcating pair $\pm iq_0$. Reversible homoclinic orbits to p-tori exist for the *truncated* normal form, but their persistence for the full equation remains a nontrivial open problem. Consequently, only approximate solutions of (6.3)–(6.4) (corresponding to nanopterons) are obtained in that case.

There remains to study the sign of s_2 as parameters (γ, T, coefficients of V) are varied. This coefficient determines (for $s_2 < 0$) the existence of homoclinic solutions for parameter values close to the curves Γ_m. Here we only describe the even potential case (hence $\alpha = 0$), and refer the reader to [20] for the general case. Figure 6.8 provides a summary of the situation. If $\beta > 0$ (hard potential), homoclinic bifurcations occur on the left side of curves Γ_m, and on the right side if $\beta < 0$ (soft potential). This comes from the fact that the multiplicative factor $(2 - \frac{q_0}{\tan(q_0/2)})$ in (6.43) changes its sign at the cusp point on Γ_m.

Numerical Computation of Waves in the High-amplitude Regime

In this section, we numerically solve system (6.27) beyond the small amplitude regime. Results are taken from reference [24].

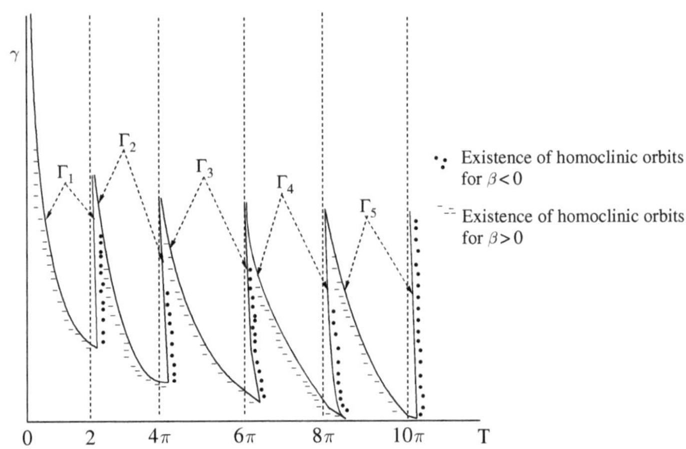

Fig. 6.8. This figure treats the case of an even potential V. In the normal form (6.40), coefficient s_2 (determined by expression (6.43)) is negative for $\beta > 0$ with (T_0, γ_0) lying on the left side of a curve Γ_m, and for $\beta < 0$ with (T_0, γ_0) lying on the right side. For $s_2 < 0$ and parameter values $(T, \gamma) \approx (T_0, \gamma_0)$ below Γ_m (*dashed regions* for $\beta > 0$, *dotted ones* for $\beta < 0$), the principal part of the normal form (6.40) admits homoclinic orbits to 0, and families of solutions homoclinic to 2-tori or periodic orbits. Families of reversible solutions homoclinic to certain periodic orbits persist for the full normal form, as stated by Theorem 6.5

We consider periodic boundary conditions $w_i(\xi + M) = w_i(\xi)$. When M is large, we end up with good approximations of spatially localized solutions, whose period is "infinite". We use a finite-difference scheme and solve the resulting non linear algebraic system of equations by an hybrid Powell method [56]. The convergence of this scheme needs a good initial guess. We compute a family of solutions depending on T (γ being fixed) by continuation from a critical value T_0 corresponding to our local analysis. The center manifold reduction provides a leading order approximation of bifurcating homoclinics, which is used as an initial guess. We refer the reader to [24] for further details on this numerical method. Another technique which can be used for the computation of solitary waves exploits a modulational instability [24, 57].

To solve the advance-delay problem, we fix $(T_0, \gamma_0) \in \Gamma_1$ and we vary the parameter $\mu = |T - T_0|$, $\gamma = \gamma_0$ being fixed. We numerically solve (6.27) and (6.29) by continuing the pulsating solitary waves (initally computed for $\mu \approx 0$) in the high amplitude regime.

We first consider the case of an even potential $V(x) = \frac{1}{2}x^2 + \frac{\beta}{4}x^4$ and compute solutions of (6.29). Figure 6.9 presents some solution profiles. The left column refers to a hard potential $V(x) = \frac{1}{2}x^2 + \frac{1}{24}x^4$, and the right one a soft one $V(x) = \frac{1}{2}x^2 - \frac{1}{4}x^4$. One can observe the difference between the oscillation frequencies of the central part of the solutions (close to q_0 for small amplitudes) and the frequency of the tail (close to q_1). The hardness of the potential determines which one of the two frequencies is the highest. Besides, we numerically matched both sides of the solution tail to a periodic solutions

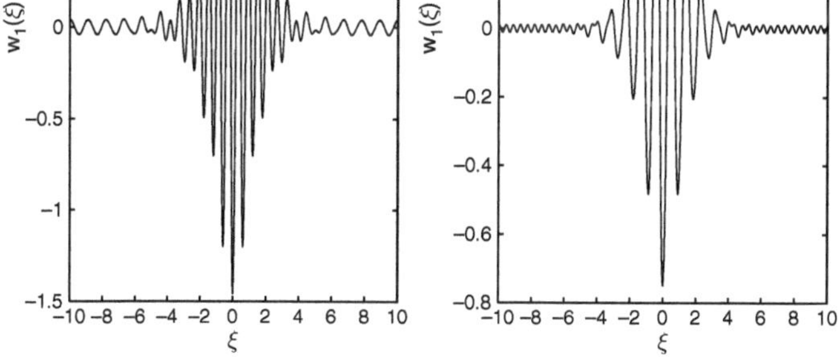

Fig. 6.9. Examples of localized solutions of (6.29) for potentials $V(x) = \frac{1}{2}x^2 + \frac{\beta}{4}x^4$. *Left figure:* $\beta = 1/6$, $\gamma = \gamma_0 \approx 0.83$, $T_0 \approx 5.59$ and $T = 5.5$. *Right figure:* $\beta = -1$, $\gamma = \gamma_0 \approx 0.9$, $T_0 \approx 6.63$ and $T = 7.4$

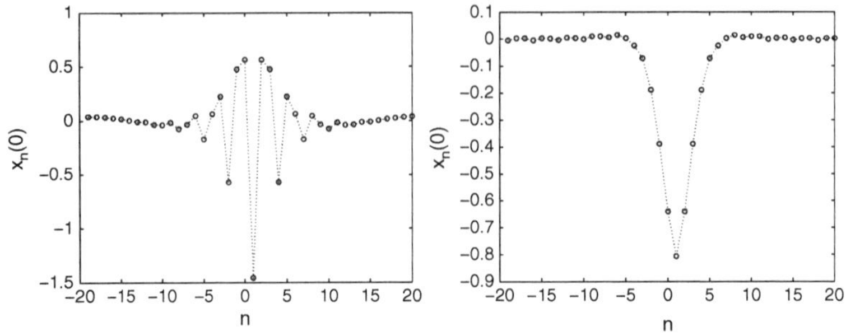

Fig. 6.10. Pulsating solitary wave solutions of (6.3)–(6.4) with $p = 2$, corresponding to profiles of Fig. 6.9. The solution is represented as a function of n, for $t = 0$. *Left figure:* $V(x) = \frac{1}{2}x^2 + \frac{1}{24}x^4$, $T = 5.5$, $\gamma = \gamma_0 \approx 0.83$, $T_0 \approx 5.59$. *Right figure:* $V(x) = \frac{1}{2}x^2 - \frac{1}{4}x^4$, $T = 7.4$, $\gamma = \gamma_0 \approx 0.9$, $T_0 \approx 6.63$

of (6.29) computed independently. When $\mu = |T - T_0|$ goes away from 0, the central hump of the solutions narrows. Its amplitude increases whereas the periodic tail becomes more visible. Figure 6.10 presents pulsating solitary wave solutions of (6.3)–(6.4) for $p = 2$, corresponding to profiles of Fig. 6.9.

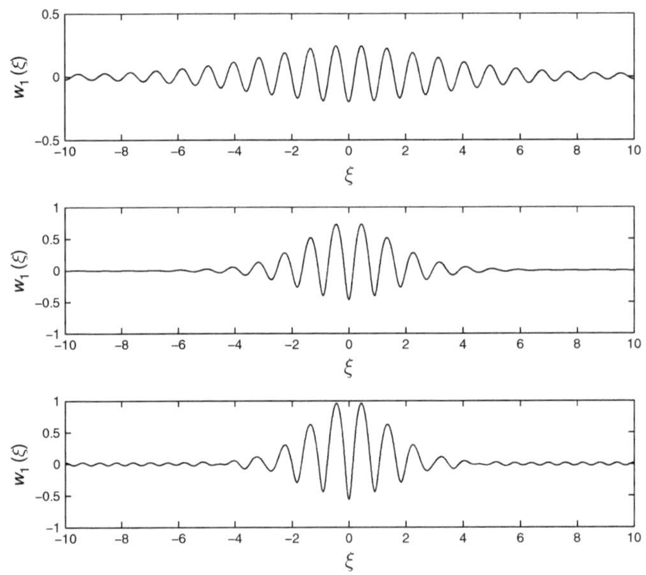

Fig. 6.11. Localized solutions of (6.27) for $V(x) = \frac{1}{2}(1 - e^{-x})^2$, $\gamma = \gamma_0 \approx 0.9$ ($(T_0, \gamma_0) \approx (6.63, 0.9) \in \Gamma_1$) and different values of T. We represent $w_1(\xi)$ (the shape of $w_2(\xi)$ is analoguous). From top to bottom we have $T = 6.7$, $T = 7.15$ and $T = 7.45$. (the value of $\mu = T - T_0$ increases from top to bottom)

The solution is represented as a function of n, for $t = 0$ (note the phase variation between nearest neighbors in the case when V is hard).

In order to illustrate the case when V is not even, we consider now the Morse potential $V(x) = \frac{1}{2}(1 - e^{-x})^2$ and solve system (6.27). We consider the point $(T_0, \gamma_0) \approx (6.63, 0.9)$ on Γ_1, fix $\gamma = \gamma_0$ and vary $\mu = T - T_0$. Figure 6.11 provides the component $w_1(\xi)$ of solutions for different values of T (the functions w_1 and $-w_2$ are different but have rather similar profiles).

For some potentials and parameter values γ, T, we find pulsating solitary waves for which an oscillating tail is not visible, at least at the scale of the central pulse. This is the case e.g. in Fig. 6.11 (middle) for the Morse potential with $T = 7.15$ and $\gamma \approx 0.9$, and in Fig. 6.1 for $V(x) = 1 - \cos(x)$ and $T = 8.1, \gamma \approx 0.9$. In the second case, however, a zoom on the solution tail reveals oscillations approximately 1000 times smaller than the central pulse amplitude.

References

1. E. Fermi, J. Pasta and S. Ulam, Technical Report LA-1940, Los Alamos National Laboratory (1955).
2. N.J. Zabusky and M.D. Kruskal, Phys. Rev. Lett. **15**, 240 (1965).
3. C.S. Gardner, J.M. Greene, M.D. Kruskal and R.M. Miura, Phys. Rev. Lett. **19**, 1095 (1967).
4. C.S. Gardner, J.M. Greene, M.D. Kruskal and R.M. Miura, Commun. Pure Appl. Math **27**, 97 (1974).
5. L.A. Kalyakin, Russian Math. Surveys **44**(1), 3 (1989).
6. G. Schneider and C.E. Wayne, in B. Fiedler, K. Gröger and J. Sprekels, eds, International Conference on Differential Equations *Appl.* 5(1), 69 (1998).
7. A. Ponno and D. Bambusi, Chaos **15**, 015107 (2005).
8. G. Friesecke and J. Wattis, Commun. Math. Phys. **161**, 391 (1994).
9. G. Friesecke and K. Matthies, Physica D **171**, 211 (2002).
10. D. Treschev, Disc. Cont. Dyn. Syst. A **11**, 867 (2004).
11. D. Smets and M. Willem, J. Funct. Anal. **149**, 266 (1997).
12. A. Pankov and K. Pflüger, Math. Meth. Appl. Sci. **23**, 1223 (2000).
13. G. Friesecke and R.L. Pego, Nonlinearity **12**, 1601 (1999).
14. G. Friesecke and R.L. Pego, Nonlinearity **15**, 1343 (2002).
15. G. Friesecke and R.L. Pego, Nonlinearity **17**, III:207, IV:229 (2004).
16. G. Iooss, Nonlinearity **13**, 849 (2000).
17. D.E. Pelinovsky and V.M. Rothos, Physica D **202**, 16 (2005).
18. G.Iooss and K.Kirchgässner, Com. Math. Phys. **211**, 439 (2000).
19. Y. Sire and G. James, C.R. Acad. Sci. Paris, **338**, Série I, 661 (2004).
20. G. James and Y. Sire, Commun. Math. Phys. **257**, 51 (2005).
21. Y. Sire, J. Dyn. Diff. Eqs. 17, 4 (2005).
22. J.P. Boyd, Nonlinearity **3**, 177 (1990).
23. J. Giannoulis and A. Mielke, Nonlinearity **17**, 551 (2004).
24. Y. Sire and G. James, Physica D **204**, 15 (2005).
25. S. Aubry and T. Crétégny, Physica D **119**, 34 (1998).

26. A.V. Savin, Y. Zolotaryuk, and J.C. Eilbeck, Physica D **138**, 267 (2000).
27. G. Iooss and G. James. Chaos **15**, 015113 (2005).
28. G. James, J. Nonlinear Sci. **13**(1), 27 (2003).
29. G. James, C.R. Acad. Sci. Paris **332**, Série I, 581 (2001).
30. G. Arioli and A. Szulkin, Ann. Sci. Math. Québec **22**, 97 (1998).
31. S. Aubry, G. Kopidakis and V. Kadelburg, Disc. Cont. Dyn. Syst. B **1**, 271 (2001).
32. G. James and P. Noble, Physica D **196**, 124 (2004).
33. P. Noble, Nonlinearity **17**, 1 (2004).
34. B. Sánchez-Rey, G. James, J. Cuevas and J.F.R. Archilla, Phys. Rev. B **70**, 014301 (2004).
35. A. Mielke, Math. Meth. Appl. Sci. **10**, 51 (1988).
36. A. Vanderbauwhede and G. Iooss, Dynamics Reported **1**, New Series, C. Jones, U. Kirchgraber and H. Walther eds, Springer Verlag, 125 (1992).
37. J. Guckenheimer and P. Holmes, *Nonlinear oscillations, dynamical systems and bifurcations of vector fields*, Springer, NY, 1983.
38. G. Iooss, *Bifurcation of maps and applications*, Math. Studies **36**, Elsevier-North-Holland, Amsterdam, 1979.
39. J. Marsden and M. McCracken, *The Hopf Bifurcation and its Applications*, Springer Verlag, NY, 1976.
40. X. Cabré, E. Fontich and R. de la Llave, Indiana Univ. Math. J. **52**, I:283, II:329 (2003).
41. X. Cabré, E. Fontich and R. de la Llave, J. Diff. Eqs., 218, 2 (2005).
42. K. Kirchgässner, J. Diff. Eqs. **45**, 113 (1982).
43. S. Flach and C.R. Willis, Phys. Rep. **295**, 181 (1998).
44. T. Kato, Perturbation theory for linear operators, Springer Verlag (1966).
45. G. Iooss and M. Adelmeyer, *Topics in bifurcation theory and applications*, Adv. Series in Nonlinear Dyn. **3**, World Sci. (1992).
46. A. Vanderbauwhede, Dynamics Reported **2**, in U. Kirchgraber and H.O. Walther, eds, John Wiley and Sons Ltd and B.G. Teubner, 89 (1989).
47. A. Mielke, Math. Ann. **277**, 121 (1987).
48. D.K. Arrowsmith and C.M. Place, *An introduction to dynamical systems*, Cambridge University Press, 1990.
49. A. Alvarez, J.F.R. Archilla, J. Cuevas and F.R. Romero, New Journal of Physics **4**, 72 (2002).
50. A. Tsurui, Prog. Theor. Phys. **48**(4), 1196 (1972).
51. S. Flach: Physica D **91**, 223 (1996).
52. S.R. Bickham, S.A. Kiselev and A.J. Sievers, Phys. Rev. B **47**, 14206 (1993).
53. S. Flach and A. Gorbach, Chaos **15**, 015112 (2005).
54. G. Iooss and M-C Pérouème, J. Diff. Eqs. **102**, 62 (1993).
55. E. Lombardi, *Oscillatory integrals and phenomena beyond all algebraic orders with applications to homoclinic orbits in reversible systems*, Lecture Notes in Mathematics **1741**, Springer-Verlag (2000).
56. M.J.D. Powell, *numerical methods for nonlinear algebraic equations*, Gordon and Breach (1970).
57. H. Feddersen, M. Remoissenet and M. Peyrard, eds, *nonlinear coherent structures in physics and biology*, Lecture Notes in Physics **393**, 159, Springer-Verlag, 1991.

7

Numerical Methods and Results in the FPU Problem

Simone Paleari[1] and Tiziano Penati[2]

[1] Dipartimento di Matematica e Applicazioni
simone.paleari@unimi.it
[2] Università degli Studi di Milano–Bicocca Dipartimento di Matematica e Applicazioni
tiziano.penati@unimi.it

7.1 Introduction

In his last work, Fermi [18], in collaboration with Pasta andUlam, studied a linear chain of equal masses connected with nonlinear springs (see Sect. 7.2.1 for details), performing one of the first numerical works ever made. Quoting from the original paper: "The ergodic behaviour of such systems was studied with the primary aim of establishing, experimentally, the rate of approach to equipartition of energy among the various degrees of freedom of the system."

They were interested in looking how long does it take for the following limit to be actually reached:

$$\overline{E}_j = \frac{1}{T}\int_0^T E_j(t)\mathrm{d}t \longrightarrow \frac{E}{N}$$

where E_j is the harmonic energy of the normal modes (see (7.3) in Sect. 7.2.1) and E is the total energy, starting with an initial condition such that, for example, $E_1 = E$.

Surprisingly, in their experiments they could not see equipartition at all. Quoting again from their original words: "Let us say here that the results of our computations show features which were, from the beginning, surprising to us. Instead of a gradual, continuous flow of energy from the first mode to higher modes, all of the problems show an entirely different behaviour. (...) Instead of a gradual increase of all the higher modes, the energy is exchanged, essentially, among only a certain few. It is, therefore, very hard to observe the rate of 'thermalization' or mixing in our problem, and this was the initial purpose of the calculation."

Their paper has been a milestone in the history of dynamical systems, and constituted the beginning of a long challenge still open nowadays.

In fact their numerical experiments gave origin to a considerable amount of work, both from the analytical point of view and from the numerical one. We

S. Paleari and T. Penati: *Numerical Methods and Results in the FPU Problem*, Lect. Notes Phys. **728**, 239–282 (2008)
DOI 10.1007/978-3-540-72995-2_7

have to say that, in our opinion, despite many important advances, the problem deserves further work. For sure it is now clear what kind of phenomena were underlying the results of the actual calculations performed by Fermi and his coworkers, but their original question has not received a complete answer yet: We still miss a satisfactory picture of the dynamics when the number N of degrees of freedom goes to infinity with finite specific energy, i.e., in the thermodynamic limit.

And since this is indeed a particularly difficult issue from the viewpoint of rigorous results, the numerical investigations, started with the work of Fermi, Pasta and Ulam, have also the fundamental role of inspiring the directions one should pursue with the analytical techniques.

In this chapter we concentrate our attention on the point of view of the numerical investigation of the dynamics of the FPU chain: it is indeed true that the FPU model has represented a formidable source of problems in that field, and many numerical methods, techniques, indicators has been used and tested with it.

We will thus take some care about the implementation, reliability and concrete use of some numerical tools; in particular we will consider the problem of the numerical integration of the equations of motion, which is the basic step for every investigation, and then we will analyse the use of Lyapunov exponents, spectral entropies, Poincaré sections and other suitably developed indicators, showing the results one can derive from them.

The picture emerging from the analysis of the simulations supports the idea of a meta-stability scenario, according to which, when the specific energy tends to zero, the phase space undergoes a sort of phase transition from a "liquid" phase to a "solid" one: several regions appear where the orbits remain trapped for very long times, and the diffusion process possibly leading to an ergodic behaviour becomes extremely slow.

7.2 The Fermi-Pasta-Ulam Problem

7.2.1 The Model

We consider a chain of N particles with nearest neighbour interaction given by a potential V. The Hamiltonian of such a system is given by

$$H(x,y) = \sum_{j=0}^{N} \left[\frac{1}{2} y_j^2 + V\big(x_{j+1} - x_j\big) \right] , \tag{7.1}$$

where $x_1, \ldots, x_N$ are the displacements of the particles with respect to the equilibrium positions, and $y_1, \ldots, y_N$ are the corresponding momenta; it is possible to impose periodic or fixed boundary conditions, and we will always consider the latter case, i.e., $x_0 = x_{N+1} = 0$.

In this quite general context, the FPU α,β-model is given by the following choice of the potential:

$$V(s) = \frac{s^2}{2} + \alpha\frac{s^3}{3} + \beta\frac{s^4}{4} ; \tag{7.2}$$

as usual we will indicate as the α-model the case $\beta = 0$, and as the β-model the case $\alpha = 0$.

As it is well known it is possible to introduce the normal modes by

$$x_j = \sqrt{\frac{2}{N+1}}\sum_{k=1}^{N} q_k \sin\frac{jk\pi}{N+1} , \quad y_j = \sqrt{\frac{2}{N+1}}\sum_{k=1}^{N} p_k \sin\frac{jk\pi}{N+1} ,$$

(q_k, p_k) being the new coordinates and momenta. The quadratic part of the Hamiltonian in the normal coordinates is given the form

$$H_2 = \sum_{j=1}^{N} E_j , \quad E_j = \frac{1}{2}(p_j^2 + \omega_j^2 q_j^2) , \tag{7.3}$$

where E_j is the harmonic energies and ω_j the harmonic frequencies

$$\omega_j = 2 \sin\left(\frac{j\pi}{2(N+1)}\right) .$$

7.2.2 The First Experiments and Conjectures

In Fig. 7.1 we reproduce a simulation similar to those obtained by Fermi and collaborators. In a α,β chain of 31 particles, with the energy initially placed on the first harmonic mode, it is possible to see a sort of recurrent dynamic: at the end of the simulation shown, after a temporary exchange of energy with some nearby modes, the initial situation is almost completely recovered, with a kind of memory effect. Further investigation has shown that the sharing of energy, if it happens, takes a very long time that, at low energies, becomes unobservable even with the most powerful computers.

Before this experimental discovery, the general expectation about these phenomena was quite different, since the common belief was that a simple nonlinear perturbation of an integrable system is enough to destroy its structure and lead to ergodicity, and thus making sense to the application of Statistical Mechanics. In the attempt of understanding the mechanisms leading to these phenomena, also in order to check their possible persistence in the thermodynamic limit, different conjectures have been proposed.

Izrailev and Chirikov [32] suggested, in view of the KAM theory that appeared in those years, the existence of an energy threshold, below which the dynamics exhibits a recurrent behaviour due to the persistence of most of the tori of underlying integrable structure. The choice of the initial conditions in

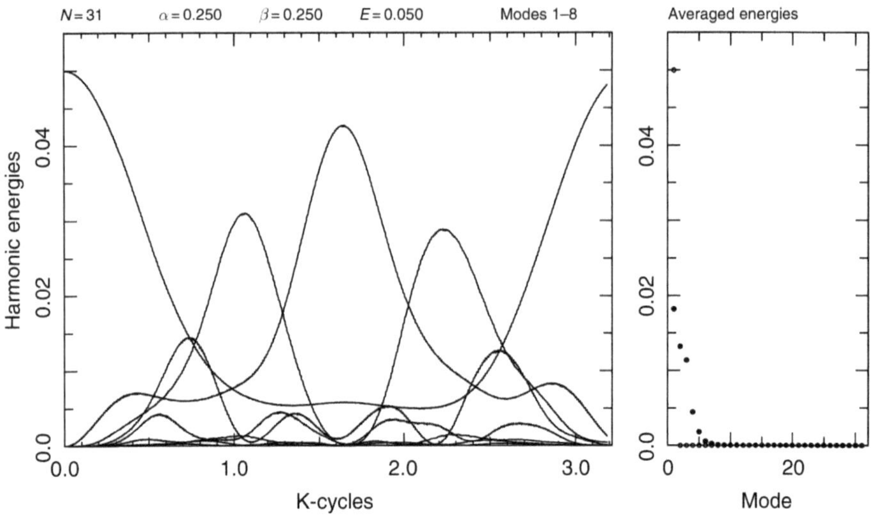

Fig. 7.1. FPU chain (parameters shown in the figure) with condition close to those used in the original experiments by Fermi, Pasta and Ulam. *Left panel*: time evolution of the harmonic energy of the first eight normal modes. It is possible to see a sort of quasi-periodic motion, with energy almost completely regained by the first mode at the end of the simulation. *Right panel*: initial conditions (*empty circles*) showing all energy given at time zero to the first mode, and averages over the whole simulation (*full circles*) showing the spread of energy limited to a few low-frequency modes. Time in natural units of the fastest oscillator (from [37])

the original experiments was simply unfortunate, since, according to them, above the threshold the system should be ergodic, with a rapid sharing of energy towards equipartition. Moreover, due to the mechanism of the resonance overlapping, when the number N of particles grows very large, such a threshold should vanish (as indeed happens for the available estimates of KAM theory), thus making the FPU phenomena non-relevant in the thermodynamic limit.

It is interesting to remark that the applicability of the KAM theory has been proved only recently by Rink [40], although he does not discuss the dependence of the threshold on N.

A different conjecture was proposed by Bocchieri et al. [9]. According to these authors the existence of a specific energy ($\epsilon = E/N$) threshold is the relevant fact: such a threshold should remain positive in the thermodynamic limit, giving important consequences about the foundations of Statistical Mechanics (see, e.g., [11, 13, 25, 26]).

This last approach could be recast, in a weakened form, in the framework of Nekhoroshev exponential stability, stating that, for very low specific energies, equipartition might be reached, but only in times that increase as $\exp(\epsilon^{-a})$, with some positive a, i.e. times which could become longer than the lifetime of physical system under analysis (see, e.g., [23, 24]).

7.2.3 The Meta-stability Scenario

As a further evolution of the original conjecture by Bocchieri et al. in the spirit of a suitable weak Nekhoroshev framework, the picture emerging from the more recent numerical results, is clearly characterized by the presence of two well distinct time scales in the FPU dynamics.

We will show with several experiments that, for specific energy small enough, a rather large class of initial data gives rise to a relatively rapid creation of what we call "natural packets", i.e. a cluster of, not necessarily consecutive, modes which share the most of the energy. The dynamics within the packet may well be chaotic, but it happens that it hardly exchanges energy with the rest of the chain. In fact we also show experimental evidence that these dynamical structures survive for exponentially long times, thus giving the second timescale. Only after these longer times one can observe a proper equipartition regime.

We stress that all these timescales appear to be with respect to the inverse of the specific energy, and that we also have some evidence of a possible persistence of these phenomena at the thermodynamic limit.

7.2.4 Further Comments

We conclude this section with some remarks of different nature.

To start with, it is worthwhile to stress that the main point of the FPU question brings intrinsically a great difficulty from the numerical point of view. In fact one has to deal with different infinite limits: ergodic properties and the equipartition condition are defined by time averages over the whole orbit, and the thermodynamic limit requires a very large number of particles to be reasonably estimated. These points are clearly critical in actual computations, and make several aspects of the investigation very hard and delicate at the same time.

Another comment is the following. The natural way to deal with these problems is the perturbative one. And the first choice among the integrable systems the FPU is close to is that of the uncoupled linear oscillators; this leads any numerical and analytical approach to the problem to be strongly based on the normal modes decomposition. Indeed the FPU question itself concerning equipartition is formulated in terms of harmonic energies. But there are other, in these cases nonlinear, integrable systems in a neighbourhood of the FPU model: the Toda chain, for example, or one could consider a resonant Birkhoff normal form of the FPU. We exploited this fact in some experiment, although in a non systematic way; we think that this point should deserve more attention.

As a last remark we mention that most of the studies performed up to now concerning the FPU problem were concentrated on the original one dimensional case. The natural further step is to consider two- or three-dimensional lattices in order to investigate the role of the dimensionality and to obtain more relevant models for Statistical Mechanics. Some preliminary explorations (see, e.g., [2]) on different types of two-dimensional lattices have been recently performed, but it is still not clear whether the FPU phenomena persists; we will not investigate this point.

7.3 Numerical Integration of Hamiltonian Systems

We begin our analysis of the numerical methods applied in the FPU problem with the first and fundamental one: the numerical integration of the equations of motion and its reliability.

The point is particularly relevant since we are dealing with a peculiar class of systems, the Hamiltonian ones, and since we are naturally lead, by the fundamental questions around the FPU problem, to follow the orbits for very long times.

We will thus introduce in this section the symplectic integrators, which are considered the most suitable ones for these purposes, and discuss their properties. Many parts of this section are widely and freely inspired by some unpublished lecture notes by Prof. Benettin and a book by Hairer et al. [31].

7.3.1 A Simple Symplectic Algorithm

To begin with, let us consider the following class of Hamiltonian functions

$$H(q,p) = T(p) + V(q) \ , \tag{7.4}$$

given by the sum of a first term depending only on the momenta and a second one depending only on the coordinates. If we take the usual kinetic term $T(p) = \frac{1}{2}p^2$, the equations of motion arising from 7.4 are $\ddot{q} = -V'(q) =: f(q)$, or as an Hamiltonian system

$$\begin{cases} \dot{q} = p \\ \dot{p} = f(q) \end{cases} . \tag{7.5}$$

In the comments that follow, for the sake of simplicity, we will consider only one degree of freedom systems.

Leap-Frog Algorithm

We illustrate here a very classical and simple numerical integrator, introduced for the first time, up to our knowledge, by Verlet [44]. In order to derive it one simply develops the solution $q(t)$ of the previous equations of motion around t in the points $t \pm \tau$, and using 7.5 one gets:

$$q(t+\tau) = q(t) + \tau p(t) + \frac{\tau^2}{2} f(q(t)) + \frac{\tau^3}{6} f'(q(t))p(t) + \mathrm{O}(\tau^4) ,$$
$$q(t-\tau) = q(t) - \tau p(t) + \frac{\tau^2}{2} f(q(t)) - \frac{\tau^3}{6} f'(q(t))p(t) + \mathrm{O}(\tau^4) ;$$

adding these two equations, and introducing $\Delta q(t) := q(t) - q(t-\tau)$, we get

$$\Delta q(t+\tau) = \Delta q(t) + \tau^2 f(q(t)) + \mathrm{O}(\tau^4) , \tag{7.6}$$
$$p(t) = \frac{1}{\tau}\Delta q(t) + \frac{\tau}{2} f(q(t)) + \mathrm{O}(\tau^2) . \tag{7.7}$$

Leap-Frog Step

Once the time has been discretized with step τ, and with the obvious notation $q_n = q(t_0 + n\tau)$, the algorithm is given by the following relations:

$$\begin{cases} \Delta q_{n+1} &= \Delta q_n + \tau^2 f(q_n) \\ q_{n+1} &= \Delta q_{n+1} + q_n \\ p_{n+1} &= \frac{1}{\tau}\Delta q_{n+1} + \frac{\tau}{2} f(q_{n+1}) \end{cases} . \tag{7.8}$$

The first remark is that it is necessary to start up the algorithm, giving an initial value for Δq by means of (7.7)

$$\Delta q_0 = \tau p_0 - \frac{\tau^2}{2} f(q_0) .$$

It is then worthwhile to remark that the first two relations of (7.8) concern only the configurations and can be carried on over the integration independently of the third one. It is thus possible to avoid the computation of the velocities, unless it is necessary for other reasons than the integration.

Symplecticity of the Algorithm

The *leap-frog* algorithm can be written as a map

$$\Psi_\tau : \begin{cases} q_{n+1} = q_n + \tau p_n + \frac{\tau^2}{2} f(q_n) \\ p_{n+1} = p_n + \frac{\tau}{2}[f(q_n) + f(q_{n+1})] \end{cases} , \tag{7.9}$$

which turns out to be symplectic, i.e. the *leap-frog* is a symplectic integrator; we will see the great importance of this property in Sect. 7.3.3. The proof of this fact is obtained showing, by easy calculations, that the symplectic 2-form $\omega = \mathrm{d}q \wedge \mathrm{d}p$ is preserved under the action of Ψ_τ

$$\mathrm{d}q_{n+1} \wedge \mathrm{d}p_{n+1} = \mathrm{d}\Psi_1(q_n, p_n) \wedge \mathrm{d}\Psi_2(q_n, p_n) = \mathrm{d}q_n \wedge \mathrm{d}p_n .$$

Local Error Estimate

In order to estimate the local error produced at every step, one can simply compare the map (7.9) with a Taylor expansion at time τ of the Hamiltonian flow

$$\Phi_H^\tau : \begin{cases} q_{n+1} = q_n + \tau p_n + \frac{\tau^2}{2} f(q_n) + \mathrm{O}(\tau^3) \\ p_{n+1} = p_n + \tau f(q_n) + \frac{\tau^2}{2} f'(q_n) p_n + \mathrm{O}(\tau^3) \end{cases} ;$$

we thus get

$$\|\Phi_H^\tau(q_n, p_n) - \Psi(q_n, p_n)\| \le \mathrm{O}(\tau^3) ,$$

and according to Definition 7.1 of Sect. 7.3.2, Ψ_τ turn out to represent an algorithm of order 2.

7.3.2 Symplectic Algorithms of Splitting Type

We will consider now a special class of symplectic algorithms, the *splitting* ones; and we will show how to build, starting from a symplectic-symmetric map, a new map which has a greater accuracy than the original one.

Let us start with some definitions (see, e.g. [31]):

Definition 7.1. *Consider a flow $\Phi^\tau(x)$ of a differential equation, and an integration algorithm represented by the map $\Psi_\tau(x)$; we say that Ψ_τ is of order k if there exists a function $\Delta(x)$ such that*

$$\Psi_\tau(x) = \Phi^\tau(x) + \tau^{k+1}\Delta(x) + \mathrm{O}(\tau^{k+2}) . \tag{7.10}$$

The map $\Psi_\tau^* := (\Psi_{-\tau})^{-1}$ is called adjoint map of Ψ_τ.

The following proposition gives the relation between Ψ_τ^* and the previous quantities $\Phi^\tau(x), \Delta(x)$:

Proposition 7.1. *If Ψ_τ is of order k, then*

$$\Psi_\tau^*(x) = \Phi^\tau(x) + (-1)^k\tau^{k+1}\Delta(x) + \mathrm{O}(\tau^{k+2}) . \tag{7.11}$$

Now we introduce the following

Definition 7.2. *An algorithm is called symmetric if $\Psi_\tau^* = \Psi_\tau$.*

The nice property of such algorithms is that their order is forced to be even; indeed, using (7.10) and (7.11) it follows

$$0 = \Psi_\tau(x) - \Psi_\tau^*(x) = \Delta(x)\tau^{k+1}[1 - (-1)^k] + \mathrm{O}(\tau^{k+2}) ;$$

so that $k = 2m$.

A Splitting Algorithm of Order 2

Consider again a Hamiltonian of the form (7.4), i.e. the sum of two terms one of which depending only on the coordinates, and the other one only on momenta. This is a particular and simple case of a system whose vector field can be splitted into two parts explicitly integrable. Denoting by Φ_T and Φ_V the flows of the two separated parts, consider the following map:

$$\Psi_\tau(x) = \left(\Phi_T^{\tau/2} \circ \Phi_V^{\tau} \circ \Phi_T^{\tau/2}\right)(x) .$$

Considering for simplicity the one degree of freedom case, with the usual kinetic term $T(p) = (1/2)p^2$, we can easily write the action of the map Ψ_τ, with the same notation of the previous section, as

$$\Psi_\tau : \begin{cases} q_{n+1} = q_n + \tau p_n + \frac{\tau^2}{2} f(q_{n+\frac{1}{2}}) \\ p_{n+1} = p_n + \tau f(q_{n+\frac{1}{2}}) \end{cases} . \tag{7.12}$$

This algorithm is easily seen to be symplectic, since it is obtained by means of Hamiltonian flows, and symmetric, since it is given by a symmetric composition of flows. It is a simple example of a splitting algorithm, and it turns out to be of order 2; indeed it is very similar to the classical leap-frog, which actually turns out to be $\Phi_V^{\frac{\tau}{2}} \circ \Phi_T^{\tau} \circ \Phi_V^{\frac{\tau}{2}}$.

Higher Order Splitting Algorithms

Now we will show how, starting from a low-order symplectic-symmetric algorithm, we can build up a higher order one, which still turns out to be symplectic and symmetric.

Proposition 7.2. *Let Ψ_τ be an algorithm of order k for the flow Φ, and α_1, α_2 two real numbers such that $\alpha_1 + \alpha_2 = 1$; then the map $\psi_\tau = \Psi_{\alpha_1\tau} \circ \Psi_{\alpha_2\tau}$ satisfies*

$$\psi_\tau(x) = \Phi^\tau(x) + (\alpha_1^{k+1} + \alpha_2^{k+1})\tau^{k+1}\Delta(x) + \mathrm{O}(\tau^{k+2}) .$$

This result is true also with more than two maps, for example:

$$\psi_\tau(x) = \left(\Psi_{\alpha_1\tau} \circ \Psi_{\alpha_2\tau} \circ \Psi_{\alpha_3\tau}\right)(x) , \tag{7.13}$$

where $\alpha_1 + \alpha_2 + \alpha_3 = 1$; we notice that, if and only if $k = 2m$, we can choose the three real numbers such that $\alpha_1^{k+1} + \alpha_2^{k+1} + \alpha_3^{k+1} = 0$. Thus ψ_τ has at least order $k+1$. In particular, if Ψ_τ is symmetric and $\alpha_1 = \alpha_3$, then ψ_τ is an even-order symmetric map again, so its order is at least $k+2$.

As a simple example one could take the map Ψ_τ given by (7.12); since it is of order 2, the conditions on α_i imply $\alpha_1 = \alpha_3 = 1/(2 - 2^{1/3})$ and $\alpha_2 = -2^{1/3}/(2 - 2^{1/3})$. With these choices, the map ψ_τ in (7.13) is a fourth order symplectic algorithm.

7.3.3 A Theorem by Benettin and Giorgilli

The integration algorithms described in the previous sections are quite simple (the leap-frog in particular) and efficient, with an accuracy which is always satisfactory, or even very good for the higher order versions. But the reason why they are so popular is much deeper, and it can be found, for example, in a result by Benettin–Giorgilli [6] (but see also [3, 30, 31]).

In fact, given a symplectic map, they prove the existence of an interpolating Hamiltonian, i.e. whose time one flow is close, in a suitable sense, to the given map. The relevance of this result for numerical integration is the following: if the given symplectic map is indeed a symplectic algorithm for a Hamiltonian H, then using such a map, one is following almost exactly the flow of the interpolating Hamiltonian, which in turn is close to H. As a further remark one has to say that the idea of looking for the interpolating Hamiltonian fits into the general procedure of the backward error analysis (see [31]) usually performed in numerical analysis.

To be more definite, consider a Hamiltonian function H, denote its flow at time τ as Φ_H^τ and take a symplectic integrator Ψ_τ of order s, i.e. such that

$$\|\Psi_\tau - \Phi_H^\tau\| = \mathrm{O}(\tau^{s+1}) \,. \tag{7.14}$$

The question is whether there exist a different Hamiltonian K_τ whose flow is exactly given by Ψ_τ.

Since τ, representing the time step of the integration, is usually quite small, all the maps involved are close to the identity. The idea is then to develop them in powers of τ, and to impose the relation $\Psi_\tau = \Phi_{K_\tau}^\tau$ at every order in τ, solving the system iteratively. In this procedure there are of course convergence problems, but one might hope in an asymptotic behaviour, so as to optimize the order of truncation and perhaps obtain a small remainder, in the spirit of Nekhoroshev estimates. And this is exactly what is possible to prove:

Theorem 7.1 (Benettin–Giorgilli [6]). *There exists τ^*, depending on the constants of analyticity of Ψ_τ, and K_τ such that $\forall \tau \le \tau^*$, one has*

$$\|\Psi_\tau - \Phi_{K_\tau}^\tau\| = \mathrm{O}\left(\tau \mathrm{e}^{-\tau^*/\tau}\right) \,. \tag{7.15}$$

From the numerical point of view, the previous statement tells us that, if the step is small enough, the approximation error is smaller than the roundoff error.

Moreover, using (7.14) and (7.15) one gets

$$\|\Phi_{K_\tau}^\tau - \Phi_H^\tau\| = \mathrm{O}(\tau^{s+1}) \,,$$

from which it is possible to deduce

$$K_\tau = H + \tau^s K' \,; \tag{7.16}$$

the last relation clearly illustrates the main characteristic of symplectic integrators: they conserve almost exactly the energy of a Hamiltonian which is τ^s close to the one we are numerically integrating.

A Simple Example

In order to give an idea of the actual effect of a symplectic integrator, we consider a simple concrete case, and we try to give at least the leading term of K' in (7.16). We consider for simplicity the splitting algorithm of order two introduced before [see formula 7.12] and we expand it with respect to powers of τ

$$\Psi_\tau = \psi_0 + \tau\psi_1 + \tau^2\psi_2 + \tau^3\psi_3 + \cdots ,$$

where ψ_0 is obviously the identity, and the first following terms are

$$\begin{aligned}\psi_1(q,p) &= (p, f(q)) ,\\ \psi_2(q,p) &= \frac{1}{2}\left(f(q), pf'(q)\right) ,\\ \psi_3(q,p) &= \frac{1}{2}\left(pf'(q), \frac{1}{2}p^2 f''(q)\right) .\end{aligned}$$

We will follow the initial steps of the algorithm described in [6] to compute the first correcting term, call it k_3, of the modified Hamiltonian K_τ. Working on the corresponding field F_τ, the easiest terms, in a power expansion in τ, are f_1 and f_2: in fact

$$f_1 = \psi_1 , \qquad f_2 = \psi_2 - \frac{1}{2}L_{\psi_1}\psi_1 ,$$

where L_G denotes the Lie derivative along G.

One can easily verify that $f_2 = 0$, as expected from the statement of the main theorem; thus, up to constants, $k_2 = 0$. Concerning the subsequent term

$$f_3 = \psi_3 - \frac{1}{12}L_{\psi_1}\psi_2 ,$$

we perform Lie derivatives

$$\left(L_{\psi_1}\psi_2\right)(q,p) = \left(L_{\psi_1}f(q), L_{\psi_1}pf'(q)\right) = \left(pf'(q), p^2f''(q) + f(q)f'(q)\right) ,$$

and then, adding ψ_3, we get

$$f_3 = \left(\frac{1}{12}pf'(q), -\frac{1}{24}p^2f''(q) - \frac{1}{6}f(q)f'(q)\right) ,$$

whose Hamiltonian is

$$k_3 = \frac{1}{24}p^2f'(q) + \frac{1}{12}f^2(q) .$$

The interpolating Hamilton function thus becomes

$$K_\tau = K + \tau^2 k_3 + \mathrm{O}(\tau^3) .$$

7.3.4 Stability of the Algorithm

Up to now we checked the conservation of energy during the numerical integration of our systems. To conclude our brief analysis of symplectic algorithms, we comment on the effect they may have on other aspects of the dynamics. In particular, considering a system with an elliptic equilibrium point at the origin, we concentrate on the persistence of its nature, and on the effects on its normal modes frequencies.

For this purpose we recall that we ended last paragraph with the computation of the first correcting term $\tau^2 k_3$ of the modified Hamiltonian K_τ in the case of the integrator given by the map (7.12). We want to stress that this term, as the others k_j, may be quadratic in the set of variables (q,p), thus modifying the quadratic part of K and the spectrum of frequencies of the small oscillations. To go on with the analysis we must specify the original Hamiltonian H we are studying: in order to have an elliptic equilibrium point at the origin we consider a set of oscillators. In a general nonlinear case it may be not an easy task to study the new frequencies, modified by the integrator, via the usual diagonalization of the quadratic part of the Hamiltonian. We thus further restrict to the simplest example, i.e. that of uncoupled oscillators. In this case the action of the algorithm in normal coordinates may be represented by the following block matrix

$$\Psi_\tau(q,p) = \begin{pmatrix} \mathrm{Id} - \frac{\tau^2}{2}\Omega_2 & \tau\mathrm{Id} - \frac{\tau^3}{4}\Omega_2 \\ -\tau\Omega_2 & \mathrm{Id} - \frac{\tau^2}{2}\Omega_2 \end{pmatrix} \cdot \begin{pmatrix} q \\ p \end{pmatrix} ,$$

where Ω_2 is the diagonal matrix of the ω_j^2. Reordering rows and columns with respect to the oscillator's coordinates $(q_1, p_1, \ldots, q_n, p_n)$ one gets the matrix in block diagonal form, with blocks A_j

$$A_j := \begin{pmatrix} 1 - \frac{\tau^2}{2}\omega_j^2 & \tau - \frac{\tau^3}{4}\omega_j^2 \\ -\tau\omega_j^2 & 1 - \frac{\tau^2}{2}\omega_j^2 \end{pmatrix} .$$

The stability of A_j is related to the signum of $\tau^2\omega_j^2 - 4$: if $|\tau\omega_j| \geq 2$ the corresponding dynamic becomes hyperbolic. The minimal requirement to preserve the qualitative behaviour of the system is to ask for $\tau \max_{j=1\ldots,n}\{\omega_j\} < 2$.

The next point we try to investigate, once we are sure that we maintain the nature of the equilibrium, is how the resonance relations are affected by the integrator. A complete analysis of these facts is beyond the scope of this paper, so we will limit ourselves to some considerations. Owing to the analogy between A_j and the exponential map $\mathrm{e}^{\Omega_2 \tau}$, we first define the rotation angles and the new frequencies

$$\theta_j := \arccos\left(1 - \frac{1}{2}\tau^2\omega_j^2\right) , \tag{7.17}$$

$$\tilde{\omega}_j := \frac{1}{\tau}\theta_j ; \tag{7.18}$$

remark that, due to the restriction $\tau \max_{j=1\dots,n}\{\omega_j\} < 2$, the definition of θ_j makes sense.

If $\tau\omega_j \ll 1$, then we easily get convinced that the angles θ_j are very close to the original values $\tau\omega_j$ and the old frequencies are a first order approximation of the new ones; indeed, from a Taylor expansion around zero, we have

$$\theta_j = \tau\omega_j + \frac{1}{6}\tau^3\omega_j^3 + \text{h.o.t.}\,, \tag{7.19}$$

$$\tilde{\omega}_j = \omega_j + \frac{1}{6}\tau^2\omega_j^3 + \text{h.o.t.}\,. \tag{7.20}$$

We consider now the low part of the FPU spectrum, since in many numerical experiments initial data are taken with all the energy in one or few low modes. It is worthwhile to remark that, for large N the low frequencies are almost resonant, being approximately integer multiples of the lowest one, as one can see by the following development

$$\omega_j \sim j\frac{\pi}{N+1} - \frac{1}{24}\left[\frac{j\pi}{N+1}\right]^3 \qquad 1 \le j \ll N\,. \tag{7.21}$$

These Taylor expansion shows the existence of almost resonances $|j\omega_1 - \omega_j| = \mathrm{O}(j^3/N^3)$, which may appear as small denominators in a suitable perturbation approach; the new frequencies $\tilde{\omega}_j$ still satisfy the same property

$$|j\tilde{\omega}_1 - \tilde{\omega}_j| = \frac{j(j^2-1)}{6}\left(\frac{\pi}{N+1}\right)^3\left(\frac{1}{4} - \tau^2\right) + \text{h.o.t.} = \mathrm{O}\left(\frac{j^3}{N^3}\right)\,,$$

due to the smallness of $\tau^2 \ll 1/4$.

As a final comment we could say that in dealing with numerical integration one has to be always very careful. The use of higher order symplectic integrators strongly improves the conservation of energy, and the natural consequence could be the choice of a larger integration step to decrease the computational times maintaining a reasonable accuracy. But, as we showed in this section, this operation might be dangerous, since it may change other important features of the system, for example producing relevant deformation of the spectrum that affect the resonance relations.

7.4 Natural Packets and Time Scales

In the spirit of the original paper, as we discussed in Sect. 7.2, the main point in the FPU problem is to estimate the rate of the relaxation to equipartition. From the point of view of numerical investigation, once one is confident with the simulations performed, as we discussed in Sect. 7.3, the issue is how to extract the useful informations, i.e. to find a reliable set of indicators which are able to detect the approach towards equipartition in a qualitative and/or quantitative way.

In this section we consider a sort of indicator, recently introduced by Giorgilli and collaborators in [7], and then further generalized in [38], by means of which one clearly recognizes the existence of two timescales and the meta-stability of clusters of normal modes in the FPU dynamics.

7.4.1 Ordered Modes Clusters

To introduce the main idea, the first remark is the following. If one aims at seeing the slower and slower rate of approach to equilibrium as the energy, or specific energy, goes to zero, one should be able to detect a non-complete sharing of the energy among the normal modes. In different words, it should be possible, for example, to consider a population of modes which does not exchange energy with the other ones. A first attempt of this kind has been done in [24], where a model with alternating masses was considered: it has been possible to show that the normal modes are split into two populations of acoustic and optical modes, with an extremely slow sharing of energy between the two sets. In the original FPU model it has never been possible to prescribe an analogous subdivision: the point enlightened in [7] is that such a task is actually performed by the dynamics itself according to the initial conditions considered, producing a sort of "natural" splitting, which depends on the region of the phase space considered.

This can be seen as follows. A naive, but quite effective, way to check the degree of equipartition reached by the system is to look directly at the distribution of energy among the modes, like in the right panel of Fig. 7.1, as time grows. For sufficiently low energy one clearly observes that the energy, initially given only to the first mode, spreads quite rapidly to a few low frequency modes, involving in sequence the first two modes, then the first three ones, then the first four ones, and so on until this process stops. Thus, according to a first timescale, the dynamic creates a "natural packet" of modes, and the system enters in a kind of meta-stable state, characterized by an extremely slow flow of energy towards the higher modes, until equipartition is possibly reached on a second and much longer time scale.

In order to put into evidence this phenomenon, and to give quantitative estimates one proceeds as follows: the natural quantities to consider are the time evolutions of packets of modes instead of that of the single modes. One thus introduces

$$E_s(t) := \sum_{j=1}^{s} \bar{E}_j(t) , \tag{7.22}$$

where $\bar{E}_j(t) := \frac{1}{t}\int_0^t E_j(s)\mathrm{d}s$ and E_j is defined in formula (7.3); giving initially all the energy to the first mode, every packet has at the beginning the whole energy, i.e., $E_s(0) = E$ for every s. To estimate the evolution towards equipartition of the system, the notion of a critical time t_s is introduced. It is defined as the first time at which the sth packet has lost a fixed fraction γ

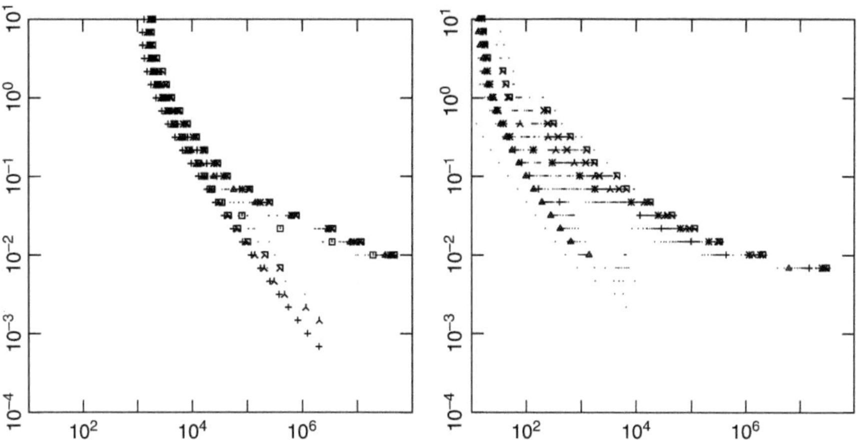

Fig. 7.2. Natural packet phenomenon. β-model, with $\beta = 1/10$, $N = 255$; the threshold γ is fixed to 0.1. *Abscissa*: critical time t_s for every packet (see footnote 1); ordinate: specific energy. *Left panel*: initial energy on mode 1. *Right panel*: initial energy equally distributed on a packet of modes from 4 to 20 (from [38])

of the energy Δ_s it has to lose to reach equipartition. In more precise terms, since at equipartition, say when $t \to \infty$, $E_s(\infty) = s\frac{E}{N}$, one has

$$t_s := \min\left\{t : \frac{E_s(t)}{E_N(t)} = \frac{E_s(0) - \gamma\Delta_s}{E_N(t)} = 1 - \left(1 - \frac{s}{N}\right)\gamma\right\} \tag{7.23}$$

For every s, and for every choice of the threshold γ, an "existence" time for the s-th packet is thus defined. Fig. 7.2 shows the critical times of the various packets[1] plotted against the fixed values of the specific energy. The two different timescales we previously introduced are clearly recognizable. The collection of points clustered in a straight line, i.e. giving a power-law dependence of the time with respect to the specific energy, represents the relatively rapid formation of the *natural packet*; the other points form a sort of second branch which behaves according to a longer timescale and which represents the destruction of the *natural packet* and the sharing of energy among all the modes up to the reaching of equipartition. The left panel refers to the initial energy being given to the first mode, while the right one refers to the case of initial energy equally distributed on a packet of low-frequency modes. Despite some differences in the actual positions of the points in the graphs, the qualitative structure with the two timescales is common to the two different initial conditions. We may also remark that for the lowest values of the energy there are no points on plot: this means that the *natural packet* is

[1] In order to avoid overcrowded pictures, here as well as in Figs. 7.4 and 7.5, for most of the packets only a single point is plotted. Symbols are used for packets $1, 3, 7, 15, \ldots, \frac{N+1}{4} - 1$ and for other 4 packets with index between $\frac{N+1}{4} - 1$ and N.

composed only by the modes initially excited, i.e. the dynamic appears to be highly frozen; in particular, for the left panel it means that the first mode does not lose more than 10% of the energy it has to lose to reach equipartition, up to the time of the calculation, 10^8. We further stress that the qualitative aspects of this phenomenon are clearly present in all our experiments, independent of the choice of the threshold γ, number of particle N, or model (α, β model, α model or β model).

In [7] a careful analysis of the packet formation has been performed, for example revealing that the *natural packet*, in the case of initial energy on the single first mode, extends up to a certain frequency which scale with specific energy as $\omega \sim \epsilon^{1/4}$, with a dependence on N only through $\epsilon = \frac{E}{N}$ in the range 7–1 023 (see [7], Fig. 7.7).

For what concerns the second branch, these pictures strongly suggest a timescale longer than as a power law with the inverse of specific energy. It can be stated that the time represented by the second branch should be a lower bound for the equipartition time. A precise estimate is illustrated in Sect. 7.5 by means of the spectral entropy indicator, with a numerical evidence of an exponential scaling.

7.4.2 Reordered Modes Clusters

We should remark that the previous definition (7.22) of packets is strictly related with the peculiar evolution one observes as the energy is initially given to the first (few) mode(s): actually a spread, which is not merely a transfer, to the successive modes.

It happens that, when the energy is initially placed in a different part of the spectrum, its sharing is no more directed towards neighbouring modes in a prescribed direction. For example it is not true that starting with the last mode excited, the flow of energy has a symmetric behaviour with respect to the previous case. Indeed, as it is shown in the right panel of Fig. 7.3 for the α, β chain with $N = 511$ and initial energy on the mode 383, the first sharing of energy happens within small clusters of modes in several parts of the spectrum. Many other combinations are possible with different choices of initially excited modes, and the produced phenomena are very rich.

In order to obtain pictures like those shown in Sect. 7.4.1, one clearly has to change some definitions. The idea is then not to impose a prescribed ordering, but to leave also such a task to the dynamic itself. At every time, the s-th packet is composed by the s most energetic modes. With such a rearrangement it is possible to produce again pictures like those of Fig. 7.2: the qualitative aspects of the natural packet phenomenon seem to persist in many different cases, with initial energy concentrated in one (or few consecutive) mode(s) placed in different parts of the spectrum.

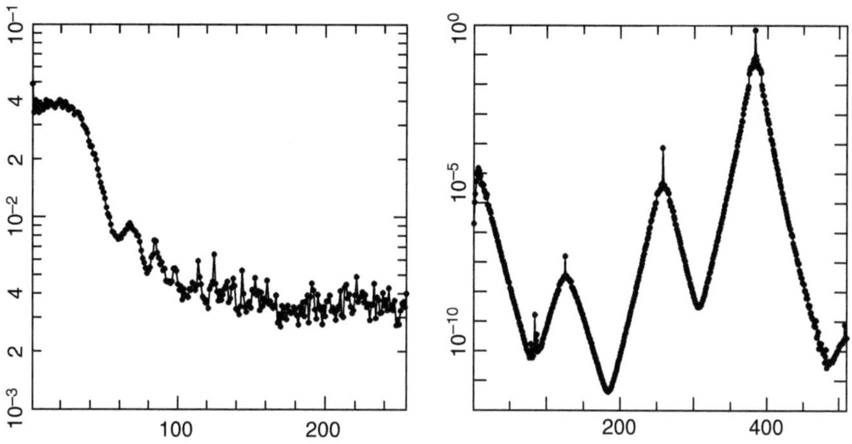

Fig. 7.3. Time averaged distributions of harmonic energies, FPU α,β-model ($\alpha = \beta = 1/4$), up to time $t = 10^6$. *Left panel*: $N = 255$, $\epsilon = 0.01$, energy initially given to the first mode. *Right panel*: $N = 511$, $\epsilon = 0.003$, energy initially given to mode 383. In the left panel it is clear the structure of a packet of consecutive low-frequency modes sharing most of the energy. In the right panel the energy is spread in several clusters of modes placed in different parts of the spectrum, thus justifying the idea of reordered packets

High-Frequency Initial Data

In Fig. 7.4 we show the results of these calculations for $N = 127$, with initial energy in the mode 127, left panel, and in a packet of modes from 120 to 127, right panel. We can see the two different branches, the one concerned with the formation of a packet involving a certain number of modes, again with a power law scaling with the specific energy, and the second one related to the further sharing with all the modes. Thus it is natural, also for these types of initial conditions, to perform some more precise estimates of the second time scale, as we illustrate in Sect. 7.5.

Other Initial Data

Figure 7.5 shows again these phenomena with a different number of particles ($N = 511$) and initial energy concentrated on mode 511, left panel, and on mode 383, right panel. Despite some minor differences, like the actual time at which we see the branching between the different time scales, the qualitative aspects of the natural packet phenomena seem to persist.

Concerning the initial data, it is possible to say that the figure in the right panel represents a sort of intermediate situation between those of Figs. 7.2

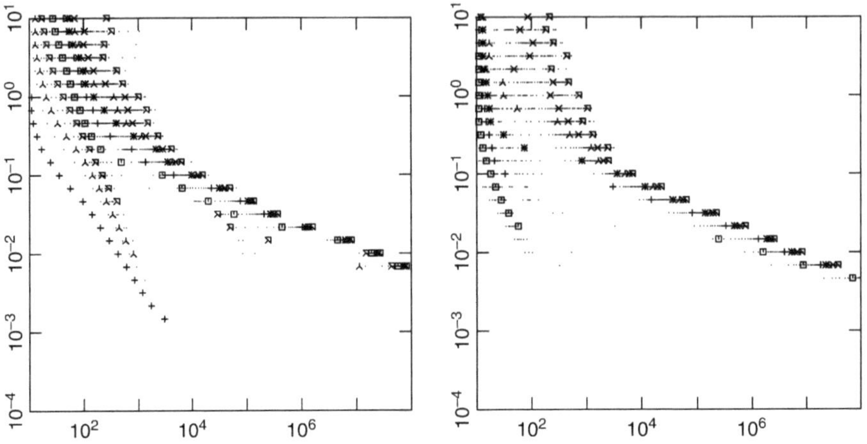

Fig. 7.4. Natural packet phenomenon. α, β-model, with $\alpha = \beta = 1/4$, $N = 127$; the threshold γ is fixed to 0.1. *Abscissa*: critical time t_s for every packet (see footnote 1); ordinate: specific energy. *Left panel*: initial energy on mode 127. *Right panel*: initial energy equally distributed on a packet of modes from 120 to 127 (from [38])

and 7.4. In the figure in the left panel it is possible to see a sort of intermediate branch: one could think of a first meta-stable state, related to the natural packet, followed by a second one which involves more modes, whose destruction eventually leads to equipartition.

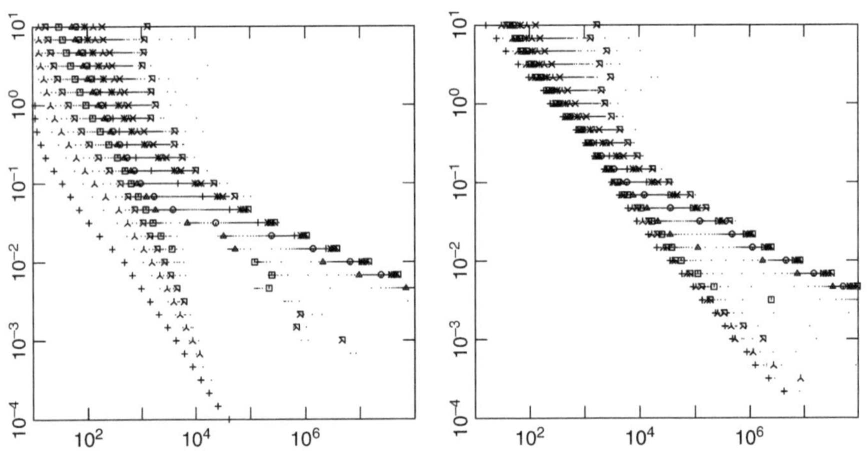

Fig. 7.5. Natural packet phenomenon. α, β-model, with $\alpha = \beta = 1/4$, $N = 511$; the threshold γ is fixed to 0.1. *Abscissa*: critical time t_s for every packet (see footnote 1); ordinate: specific energy. *Left panel*: initial energy on mode 511. *Right panel*: initial energy on mode 383 (from [38])

7.4.3 Discussion

Some remarks are in order. From the numerical point of view, an important aspect is that the calculation of the relevant data do not constitute a great computational overhead; indeed, like in the case of the spectral entropy (see Sect. 7.5), besides the numerical integration of the orbits we are interested in, one has to perform the usual Fourier transform (by means of FFT algorithm) to get the normal modes and very few further computations. The situation is quite different, for example from the case when one uses the Lyapunov exponents (see Sect. 7.6).

The observation that the dynamics creates packets of modes is probably already present in other previous papers (see, e.g., [15]), but up to our knowledge it has never been revealed before in a so clear and effective way. It gives both qualitative pictures and quantitative estimates of the dynamics, putting into evidence the existence of different timescales, and the meta-stability of the natural packets. Moreover the so obtained splitting of the system in slowly interacting part is not a priori prescribed, but emerges naturally from the evolution of the phase space region chosen for the initial conditions; the metastability could be explained, as we do in [27], in terms of resonant states: indeed for low modes, the frequencies involved in the packet satisfy the almost resonances considered in (7.21).

This indicator, initially conceived for initial conditions involving the lowest part of the spectrum, appeared robust enough to be generalized to cover other cases, with energy given to modes of different frequencies. This is particularly relevant, since the existence of the natural packet phenomenon with different classes of initial data confirms that these mechanisms are not restricted only to particular regions of the phase space. We will stress again this point discussing the results of Sect. 7.5.

A further comment, again related to the relevance of these phenomena for Statistical Mechanics, comes from the indication that some quantitative aspects of the natural packets as well as their existence and meta-stability appear to be independent from N, suggesting their possible persistence in the thermodynamic limit (concerning this point see also Sect. 7.5, and Fig. 7.6 in particular).

As a final remark concerning the interplay between analytical and numerical results, and the inspiring role of the latter, we should mention that, a first rigorous result for the natural packet formation is now available [1]: the statement says that, for an α-model with periodic boundary conditions, with low-frequency initial excitation, for sufficiently large N and sufficiently small ϵ, the dynamics remains close, for timesscaling as a suitable inverse power of ϵ, to a suitable solution of a pair of Korteveg–deVries equations, which constitute a resonant normal form of the system; in terms of normal modes, such a solution appears as a natural packet.

7.5 Spectral Entropy

In this section we will describe this now by classical indicator and some of the results we obtained by means of it.

It has been used in many papers (see, e.g., [8, 17, 33, 34, 38, 43]), and many conclusions have been drawn thanks to it, but we defer from other contributions in this volume for more details on the literature.

7.5.1 The Numerical Indicator

This indicator too is based, like the one described in Sect. 7.4 and many others, on the normal modes structure of the linear part of the system. From the point of view of its numerical calculation, apart from the Fourier transform needed by the normal modes conversion, it is absolutely not expensive.

The definition is the following:

$$S := -\sum_{j=1}^{N} e_j \ln e_j \qquad e_j := \frac{E_j}{\sum_k E_k} ,$$

where E_j are the usual harmonic energies defined in (7.3); the formulas justify the name of spectral entropy for S, which is a quantity ranging from 0, when all the harmonic energy is confined in a single oscillator, and $\ln N$, in exact equipartition. In order to have a normalized indicator, the following quantity is introduced:

$$n_{\text{eff}} := \frac{e^S}{N} ,$$

which varies in $[1/N, 1]$ and can be interpreted as an effective fraction of modes involved in the dynamic. It is possible to consider an analogous quantity, see, for example, [43], defined as an effective fraction of oscillators involved in the dynamic, but we will concentrate on n_{eff}.

S, and thus n_{eff} too, are defined on the phase space, and so it is natural to evaluate them on the flow, to check the "level of equipartition" of the orbit during its evolution. Starting with one or a few modes excited, it is possible to see the indicator growing from its initial value close to zero to a saturating value corresponding to equipartition, if the energy given is sufficiently high and the time evolution computed sufficiently long.

In principle one could try to detect, in the time course of n_{eff}, the formation and the meta-stability of the natural packets introduced and described in Sect. 7.4: even though it is sometime possible a posteriori to recognize them, it turns out that this indicator is too rough for this purpose. It is in fact not reasonable for a single number, as spectral entropy is, to contain too many detailed informations.

It is instead less ambiguous to consider the time at which it saturates, or even better the time at which it overcomes a certain threshold c close to its saturating value:

$$t_e := \min \{t : n_{\text{eff}}(t) = c\} \tag{7.24}$$

Concerning this point one has to remark that the quantities E_j, and thus also S and n_{eff}, fluctuate quite a lot, especially in a genuine equipartition regime; it some averaging is thus useful, for example over time, or over different orbits in the same class of initial conditions. If one performs a time average of E_j before computing S, then n_{eff} will saturate at 1 at equipartition; any kind of averages performed after the computation of S will results in a lower saturating level.

Independently of the particular way one calculates it, the above defined time t_e should be a reliable lower bound for the equipartition time, i.e. it could be used to estimate the second time scale whose existence is put into evidence in Sect. 7.4 by means of the natural packet phenomenon. And it is the "lower bound" nature of t_e that makes it useful if one aims, as we do, at showing a strong freezing of the dynamics when (specific) energy goes to zero.

7.5.2 Exponentially Long Times to Equipartition

As we said, there all a lot of papers (see, e.g., [7, 8, 15, 16, 34, 39], without any claim of completeness) dealing with quantitative estimates for the equipartition times, many of them using the spectral entropy indicator.

We recall only the papers [17, 43], where power law scalings with the inverse of specific energy of the kind $T_{\text{eq}} \sim \epsilon^{-3}$ and $T_{\text{eq}} \sim \epsilon^{-2}$ are found for the energy initially placed respectively in the low and high parts of the spectrum; and we quote also [39], probably the first one to put into evidence the possibility of exponentially long times.

Since the pictures of Sect. 7.4 suggest for the second time scale a stronger than power law dependence with respect to $1/\epsilon$, it has been natural to further investigate this point. In what follows we briefly describe the corresponding results.

Low-Frequency Initial Data

In order to compare the results with those in [17], the FPU β-model has been considered, with $\beta = 0.1$, initial data with energy equally distributed on a packet of modes $\left[\frac{N+1}{64}, \frac{5(N+1)}{64}\right]$, i.e., with fixed frequency range. The specific energy has been varied in the range $[0.0089, 7.7]$ for $N \in \{255, 511, 1023\}$. For every one of these conditions, 25 different orbits have been integrated, changing randomly the phases of the oscillators: the indicator used is the average over these different orbits, which saturates at values slightly less than 0.7. In order to see the scaling of the relaxation time with the specific energy, we plotted, for every ϵ the time at which n_{eff} overcomes a fixed threshold, that we choose equal to 0.5.

The results are illustrated in the left panel of Fig. 7.6. There is a numerical evidence supporting the exponential scaling of relaxation times to

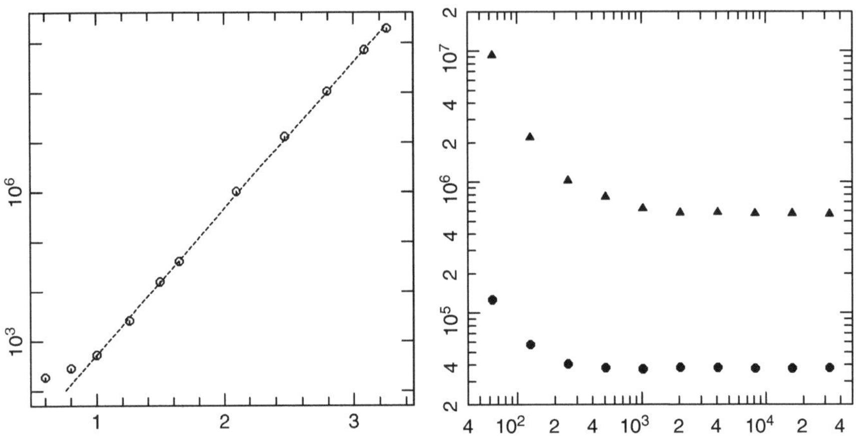

Fig. 7.6. Exponentially long relaxation times and their thermodynamic limit. β-model with $\beta = 1/10$. *Left panel*: time needed by n_{eff} to overcome the threshold 0.5 versus the specific energy power to $-1/4$, for fixed $N = 255$ and ϵ in the range $[0.0089, 7.7]$, in semi-log scale; the *straight line* is the best fit using all the points with $\epsilon \leq 1$. *Right panel*: time needed by n_{eff} to overcome the threshold 0.5 versus the number of particles N, for a couple of specific energies (0.0516 for *triangles* and 0.137 for *circles*) and N in the range $[63, 32767]$. After an initial transient, there is clearly no dependence on N (from [8, 38])

equipartition with respect to specific energy, with a possible law of the type $T \sim \exp(\epsilon^{-1/4})$. We notice that in our calculations we explore a slightly lower specific energy range than in [17].

In the left panel of Fig. 7.6, results for $N = 255$ are shown; we remark that the same exponential estimate has been obtained also for $N = 511$ and $N = 1023$.

Thermodynamic Limit

For the case of low-frequency initial data, in order to check the persistence of these phenomena in the thermodynamic limit, we repeated the same calculation as before varying $N \in [63, 32\,767]$ for a couple of fixed values of specific energies $\epsilon \in \{0.052, 0.14\}$.

In Fig. 7.6 one clearly sees that after an initial decrease of the times for the first low values of N, the subsequent points reported in the plot relative to values of N greater than 1 023 remains practically constant, thus showing another numerical indication of the possible persistence in the thermodynamic limit of the phenomena presented in [8].

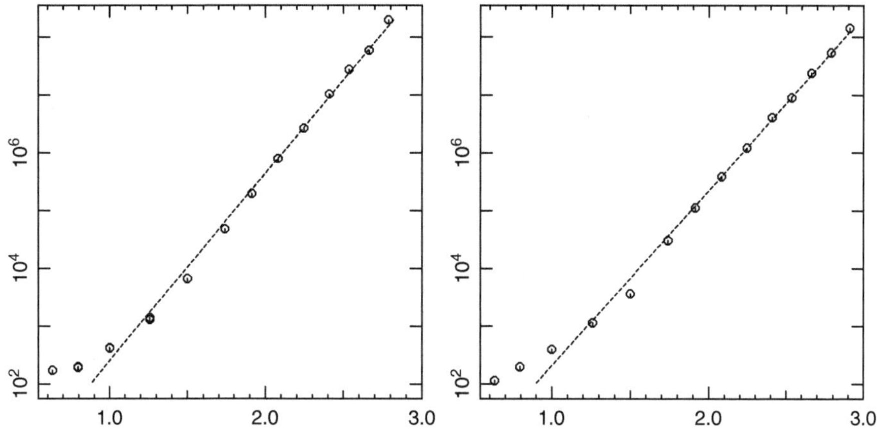

Fig. 7.7. Exponentially long times. α, β-model, with $\alpha = \beta = 1/4$, $N = 127$. Plot of time needed for n_{eff} to overcome 0.55 versus $\epsilon^{-1/5}$; specific energies are in the range $[0.0046, 10]$, and the *straight lines* are the best fit using all the point with $\epsilon \leq 1$. *Left panel*: initial data on the mode 127. *Right panel*: initial energy in equipartition on a packet of modes from 120 to 127. Every point is obtained by averaging over 10 orbits (from [38])

High-Frequency Initial Data

Also in the case of high-frequency initial data, as explored in [43], we performed a computation similar to the one previously illustrated, for the α, β-model. The only difference is in the number of different orbits used to average n_{eff}, lowered to 10 to reduce the computational time. The results are shown in Fig. 7.7. The fitting with an exponential law describes in a quite good way the experimental data. With respect to the case of initial low-frequency excitation, the scaling $T_{\text{eq}} \sim \exp(\epsilon^{-1/5})$ has a different exponent for the specific energy; we have no theoretical explanation for this particular number, but such a difference might not be surprising. In fact, if one conjectures that a resonance mechanism is actually responsible for the long-time freezing of the dynamics, it is clear that different parts of the spectrum may be characterized by different type of resonances.

7.5.3 Discussion

Concerning the indicator itself, we have to stress again that it is quite easy and cheap to calculate. Moreover, it is constituted by a single number, which is at the same time a strength and a weakness: In fact it may summarize in a single quantity the level of equipartition of an orbit, and this fact is really useful. But on the other side, for the same reason, one cannot expect that it may contain a rich amount of informations. A kind of opposite situation, say, happens for the critical times t_s defined in formula (7.23) to describe the natural packets phenomenon: they are many numbers and they contain more informations, and thus it requires some care in order to obtain a global picture, as in the figures representing the two timescales and the meta-stability of natural packets.

A different remark about the indicator concerns the way one calculates it; as we noted after the definition of t_e in formula (7.24), a time average of the harmonic energies before the computation of S clears out their fluctuations, while a different way of calculation partially maintain these informations. And it is in fact possible to show that if one computes S in the two ways, their difference is related to the mean square deviation of the harmonic energies. According to us this point should deserve more attention, because one should wonder if equipartition is really the relevant condition for Statistical Mechanics; in fact even a set of uncoupled oscillators may indeed be put in a condition of equipartition. It is probably more relevant to investigate the fluctuations, by means of the mean square deviations, or by means of some quantity related to a normalized total variations of the harmonic energies. We are presently trying to make some steps in this direction.

Another important point in the discussion of these results, in connection with their relevance for Statistical Mechanics, is the following: it is often said that, even if the "FPU behaviour" survives in the thermodynamic limit, the fraction of the phase space associated with it is negligible, and that the choice of the initial energy on a single mode, the first one, is not physical. Without entering in the discussion of what is physically relevant or not, we could now say that, in view of the results presented in this section, the exponentially long times and the existence of meta-stable packets of modes are phenomena not so isolated in particular and small regions of the phase space.

Concerning the comparisons with the results of [17] and of [43], we remark that in both cases we explore a lower specific energy range. And indeed in the latter paper the authors do not exclude, for very low energies, that the equipartition times could increase more rapidly than as a power law.

7.6 Lyapunov Exponents

In this section we will be concerned with a class of numerical tools which are based on the local behaviour around an orbit studied by means of the corresponding variational equation. This idea is of course the natural generalization of the linear stability analysis in a neighbourhood of an equilibrium point; the same analysis is indeed well defined for a periodic orbit, which is a stationary point of a suitable associated map, and the goal is to obtain a kind of generalized eigenvalues and eigenvectors for every orbit.

Most of this section will be devoted to the so called Lyapunov Characteristic Exponents (LCEs), their definition and properties, the algorithm one uses to calculate them, and some results we obtained. Then some variants will be considered and analysed.

7.6.1 Maximal LCE

In order to introduce the (maximal) LCE, we recall that our aim is to study the local behaviour around a trajectory, and to check a possible exponential divergence of nearby orbits.

Consider a dynamical system, represented by its flow Φ, on a Riemannian manifold M and denoted by $d(\cdot,\cdot)$ the metric and by $\|\cdot\|$ the norm on the tangent space; given a point $x_0 \in M$ on the orbit we are interested in, consider a regular curve passing through x_0

$$\begin{cases} y : (-\epsilon, \epsilon) \subset \Re \to M \\ y(0) = x_0 \end{cases}$$

transversal to the flow by x_0.

Every point $y(s)$ on the curve y evolves through the flow of the system, and may be seen as an initial data for a nearby orbit. We are thus interested in the deformation, induced by the dynamic, of the distance of the two points $y(0) \equiv x_0$ and $y(s)$, in the limit for s going to zero:

$$\gamma(t, x_0, \xi_0) := \lim_{s\to 0} \frac{\mathrm{d}(\Phi^t(y(s)), \Phi^t(x_0))}{\mathrm{d}(y(s), x_0)} , \tag{7.25}$$

where $\xi_0 := \frac{\mathrm{d}y}{\mathrm{d}s}(0)$. Taking the first-order approximation of both $y(s)$ and $\Phi^t(y(s))$, we obtain

$$\begin{aligned} y(s) &= x_0 + s\xi_0 + \text{h.o.t} , \\ \Phi^t(y(s)) &= \Phi^t(x_0) + s\langle D_x\Phi^t(x_0), \xi_0\rangle + \text{h.o.t} . \end{aligned}$$

Let us now define

$$\xi(t, x_0, \xi_0) := \langle D_x\Phi^t(x_0), \xi_0\rangle ;$$

actually it clearly turns out that $\xi(t, x_0, \xi_0)$ is the solution of the variational equation

$$\dot{\xi}(t) = J_xF\left(\Phi^t(x_0)\right)\xi(t) , \qquad \xi(0) = \xi_0 ,$$

where the flow $\Phi^t(x_0)$ is the solution of the Cauchy problem $\dot{x}(t) = F(x(t))$, $x(0) = x_0$.

Standard Definition

It is possible to rewrite the formula (7.25) as

$$\gamma(t, x_0, \xi_0) = \frac{\|\xi(t)\|}{\|\xi(0)\|} , \tag{7.26}$$

dropping the dependence on the unchanged variables. Being $\gamma(t, x_0, \xi_0)$ a measure of the expansion or compression of the tangent vector ξ_0 after a time t, in the spirit of the linear analysis of a stationary point, we take its logarithm to detect exponential behaviour, normalize with t to obtain an average coefficient, and take the limit to consider the effect on the whole orbit: we thus define the (maximal) LCE as

$$\chi(x_0,\xi_0) := \lim_{t\to\infty} \frac{1}{t} \ln(\gamma(t,x_0,\xi_0)) \ . \tag{7.27}$$

Alternative Definitions

It is possible to provide an integral definition of LCE: indeed using (7.26) one has

$$\ln(\gamma(t,x_0,\xi_0)) = \ln\|\xi(s)\| - \ln\|\xi(0)\| = \int_0^t \frac{\mathrm{d}}{\mathrm{d}s}\ln\|\xi(s)\|\mathrm{d}s \ ;$$

and the one computes the derivative of $\ln\|\xi(s)\|$

$$\frac{\mathrm{d}}{\mathrm{d}s}\ln(\|\xi(s)\|) = \frac{1}{\|\xi(s)\|}\langle\dot{\xi}(s),\hat{\xi}(s)\rangle \ ,$$

where $\hat{\xi}(s) = \xi(s)/\|\xi(s)\|$.

A further description of the LCE, which will be used for actual computations, comes out as follows. One can write the logarithm of γ (7.26) in this form

$$\ln\gamma(t,x_0,\xi_0) = \sum_{s=1}^{t} \ln\left(\frac{\|\xi(s)\|}{\|\xi(s-1)\|}\right) = \sum_{s=1}^{t} \ln\gamma(1,\Phi^{s-1}(x_0),\xi(s-1)) \ ;$$

in this way χ might be seen as a logarithmic average of the rate of expansion in each time unit.

Summarizing we have the three equivalent definitions

$$\begin{aligned} \chi &= \lim_{t\to\infty} \frac{1}{t}\ln(\gamma(t,x_0,\xi_0)) \ , & (7.28)\\ &= \lim_{t\to\infty} \frac{1}{t}\int_0^t \frac{1}{\|\xi(s)\|}\langle\dot{\xi}(s),\hat{\xi}(s)\rangle\mathrm{d}s \ , & (7.29)\\ &= \lim_{t\to\infty} \frac{1}{t}\sum_{s=1}^{t} \ln(\gamma(1,\Phi^{s-1}(x_0),\xi(s-1))) \ . & (7.30) \end{aligned}$$

We recalled the integral definition since it is the starting point of another indicator related to the LCE, the Mean Exponential Growth factor of Nearby Orbits (MEGNO), discussed in [14]. This indicator is in fact given by $Y(t) = \frac{2}{t}\int_0^t \frac{s}{\|\xi(s)\|}\langle\dot{\xi}(s),\hat{\xi}(s)\rangle\mathrm{d}s$ and it proved to be useful in the investigation of both the regular and chaotic part of the phase space. We will nevertheless enter in the discussion of this indicator.

7.6.2 Some Rigorous Results

In the previous section we have presented three different ways to introduce the (maximal) LCE of a tangent vector ξ_0 along an orbit $\Phi^t(x_0)$, but we omitted any discussion about the existence of these limits. Usually, in literature, one overcomes the problem giving a weaker definition

$$\chi = \lim\sup_{t\to\infty} \frac{1}{t} \ln(\gamma(t, x_0, \xi_0)) .$$

Exploiting the common properties of lim sup one can prove the

Theorem 7.2. *For each $x \in M$, it holds*

(1) the quantity $\chi(x, \xi_0)$, $\xi_0 \in T_x M$, can assume at most $m \leq n$ different values

$$\chi_1(x) > \chi_2(x) > \cdots > \chi_m(x) ;$$

(2) let $L_i = \{\xi_0 \in T_x M : \chi(x, \xi_0) \leq \chi_i\}$, then

$$T_x M = L_1 \supset L_2 \supset \cdots \supset L_m$$

and

$$\xi_0 \in L_i \setminus L_{i+1} \Rightarrow \chi(x, \xi_0) = \chi_i(x), \quad i = 1, \ldots, m.$$

The idea of the proof is to show, firstly, that the sets

$$L(\theta) = \{\xi_0 \in T_x M : \chi(x, \xi_0) \leq \theta\}$$

are vector subspaces of the tangent space $T_x M$. Plainly, if $\theta' < \theta$ then $L(\theta') \subset L(\theta)$. If there exists $\xi_0 \in L(\theta)$ such that $\chi(x, \xi_0) = \theta$, then $\xi_0 \notin L(\theta')$ and $\dim L(\theta) > \dim L(\theta')$.

Looking at the statements of the theorem one realizes that a single LCE $\chi(x, \xi_0)$ is function of the whole one-dimensional subspace to which ξ_0 belongs. Thus, it is natural to generalize

Definition 7.3. *Consider a p-dimensional subspace $E \subseteq T_x M$ and $\xi_1, \ldots, \xi_p$ a generic basis of it; if the limit*

$$\chi^{(p)}(x, E) = \lim_{t\to\infty} \frac{1}{t} \ln \frac{Vol^p(D_x\Phi^t\xi_1, \ldots, D_x\Phi^t\xi_p)}{Vol^p(\xi_1, \ldots, \xi_p)} \tag{7.31}$$

exists, it is called LCE of order p.

The setting we are dealing with suggests other definitions:

Definition 7.4. *Let us take $\nu_i = \dim L_i - \dim L_{i+1}$ independent vectors in $L_i \setminus L_{i+1}$, $i = 1, \ldots, m$ which form a basis $e_1, \ldots, e_n$ in $T_x M$, and $g_1, \ldots, g_n$ any other basis of $T_x \mathcal{M}$. The first basis is called "normal" iff*

$$\sum_{i=1}^{n} \chi(x, e_i) \leq \sum_{i=1}^{n} \chi(x, g_i) .$$

Definition 7.5. *The quantity ν_i we have defined above is called "multiplicity" of $\chi_i(x)$. The collection of all the LCEs, each repeated with its own multiplicity, is called "spectrum."*

We are now ready to give a statement of one of the most classical and relevant results concerning LCE:

Theorem 7.3 (Oseledec [36, 10]). *Consider a dynamical system $(\mathcal{M}, \mu, \Phi)$:*

(1) for almost all $x \in M$ there exists the "exact" LCE (7.27); the same holds for (7.31);
(2) the spectrum

$$Sp(x) = \{\chi_1, \nu_1, \ldots, \chi_m, \nu_m\}$$

is a measurable function of x;
(3) for each "normal" basis $e_1, \ldots, e_p$ of any subspace $E \subseteq T_x M$ we have

$$\chi^{(p)}(x, E) = \sum_{i=1}^{p} \chi(x, e_i) \, . \tag{7.32}$$

Our aim is, now, to deduce from this theorem some properties concerning Hamiltonian systems. Since the flow of an Hamiltonian field is symplectic, it preserves, in particular, the $2n$-dimensional volume, and thus, according to point (3) of the previous theorem

$$\chi(x, T_x M) = \sum_{i=1}^{2n} \chi_i(x) = 0 \, .$$

But symplecticity gives indeed more; in the Hamiltonian case, one can show that the spectrum of the LCEs is symmetric:

$$Sp(x) = \{\chi_1(x), \ldots, \chi_n(x), -\chi_n(x), \ldots, -\chi_1(x)\} \, .$$

In the autonomous case the total energy is preserved, so the orbit lies on a $2n$–1 manifold E. If this hyper-surface is compact and the orbit we are following does not belong to a stable or unstable manifold of a fixed points, it happens that the two central LCEs are equal to 0; this means that there exist two tangent directions along which no exponential divergence is possible. One of these two directions has component orthogonal to the energy surface E, and the other one is exactly the tangent field; and the presence of another independent integral of motion implies other two zero LCEs.

7.6.3 Numerical Computation

As we showed in the previous sections, from the theoretical point of view the LCE is a very good indicator of hyperbolicity of a motion, but some care is needed in its actual computation. It is clear that, considering for example the standard definition (7.27) for the maximal LCE, the true quantities one may compute are

$$\chi(t, x_0, \xi_0) := \frac{1}{t} \ln(\gamma(t, x_0, \xi_0)) \; ; \tag{7.33}$$

and the definition of χ, as a limit for t going to ∞ of $\chi(t)$ puts immediately into evidence the point: if the system has not a strong and say uniform hyperbolicity, the convergence to the limiting value may be slow, and even non monotone if the trajectory is visiting regions of the phase space with different behaviours. It thus may be necessary to follow the flow for long time, further increasing the computational load given by the integration of both the orbit and its tangent dynamic.

Moreover, the detection of weak hyperbolicity is delicate also because even in the case of completely integrable systems, after a long-time integration, the dynamic may exhibit a small exponential divergence of nearby orbits, due to the errors during integration. The orbit we are following via a symplectic algorithm is a pseudo-orbit of a small perturbation of the original system, and thus generically no more integrable.

In what follows we briefly consider the reliability of the numerical integration of the variational equations, in the spirit of Sect. 7.3.3, and we illustrate the algorithm used to compute all the LCEs.

Numerical Integration of Variational Equations

In the same spirit of Sect. 7.3, one might ask about the reliability of the numerical integration of the tangent dynamic. If the underlying flow is given by a natural Hamiltonian $H = T + V$, it turns out to be a non-autonomous Hamiltonian system with Hamiltonian $G(u, v, t) = (1/2)\langle v, v\rangle - (1/2)\langle A(\Phi_H^t)u, u\rangle$. Integrating the flow Φ_H with a symplectic algorithm of order k, and step τ, we evaluate A on the flow of $H_\tau = H + \tau^k H'$ up to roundoff errors (see Theorem 7.1 in Sect. 7.3.3), and the actual Hamiltonian representing the tangent dynamic is

$$G_\tau(u, v, t) = \frac{1}{2}\langle v, v\rangle - \frac{1}{2}\langle A(\Phi_{H_\tau}^t)u, u\rangle \; .$$

In order to exploit once again the good properties of symplectic integrators, we have to extend the non conservative system given by G_τ in an autonomous form considering t as a variable conjugated to the energy:

$$K(q, p, s, E) = G_\tau(q, p, s) - E \; ;$$

A leap-frog type algorithm should be represented by the following map:

$$\begin{cases} u_{n+1} &= u_n + \tau v_n + \frac{\tau^2}{2} A(s_n) u_n \\ v_{n+1} &= v_n + \frac{\tau}{2} A(s_n)(u_n + u_{n+1}) \\ s_{n+1} &= s_n + \tau \\ E_{n+1} &= E_{n+1}(E_n, s_n, u_n, v_n) \end{cases} \; ; \tag{7.34}$$

one has to check that it preserves the symplectic form $\omega = \mathrm{d}u_n \wedge \mathrm{d}v_n + \mathrm{d}s_n \wedge \mathrm{d}E_n$ in $\Re^{2n+2}$. Actually there exist a result on time-dependent canonical transformations that may be stated as follows:

Theorem 7.4. *Let $H(q,p,t)$ be a non-autonomous Hamilton function and $q = q(\overline{q},\overline{p},t), p = p(\overline{q},\overline{p},t)$ a time-dependent transformation which preserves the fundamental Poisson brackets identically in t. Then the transformation is canonical, and there exists a function $F(q,p,t)$ such that the transformed Hamiltonian is*

$$H(\overline{q},\overline{p},t) = [H(q,p,t) - F(q,p,t)]_{q=q(\overline{q},\overline{p},t),p=p(\overline{q},\overline{p},t)} \ .$$

It is easy to see that the theorem applies; and it is also possible to show that the symplecticity constraint implies $E_{n+1} = E_n + f(s_n, u_n, v_n)$.

And now, since the map (7.34) is symplectic, by Theorem 7.1, it is almost exactly the flow of a perturbed Hamiltonian system with Hamilton function

$$K_\tau = K + \tau^a K' \ .$$

This makes reliable also the long time integration of the tangent dynamic.

LCE's Algorithm

We briefly illustrate here the classical [4, 5] algorithm used to calculate all the LCEs. The first remark is that, due to the exponential increase of the length of tangent vectors we have to renormalize the vectors from time to time to avoid a computer overflow, and thus we use definition (7.30). The second remark is that all the tangent vectors tend to align along the most expansive eigenvector: to avoid a numerical undetectability of the other exponents it is then also necessary to orthogonalize them via the Grahm–Schmidt procedure.

With some more detail, one fixes a relatively short time interval σ between successive orthonormalizations. We have by (7.30)

$$\chi_1 = \lim_{n\to\infty} \frac{1}{n\sigma} \sum_{j=1}^{n} \ln \alpha_j \ , \qquad \alpha_j = \|\langle D_x \Phi^\sigma(\Phi^{(j-1)\sigma}(x_0)), \hat{\xi}_{j-1}\rangle\| \ ,$$

where the hat over a vector denotes its versor.

To obtain the p_{th} exponent, we use Definition 7.3 and the third point of Oseledec theorem. Let us illustrate it for $p = 2$.

$$Vol^2(D_x\Phi^\sigma(x_{j-1})\xi_{j-1}, D_x\Phi^\sigma(x_{j-1})\eta_{j-1}) = \alpha_j\beta_j \ ,$$

where α_j has been already defined above and

$$\beta_j = \|\eta_j^\perp\| \ , \qquad \eta_j^\perp = \eta_j - \langle\hat{\xi}_j, \eta_j\rangle\hat{\xi}_j \ .$$

Using (7.32) and (7.31)

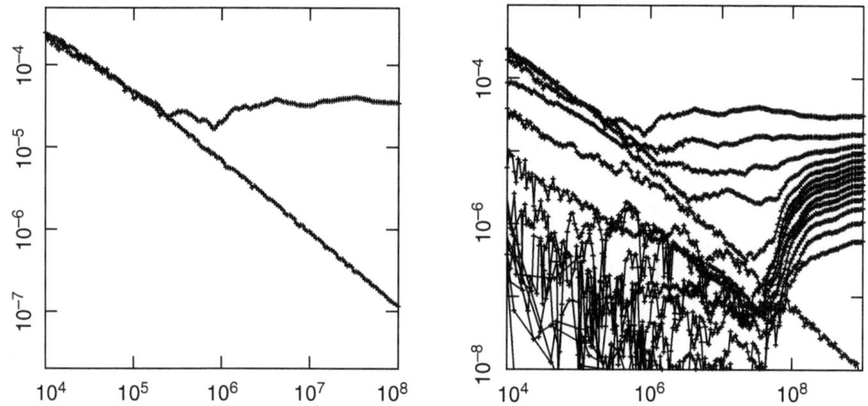

Fig. 7.8. Lyapunov exponents; $\chi(t)$ versus time. FPU α-model ($N = 15$, $\alpha = 1/4$, $\epsilon = 0.0398$). *Left panel*: comparison between FPU and Toda and definition of the trapping time. *Right panel*: all the LCEs for the FPU

$$\chi_1+\chi_2 = \chi^{(2)} = \lim_{n\to\infty} \frac{1}{n\sigma} \sum_{j=1}^{n} \ln\, \alpha_j\beta_j = \lim_{n\to\infty} \frac{1}{n\sigma} \sum_{j=1}^{n} \ln\, \alpha_j + \lim_{n\to\infty} \frac{1}{n\sigma} \sum_{j=1}^{n} \ln\, \beta_j \ ,$$

so that

$$\chi_2 = \lim_{n\to\infty} \frac{1}{n\sigma} \sum_{j=1}^{n} \ln\, \beta_j \ .$$

We conclude this part remarking that in actual computations we always used the energy norm instead of the Euclidean one, since it fits better with the problem and gives cleaner numerical results.

7.6.4 Some Results and Discussion

We present here few examples of computations of LCEs. The first one is inspired by [12], where a comparison between the maximal LCE of the FPU and of the Toda model is performed. The Toda chain is a system of the form 7.1 too, but with a potential

$$V(s) = \frac{e^{2\alpha s}}{4\alpha^2} \ ,$$

whose Taylor development coincides up to the third order with the FPU one (see (7.2) and exploit boundary conditions); such a model turns out to be integrable. Because of this property its LCEs are all zero, and thus the actual quantities $\chi_{\mathrm{Toda}}(t)$ one calculates [see formula (7.33)] go to zero as $(\ln\ t)/t$. Exploiting such a fact they define the "trapping time" t_τ as the first time at which the time evolution of the maximal $\chi_{\mathrm{FPU}}(t)$ detaches from that of the Toda chain, provided the two system have been considered with the same initial datum.

A computation of this kind is shown in Fig. 7.8: the quantity described above can be identified quite clearly. On the other hand its dynamical meaning appears to be not so well defined. The interpretation given in [12], which states that the two systems share almost the same evolution up to t_τ, and after that FPU one reaches a chaotic regime, is for sure too strong. In fact one has to remember that the LCEs give local indications related to a single orbit: if the system is ergodic, and thus it is possible to explore almost all the phase space following one trajectory, then the local information turns out to be essentially global. Moreover the precise time t_τ is directly related to the strength of the maximal LCE, i.e. it represents the time needed to detect it in the actual computation. All this considerations are confirmed by the experiments with other indicators, like those in Figs. 7.11 and 7.12, where evidence of a different behaviour is shown before the trapping time, and those in Figs. 7.9, 7.13 and 7.14, where evidence of a non-complete chaotic behaviour after the trapping time is given.

Another observation one can have from these experiments is a confirmation of what we said defining the quantity $\chi(t)$ in formula 7.33: in fact looking at the maximal LCE in Fig. 7.12 it is very clear that the local contribution constituted by $\gamma(t)$ may well change during the evolution, depending on the different regions the orbit is visiting; the value of $\chi(t)$ one obtains during the computation, even if it seems stabilized, is not necessarily a good approximation of the true exponent.

A further indication of the existence of different parts of the phase space comes from the computation of all the LCEs, as shown in the right panel of Fig. 7.8: although the first three exponents exhibit a substantially stable behaviour, the other ones experience a sudden increase for time slightly less than 10^8. This computation confirms that also the smaller LCE contains useful information concerning the hyperbolicity around the orbit. Unfortunately these computations are really heavy from the numerical point of view.

We remark once again, looking at Fig. 7.9, the locality of the information carried by the LCE: in fact the distribution of harmonic energies still shows a lack of equipartition at times equal to 10^8.

In the previous discussion of the results one can obtain using the LCEs, we stressed some of their limits: their local nature in space and time, the effect of the averaging process which forces to longer computations to detect small exponents, the numerical difficulties in the computing of all the exponents in order to have a better image of exponential divergence. In the next section we present some different indicators, again based on the tangent dynamic, which try to solve some of these problems.

7.6.5 Variations on the Theme

With respect to the LCEs, which are good indicators for irregular and stochastic regions of the phase space, we will discuss now some variants, developed

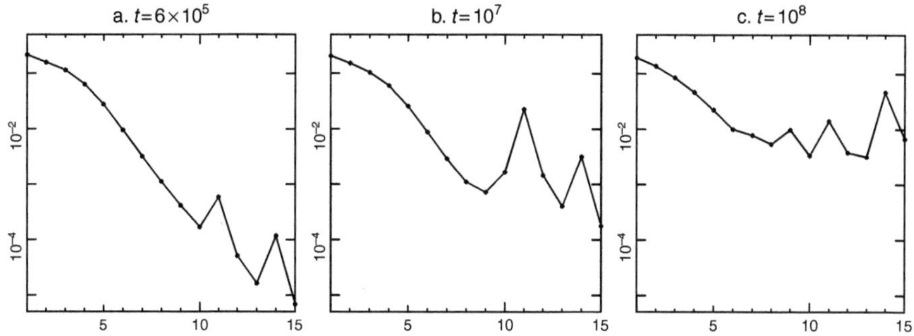

Fig. 7.9. Time averaged distributions of harmonic energies at different times, for the same orbit of Fig. 7.8. Abscissæ: the mode number k. Ordinates: the energy of the modes averaged up to time t reported in the label

during the last decade; they appear to better identify also regular orbits, being able to distinguish between resonant (islands) or non-resonant (KAM tori) regular orbits too. Moreover they have the advantage of requiring a shorter evolution time for their computation, since they typically avoid the averaging over time; nevertheless, in the presence of weak chaos and small LCEs, they also require a time which is inversely proportional to the LCEs to see their effect.

Smaller Alignment Index: SALI

In this paragraph we will consider the *smaller alignment index* (SALI), which has been introduced and developed by Skokos [41, 42] starting from 2001. This method can in principle reveal an exponential divergence of nearby orbits earlier than the maximum LCE. The main idea is to perform integration of variational equations for two different initial vectors ξ_0, η_0 and to obtain a simple measure of how much parallel the two vectors become during the evolution. At every fixed time-interval each vector is normalized and, after having defined

$$d_-(t) := \|\xi(t) - \eta(t)\| \,, \qquad\qquad d_+(t) := \|\xi(t) + \eta(t)\| \,,$$

we look for the minimum between d_- and d_+

$$\mathrm{SALI} := \min(d_-(t), d_+(t)) \,.$$

This is not the only choice: one could directly measure the angle between the vectors computing the inner product $\langle \xi(t), \eta(t) \rangle$. But some care should be taken, since if the two vectors are almost parallel, the SALI is small, say $\mathrm{O}(\delta)$, but $1 - \langle \xi(t), \eta(t) \rangle$ is smaller, actually $\mathrm{O}(\delta^2)$; and so the alignment may be revealed even faster than with SALI, although with more sensitivity to some noise, and the choice of the threshold could be critical.

If the orbit admits some positive LCEs, then all the solutions of the variational equations tend to be aligned to the most expansive hyperbolic eigenvector. Let us further investigate what one may deduce if such alignment of the tangent vectors actually occurs. Consider the non-isochronous Hamiltonian $H = \langle\omega(I), I\rangle$, whose solution is, in general, a quasi-periodic orbit over the torus identified by the initial values of $I = (I_1, \dots, I_N)$:

$$\Phi^t(I(0), \varphi(0)) := \begin{cases} I_j(t) = I_j(0) \\ \varphi_j(t) = \varphi_j(0) + \omega_j(I(0))t \ . \end{cases}$$

The Jacobian of the flow with respect to the phase variables I, φ represents the solution of the variational equations; in this case, if $X = (X_1^I, \dots, X_N^I, X_1^\varphi, \dots, X_N^\varphi)$ is a generic tangent vector, we have

$$\langle D_{(I,\varphi)}\Phi^t, X\rangle = \begin{pmatrix} X^I \\ X^\varphi + t(D_I\omega)\cdot X^I \end{pmatrix} \ . \tag{7.35}$$

From the previous formula one deduces that every initial vector X will be oriented and will become tangent to the N dimensional torus: Such evolution is not exponentially fast, but linear, accordingly with the regularity of the dynamic. The second remark is that if one considers two vectors, they will both collapse on the same lower dimensional tangent space, but there is no reason for them to get aligned within this subspace; the only possibility is that of a system with only one degree of freedom, because in such a case the space tangent to the torus is one-dimensional and all the vectors within it are obviously parallel.

We may thus conclude that in the case of a regular orbit, if the dimension N is greater than 1, SALI will tend to a constant value, while in the degenerate case $N = 1$ it will go to zero as a power law.

As it has been shown in the above quoted papers, SALI has the advantage of reaching in few iterations the limit of accuracy of the computer, putting into evidence (see left panel of Fig. 7.10) a strong divergent phenomenon well before LCE reaches its proper value.

A Generalization of SALI

In the discussion of the LCEs we showed how it is possible to obtain more detailed information considering all the exponents, not only the maximal one. In the same spirit we tried to generalize the SALI indicator by computing not only the "maximal" one, but all the small alignment indexes.

To this end we considered two sets V_1 and V_2 of n vectors; for each set independently we performed the same procedure one uses to compute all the LCEs, with repeated orthonormalizations during the evolution with the tangent dynamic. Under the action of the hyperbolic directions, if present, one vector $v_1^1 \in V_1$ and one $v_1^2 \in V_2$ will become parallel, as in the original SALI

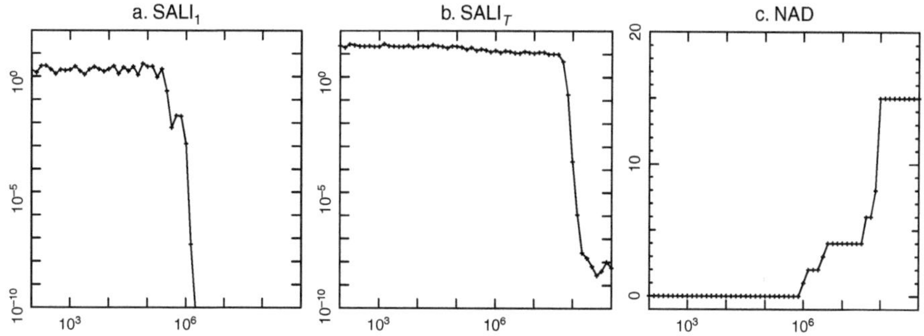

Fig. 7.10. FPU α-model, same orbit of Fig. 7.8 ($N = 15$, $\alpha = 1/4$, $\epsilon = 0.0398$). *Left panel*: the original SALI_1 versus time, showing the first alignment before 10^6. *Central panel*: the total SALI_T versus time, showing a complete alignment for times close to 10^8. *Right panel*: the NAD versus time, showing the increase, in the time interval 10^6–10^8, of the number of directions aligned

method, along the most expansive eigenvector e_1. If there exist a second positive LCE, another couple of vectors v_2^1, v_2^2 will tend to align, not along the second eigenvector, but, due to the orthogonalization, along its projection over the plane orthogonal to e_1. The time needed for each successive couple of vectors to align depends on the strength of the corresponding LCE. We notice that last couple v_n^1, v_n^2 will be forced to align due to the orthogonality to all the other vectors.

In actual computations, we performed the above described procedure with two initial sets of n orthonormal vectors: the first one oriented as the natural basis of R^n, and the second one randomly oriented. For each couple of vectors we have computed the single quantities $d_j^-(t) = \|v_j^1(t) - v_j^2(t)\|$ and $d_j^+(t) = \|v_j^1(t) + v_j^2(t)\|$, and we defined

$$\mathrm{SALI}_j := \min(d_j^-(t), d_j^+(t)) \,, \qquad \mathrm{SALI}_T = \sum_{j=1}^{n} \mathrm{SALI}_j \,.$$

As we said, SALI_1 is nothing but the SALI defined by Skokos, so it can reveal whether an exponential divergence has occurred or not.

In Fig. 7.10 we report some computations of this indicator: comparing the first two panels with Fig. 7.8 it is possible to observe a substantial agreement of the time given by SALI_1 with the time at which the maximal LCE stops going to zero, and of the time given by SALI_T with the time at which all the LCEs begin to manifest their positivity.

In the right panel we show a quantity derived from SALI_T (actually from the version obtained computing the angle $1 - \langle \xi(t), \eta(t) \rangle$) which measures the Number of Aligned Directions (NAD):

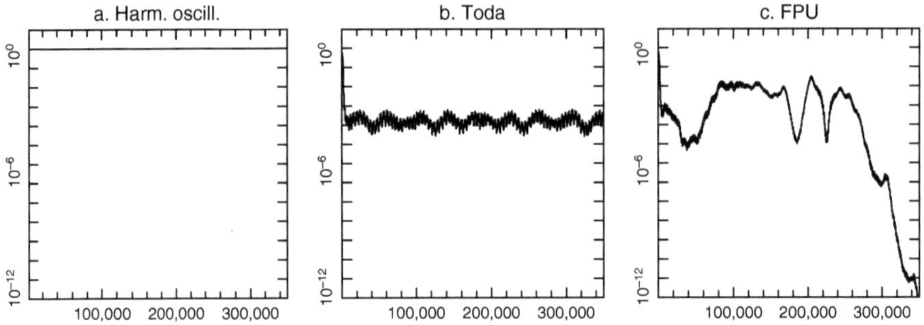

Fig. 7.11. Volumetric angle versus time described by 15 vectors up to time 3.5×10^5. $N = 15$ for all the models. *Left panel*: non-interacting harmonic oscillators. *Central panel*: Toda chain (initial datum equal to that of FPU chain). *Right panel*: FPU α-model, ($\alpha = 1/4$, $\epsilon = 0.0398$, same orbit as in Figs. 7.8, 7.9 and 7.10). The first two systems are both integrable, but the indicator seems to detect the non-isochronous nature of the Toda lattice. Moreover the non-integrable nature of the FPU is visible from the very beginning in the comparison between Toda and FPU

$$\mathrm{NAD}(t) := \sum_{j=1}^{n} [1 - \mathrm{SALI}_j + \epsilon] \ ,$$

where $[x]$ is the integer part of x, and ϵ is a small constant used to set a threshold. This indicator gives an estimate of the number of hyperbolic directions which are already explicitly visible.

We must say that Poincaré sections (see Sect. 7.7) show that a weak and local chaos has appeared before the time indicated by these experiments.

As a further comment we say that SALI_T and NAD could be easily used to detect a time of full, though local, hyperbolicity when all the expanding directions make their action visible. One should then study the scaling of that time with the (specific) energy. This task, however, turns out to be very hard from the computational point of view, because, for this indicator, the evolution of a double number of tangent vectors is required.

Volumetric Angle

Searching for other tools to detect differences, on short timescales, between the FPU model and the Toda one (as it is done by Poincaré sections), we tried to compare the different decreasing rate of a kind of solid angle between a given set of tangent vectors. In particular, considering $2 \leq p \leq n$ independent vectors $(e_1, \ldots, e_p)$, initially orthonormalized, and following their evolution along the tangent flow, at fixed intervals one computes the product of the sinus of the angle between each consecutive couple $(e_j(t), e_{j+1}(t))$ as a measure of the alignment of all the vectors. In Fig. 7.11 we can see the results of such experiment in three different systems with the same initial datum. As expected [see formula (7.35)], the isochronous integrable system (left panel) does not exhibit any alignment, since $D_I\omega = 0$. In the non-isochronous integrable system (the Toda chain, central panel) we observe a sensitive initial alignment due to the non-zero term $D_I\omega$; after this transition, the angle slightly moves around a stable value for any long time. The last example (FPU α-model, right panel) shows wide amplitude oscillations near the value reached in the previous case.

Fast Lyapunov Indicator: FLI

In a set of papers (see, e.g., [19, 20, 21, 22, 28, 29, 35]), Froeschlé and his collaborators, introduced and used the so called Fast Lyapunov Indicator (FLI).

As we discussed, the usual Lyapunov exponents have some drawbacks: Since they are defined over the whole trajectory, they require in principle long and expensive numerical computations, and moreover they may distinguish only between chaotic and non-chaotic orbits, possibly giving a "measure" of chaoticity. But the phase space structure, and thus the dynamic, is more rich and complicated, and also within the non chaotic motions there are different possibilities. With this respect, the FLI has exactly the advantage of being faster to calculate, and of being able to distinguish among regular orbits, between the resonant ones, islands and non-resonant ones, KAM tori.

The underlying idea is to consider the information contained in the initial part of the evolution of the tangent dynamic, the part which is usually considered as a sort of initial noise whose effect is thrown away through the time average in the limit for $t \to \infty$; and the second point is indeed to avoid the time average. One in fact realizes that the evolution of this indicator rapidly distinguish between chaotic, resonant and non-resonant orbits.

A possible definitions (see [19] for a discussion on other definitions) is the following: denoting by $\xi(t)$ the usual time evolution of vector along the tangent dynamic, we have:

$$\mathrm{FLI}(x_0, \xi_0, T) := \sup_{0<s<T} \ln \|\xi(s)\| ,$$

where $\xi(s)$ denotes the usual time evolution of the vector ξ_0 along the tangent dynamic.

Another great advantage of the use of FLI is that its computation on a grid of initial conditions in the phase space gives a picture of the so called Arnold web, i.e. the set of resonances of the system. From this point of view such a tool turns out to be very effective. In this way one obtains a kind of global information; and this is equivalent to that obtained by the usual Lyapunov exponent evaluated on a single orbit only when the system is ergodic, as the Oseledec theorem states. On the contrary, if the dynamic is not ergodic, the information carried on by a single trajectory is for sure local.

In some of the above quoted papers (see, e.g., [28, 29, 35]), the global geometry of the Arnold web, put into evidence by means of the FLI, allowed the authors to carefully investigate the diffusion phenomena along resonant lines: They obtain numerical evidence of Arnold diffusion, and estimate the exponential dependence of the diffusion coefficient from the perturbative parameter, typical of a Nekhoroshev regime.

Up to our knowledge this indicator has never been used in the FPU model, but we think it could be interesting to investigate its Arnold web in this way. Of course the problem, as we will discuss in the Sect. 7.7 about Poincaré sections, is connected with the use of a low-dimensional indicator, as the FLI is, in a problem where the thermodynamic limit is one of the relevant points.

7.7 Poincaré Sections

Even if it can be considered hopeless in non-low-dimensional systems, we tried anyway to investigate local chaotic behaviour by means of Poincaré sections.

As usual, we consider the Poincaré map, i.e. the discretized orbit obtained by the intersection of the whole trajectory $\gamma(t)$ with a transversal hyperplane Π $((d/dt)\gamma(t) \notin \Pi)$. In dealing with perturbations of harmonic oscillators, a proper choice for such a plane could be $\{p_j = 0\}$ (i.e. when one of the momenta is zero), provided that the corresponding action I_j is non-zero. In fact, if the projection of the orbit (q_j, p_j) has an oscillatory behaviour, the corresponding velocity must change its signum. Since we considered experiments with initial data with energy on low frequencies modes, we used as a section plane $\Pi = \{p_j = 0\}$, j being the index of one of those modes.

Due to the autonomous nature of the system, the orbit lies on a $2n - 1$ dimensional manifold E of constant energy, and the trajectory of the map on the $2n - 2$ dimensional manifold $E \cap \Pi$. In the optimal case of $n = 2$, we can injectively project over a suitable plane, obtaining a clear drawing of regular and chaotic orbits. For higher number of degrees of freedom, one has to choose the plane over which to project the higher dimensional section, and this is a delicate task. That is why, in the case of nonlinear coupling of harmonic oscillators, we have chosen the planes of normal modes (q_k, p_k): if the perturbation (energy) is not too high, the dynamics may be described through non linear modes (or linear combination of them) close to the linear ones. With this approach, we are thinking of a Birkhoff (resonant or not) normal form as a good approximation of the system for our initial data. As a matter of fact, many figures realized with a low-mode section $\{p_j = 0\}$ seem to agree with the effect of an almost resonance between the first mode and (q_j, p_j) $(\omega_j - j\omega_j = \mathrm{O}(1/N)^3)$. In particular we will always show, for uniformity purposes, sections on $\{p_3 = 0\}$ projected on (q_1, p_1), and it is well visible the underlying structure of a 1:3 resonance.

Clearly, the limit of such a qualitative study of the dynamic is the possibility of producing readable pictures; in fact, since the projection is no more injective for high n, very confused images may emerge, also in the case of a weakly chaotic orbit, and indeed even a quasi-periodic orbit on a high-dimensional torus may be difficult to be recognized after a projection on a

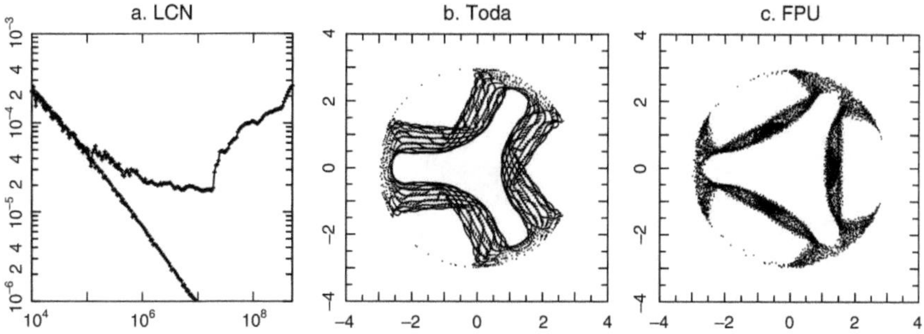

Fig. 7.12. **a** The maximal LCN versus time for the FPU α-model and the Toda lattice, computed for the orbit with the same initial starting point. Here $N = 7$, $\alpha = 1/4$, $\epsilon = 0.25$ and the phase $\varphi = 0.8995\pi$. **b** Poincaré section with the plane $p_3 = 0$, projected over the plane (q_1, p_1) of the orbit computed for the Toda lattice. There is evidence of a quasi-periodic behaviour, as expected in view of the integrability of the model. **c** Poincaré section for the orbit of the FPU system. The orbit appears be be confined in a region similar to that of the Toda model, but the figure suggests that there is a local chaotic behaviour (from [27])

plane. This explain our choice not to exceed the number of 15 particles with this kind of numerical experiments.

7.7.1 Some Results

The first results we show concern again the comparison of the FPU system with the Toda one, as in Sect. 7.6.4. In Fig. 7.12 we provide further evidence that the trapping time introduced in [12] is not a very relevant and precise dynamical indicator. In the left panel we put the comparison between the maximal LCEs for the two systems, and in the other panel the sections corresponding to the two orbits, for a time interval $[0, 5 \times 10^4]$, i.e. before the trapping time which may be estimated to be $t_\tau \sim 1.5 \times 10^5$. The clear information one obtains from the figure is that the FPU dynamics exhibits a local chaoticity well before that time: in fact, despite the points are bounded in a subregion of the plane (locality), they do not show the regularity of the corresponding section for the Toda (chaoticity).

Following the same orbit of the FPU system for longer times, as illustrated in Fig. 7.13, it is possible to see that the chaoticity remains local; in each panel, a window of 10^6 time units is considered. Since we are projecting 12 dimensions over a plane, the confinement of the points of the section is a quite strong indication that, on intermediate time scales, the evidence of an ergodic behaviour is still missing. One may also observe that the ergodicity, possibly recovered for t going to infinity, could be reached as the cumulative effect of several meta-stable states that fill up the phase space, and through which the orbit wander. It is worthwhile to remark that the pattern found in the left

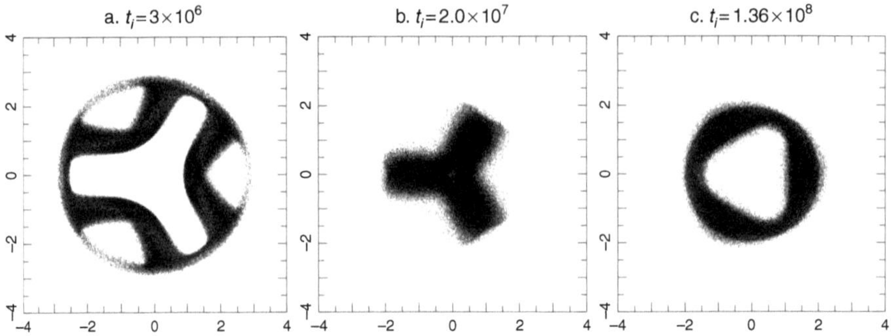

Fig. 7.13. Poincaré sections with $p_3 = 0$ for the FPU chain, projected on the plane (q_1, p_1), for the same orbit as in Fig. 7.12c, with a time window of 10^6 and initial times indicated above each panel (from [27])

panel reappears for times close to 10^9, remaining unchanged for more than 1.5×10^7 time units.

A very similar situation is again visible for $N = 15$, and the results are given in Fig. 7.14. The first panel shows the initial evolution on a short time interval of 10^5 to give a further confirmation that the system exhibits a weak stochasticity in very short times. The subsequent panels, on the other hand, illustrate again the strong confinement outside thirty dimensional cylinders emerging from the section plane.

A further remark comes from the comparison with the data shown in Figs. 7.8 and 7.10, where the LCEs and the SALI variants are used on the same orbit: it appears that even for times close to 10^8, where all the LCEs begin to be visible and the number of aligned directions reaches its maximum, the trajectory is still confined in bounded regions of the phase space.

7.7.2 Discussion

The remarkable aspect concerning the use of Poincaré sections is the problem of the dimensionality, as we already discussed at the beginning of this section. This tool is clearly a low-dimensional one, but once again it may give non sufficient but necessary conditions to reach a regime compatible with Statistical Mechanics. In fact, if one is able, as we do, to show that the projections of the points of the Poincaré map remain confined in bounded regions, then it can be safely concluded that the orbit does not visit the whole energy surface within the quite long time scale considered.

As argued in [27], analysing some features of the dynamics through the use of the Lyapunov exponents (see Sect. 7.6) and the use of Poincaré sections, one might imagine a scenario in which, lowering the specific energy, the phase space contains many subregions where the dynamics is frozen for very long times: the equipartition could be seen only on the infinite time scale, as a

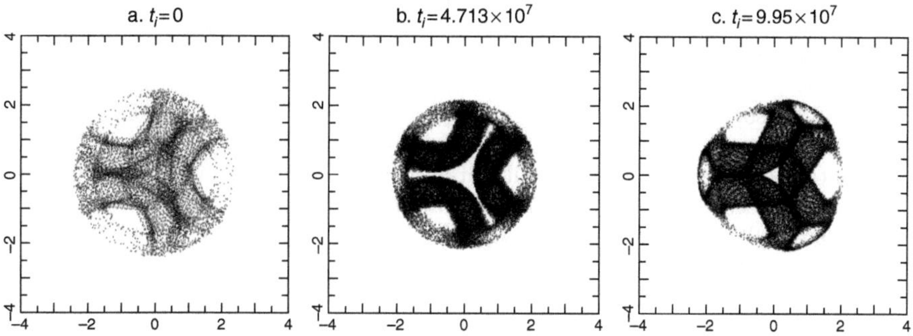

Fig. 7.14. Poincaré sections with $p_3 = 0$ for the FPU chain, projected on the plane (q_1, p_1), for the FPU α-model ($N = 15$, $\alpha = 1/4$, $\epsilon = 0.0398$), actually the same orbit as in Figs. 7.8, 7.9 and 7.10. In the left panel a time window of 10^5 is considered, in order to show the local chaoticity from the very beginning of the evolution. The other panel represent time window of 5×10^5, and show the confinement still evident after long times

results of the wandering, that happens from time to time, of the orbits among the different regions. In this respect, the presence of a third branch in the left panel of Fig. 7.5, in the Sect. 7.4 concerning the natural packets, may be seen as a further element supporting this interpretation, once it is considered as an intermediate meta-stable state between the natural packet and the reach of equipartition.

Acknowledgements

We warmly thank Antonio Giorgilli for having introduced us to the study of the FPU problem and for his constant support; we also thank Luigi Galgani for several useful discussions and Giancarlo Benettin for inspiring discussions and his notes on symplectic integrators.

References

1. D. Bambusi and A. Ponno, On meta-stability in FPU, Comm. Math. Phys, **264**, 539–561 (2006).
2. G. Benettin, Time scale for energy equipartition in a two-dimensional FPU model, Chaos, **15**, 015108, 10 (2005).
3. G. Benettin and F. Fassò, From Hamiltonian perturbation theory to symplectic integrators and back, Proceedings of the NSF/CBMS Regional Conference on numerical analysis of Hamiltonian differential equations (Golden, CO, 1997), **29**, 73–87 (1999).
4. G. Benettin, L. Galgani, A. Giorgilli and J.-M. Strelcyn, *Lyapunov characteristic exponents for smooth dynamical systems and for Hamiltonian systems; a method for computing all of them, part 1: theory*, Meccanica, 9–20 (1980).

5. G. Benettin, L. Galgani, A. Giorgilli and J.-M. Strelcyn, *Lyapunov characteristic exponents for smooth dynamical systems and for Hamiltonian systems; a method for computing all of them, part 2: numerical applications*, Meccanica, 21–30 (1980).
6. G. Benettin and A. Giorgilli, On the Hamiltonian interpolation of near-to-the-identity symplectic mappings with application to symplectic integration algorithms, J. Statist. Phys., **74**, 1117–1143 (1994).
7. L. Berchialla, L. Galgani and A. Giorgilli, Localization of energy in FPU chains, Discrete Contin. Dyn. Syst., **11**, 855–866 (2004).
8. L. Berchialla, A. Giorgilli and S. Paleari, Exponentially long times to equipartition in the thermodynamic limit, Phys. Lett. A, **321**, 167–172 (2004).
9. P. Bocchieri, A. Scotti, B. Bearzi and A. Loinger, Anharmonic chain with Lennard-Jones interaction, Phys. Rev. A, **2**, 2013–2019 (1970).
10. L.A. Bunimovich, S.G. Dani, R.L. Dobrushin, M.V. Jakobson, I.P. Kornfeld, N.B. Maslova, Y.B. Pesin, Y.G. Sinai, J. Smillie, Y.M. Sukhov and A.M.Vershik, *Dynamical systems, ergodic theory and applications*, vol. 100 of Encyclopaedia of Mathematical Sciences, Springer-Verlag, Berlin, revised edition, 2000. Edited and with a preface by Sinai, Translated from the Russian, Mathematical Physics, I.
11. A. Carati, L. Galgani, A. Ponno and A. Giorgilli, The Fermi-Pasta-Ulam problem, Nuovo Cimento B (11), **117**, 1017–1026 (2002).
12. L. Casetti, M. Cerruti-Sola, M. Pettini and E.G.D. Cohen, The Fermi–Pasta–Ulam problem revisited: stochasticity thresholds in nonlinear Hamiltonian systems, Phys. Rev. E (3), **55**, 6566–6574 (1997).
13. C. Cercignani, L. Galgani and A. Scotti, Phys. Lett. A, **38**, 403 (1972).
14. P. Cincotta, C. Giordano and C. Simó, Phase space structure of multi-dimensional systems by means of the mean exponential growth factor of nearby orbits, Phys. D, **182**, 151–178 (2003).
15. J. De Luca, A. Lichtenberg and M.A. Lieberman, Time scale to ergodicity in the fermi-pasta-ulam system, Chaos **5**, 283–297 (1995).
16. J. De Luca, A. Lichtenberg and S. Ruffo, Energy transition and time scale to equipartition in the Fermi–Pasta–Ulam oscillator chain, Phys. Rev. E (3), **51**, 2877–2884 (1995).
17. J. De Luca, A. Lichtenberg and S. Ruffo, Finite times to equipartition in the thermodynamic limit, Phys. Rev. E (3), **60**, 3781–3786 (1999).
18. E. Fermi, J. Pasta and S. Ulam, Studies of nonlinear problems, in Collected papers (Notes and memories). Vol. II: United States, 1939–1954, 1955. Los Alamos document LA-1940.
19. M. Fouchard, E. Lega, C. Froeschlé and C. Froeschlé, On the relationship between fast Lyapunov indicator and periodic orbits for continuous flows, Celestial Mech. Dynam. Astronom., **83**, 205–222, (2002). Modern celestial mechanics: from theory to applications (Rome, 2001).
20. C. Froeschlé, M. Guzzo and E. Lega, Graphical evolution of the Arnold web: from order to chaos, Science, **289**, 2108–2110 (2000).
21. C. Froeschlé and E. Lega, On the structure of symplectic mappings. The fast Lyapunov indicator: a very sensitive tool, Celestial Mech. Dynam. Astronom., **78** (2000), 167–195 (2001). New developments in the dynamics of planetary systems (Badhofgastein, 2000).
22. C. Froeschlé, E. Lega and R. Gonczi, Fast Lyapunov indicators. Application to asteroidal motion, Celestial Mech. Dynam. Astronom., **67**, 41–62 (1997).

23. F. Fucito, F. Marchesoni, E. Marinari, G. Parisi, L. Peliti, S. Ruffo and A. Vulpiani, Approach to equilibrium in a chain of nonlinear oscillators, J. Physique, **43**, 707–713 (1982).
24. L. Galgani, A. Giorgilli, A. Martinoli and S. Vanzini, On the problem of energy equipartition for large systems of the Fermi-Pasta-Ulam type: analytical and numerical estimates, Phys. D, **59**, 334–348 (1992).
25. L. Galgani and A. Scotti, Phys. Rev. Lett., **28**, 1173 (1972).
26. L. Galgani and A. Scotti, Recent progress in classical nonlinear dynamics, Riv. Nuovo Cimento (2), **2**, 189–209 (1972).
27. A. Giorgilli, S. Paleari and T. Penati, Local chaotic behaviour in the FPU system, Discrete Contin. Dyn. Syst. Ser. B, **5**, 991–1004 (2005).
28. M. Guzzo, E. Lega and C. Froeschlé, On the numerical detection of the effective stability of chaotic motions in quasi-integrable systems, Phys. D, **163**, 1–25 (2002).
29. M. Guzzo, E. Lega and C. Froeschlé, First numerical evidence of global Arnold diffusion in quasi-integrable systems, Discrete Contin. Dyn. Syst. Ser. B, **5**, 687–698 (2005).
30. E. Hairer, Backward analysis of numerical integrators and symplectic methods, Ann. Numer. Math., **1**, 107–132 (1994). Scientific computation and differential equations (Auckland, 1993).
31. E. Hairer, C. Lubich and G. Wanner, *Geometric numerical integration*, vol. 31 of Springer Series in Computational Mathematics, Springer-Verlag, Berlin, 2002. Structure-preserving algorithms for ordinary differential equations.
32. F.M. Izrailev and B.V. Chirikov, Stochasticity of the simplest dynamical model with divided phase space, Sov. Phys. Dokl., **11**, 30 (1966).
33. H. Kantz, Vanishing stability thresholds in the thermodynamic limit of nonintegrable conservative systems, Phys. D, **39**, 322–335 (1989).
34. H. Kantz, R. Livi and S. Ruffo, Equipartition thresholds in chains of anharmonic oscillators, J. Statis. Phys., **76**, 627–643 (1994).
35. E. Lega, M. Guzzo and C. Froeschlé, Detection of Arnold diffusion in Hamiltonian systems, Phys. D, **182**, 179–187 (2003).
36. V.I. Oseledec, A multiplicative ergodic theorem. Characteristic Ljapunov, exponents of dynamical systems, Trudy Moskov. Mat. Obšč., **19**, 179–210 (1968).
37. S. Paleari and T. Penati, *Relaxation time to equilibrium in Fermi–Pasta–Ulam system*, in Symmetry and perturbation theory (Cala Gonone, 2004), World Science Publishing, River Edge, NJ, 2004, pp. 255–263.
38. S. Paleari and T. Penati, *Equipartition times in a Fermi–Pasta–Ulam system*, Discr. Contin. Dyn. Syst., (2005). Dynamical systems and differential equations (Pomona, CA, 2004).
39. M. Pettini and M. Landolfi, Relaxation properties and ergodicity breaking in nonlinear Hamiltonian dynamics, Phys. Rev. A (3), **41**, 768–783 (1990).
40. B. Rink, Symmetry and resonance in periodic FPU chains, Comm. Math. Phys. **218**, 665–685 (2001).
41. C. Skokos, Alignment indices: a new, simple method for determining the ordered or chaotic nature of orbits, J. Phys. A, **34**, 10029–10043 (2001).
42. C. Skokos, C. Antonopoulos, T.C. Bountis and M.N. Vrahatis, Detecting order and chaos in Hamiltonian systems by the SALI method, J. Phys. A, **37**, 6269–6284 (2004).

43. K. Ullmann, A. Lichtenberg and G. Corso, Energy equipartition starting from high-frequency modes in the Fermi–Pasta–Ulam β oscillator chain, Phys. Rev. E (3), **61**, 2471–2477 (2000).
44. L. Verlet, Computer "experiments" on classical fluids. I.Thermodynamical properties of Lennard-Jones molecules, Phys. Rev., **159**, 98–103 (1967).

8

An Integrable Approximation for the Fermi–Pasta–Ulam Lattice

Bob Rink*

Mathematics Department, Imperial College London.
b.rink@ic.ac.uk.

Abstract. This contribution presents a review of results obtained from computations of approximate equations of motion for the Fermi–Pasta–Ulam lattice. These approximate equations are obtained as a finite-dimensional Birkhoff normal form. It turns out that in many cases, the Birkhoff normal form is suitable for application of the KAM theorem. In particular, this proves Nishida's 1971 conjecture stating that almost all low-energetic motions of the anharmonic Fermi–Pasta–Ulam lattice with fixed endpoints are quasi-periodic. The proof is based on the formal Birkhoff normal form computations of Nishida, the KAM theorem and discrete symmetry considerations.

8.1 Introduction

The Fermi–Pasta–Ulam (FPU) lattice is the famous discrete model for a continuous nonlinear string, introduced by Fermi, Pasta and Ulam [9]. It consists of a number of equal point masses that nonlinearly interact with their nearest neighbors. Assuming the lattice consists of a finite number of particles N and satisfies periodic boundary conditions, the physical variables of the FPU lattice are the positions q_j ($j \in \mathbb{Z}/N\mathbb{Z}$) of the particles, see Fig. 8.1, and their conjugate momenta p_j ($j \in \mathbb{Z}/N\mathbb{Z}$).

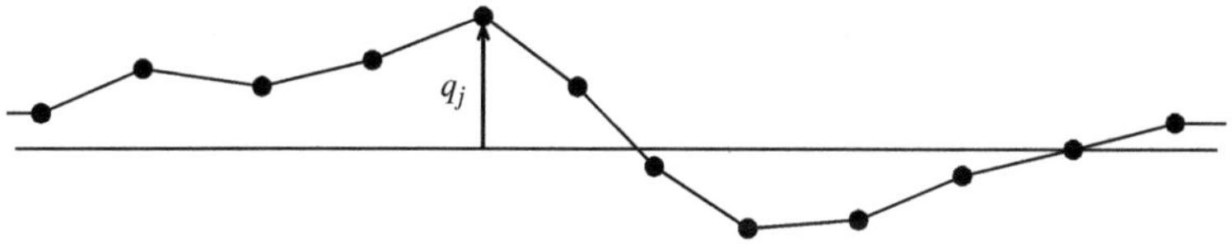

Fig. 8.1. Schematic picture of the FPU lattice

* The author is supported by an EPSRC postdoctoral fellowship and an MSRI general membership.

B. Rink: *An Integrable Approximation for the Fermi–Pasta–Ulam Lattice*, Lect. Notes Phys. **728**, 283–301 (2008)
DOI 10.1007/978-3-540-72995-2_8

Positions and momenta are elements of the $2N$-dimensional space of q_js and p_js, the cotangent bundle $T^*\mathbb{R}^N$. Equipped with the canonical symplectic form $\mathrm{d}q \wedge \mathrm{d}p := \sum_{j=1}^N \mathrm{d}q_j \wedge \mathrm{d}p_j$ this is a symplectic manifold and a Hamiltonian function $H : T^*\mathbb{R}^N \to \mathbb{R}$ generates the Hamiltonian vector field X_H implicitly defined by the relation $\mathrm{d}q \wedge \mathrm{d}p(X_\mathrm{H}, \cdot) = \mathrm{d}H$. That is the integral curves of X_H are the solutions of the system of ordinary differential equations

$$\dot{q}_j = \frac{\partial H}{\partial p_j} \ , \ \dot{p}_j = -\frac{\partial H}{\partial q_j} \ , \ j \in \mathbb{Z}/N\mathbb{Z} \ .$$

For the FPU lattice, the Hamiltonian function is the sum of the kinetic energies of all the particles and the interparticle potential energies:

$$H = \sum_j \frac{1}{2} p_j^2 + W(q_{j+1} - q_j) \ , \tag{8.1}$$

in which $W : \mathbb{R} \to \mathbb{R}$ is traditionally a potential energy density function of the form

$$W(x) = \frac{1}{2!} x^2 + \frac{\alpha}{3!} x^3 + \frac{\beta}{4!} x^4 \ . \tag{8.2}$$

The parameters α and β measure the nonlinearities in the forces between the particles in the lattice.

Fermi, Pasta and Ulam were interested in the statistical properties of the nonlinear FPU lattice. In fact, they expected that it would attain a thermal equilibrium, as is expected in statistical mechanics. This means that the initial energy of the lattice should be redistributed and, averaged over time, equipartitioned among all the Fourier modes of the lattice, see [18]. They performed a numerical experiment to investigate how and at what time-scale this would occur. The astonishing result of their integrations was that there was no sign of energy equipartition at all, see [9] and [18]: energy that was initially put in one Fourier mode was shared by only a few other modes. Moreover, within a rather short time nearly all the energy in the system returned to the initial mode. This recurrent behavior has been observed in experiments on the FPU lattice with quite small as well as very large numbers of particles, on short and long time-scales, and we are led to believe that at low energy the FPU lattice behaves more or less quasi-periodically. This observation was a big surprise. On the other hand, when the initial energy of the lattice is larger then a certain threshold, equipartition indeed occurs.

For a theoretical understanding of the FPU experiment, one has often tried to link the FPU lattice to a completely integrable system. These are dynamical systems possessing a complete set of integrals of motion and therefore they display the regular type of behavior that was observed in the FPU experiment. More precisely, it is well known [2] that periodic and quasi-periodic motion is typical in completely integrable finite-dimensional Hamiltonian systems due to the theorem of Liouville–Arnol'd. The FPU lattice is not completely integrable, but one can nevertheless remark the following.

First, it turns out that the special FPU lattice for which

$$W(x) = \frac{1}{a^2}\mathrm{e}^{ax} - \frac{1}{a^2}(1+ax) = \frac{1}{2!}x^2 + \frac{a}{3!}x^3 + \frac{a^2}{4!}x^4 + \frac{a^3}{5!}x^5 + \dots$$

is in fact completely integrable. This lattice is called the Toda lattice, and it possesses a Lax pair representation as was shown by Flaschka in [10]. For the general FPU lattice such a thing is definitely not true.

On the other hand, it is not difficult to derive integrable partial differential equations for the asymptotic evolution of long low-amplitude waves in FPU chains with a large number of particles. The first theoretical understanding of the FPU experiment therefore came when Zabusky and Kruskal [30] formally derived that the evolution of long *unidirectional* waves of low amplitude is at lowest order governed by a Korteweg–de Vries (KdV) equation. For instance, one may set $\varepsilon := \frac{1}{N} \ll 1$ and assume the existence of a smooth function $u^{\mathrm{L}} = u^{\mathrm{L}}(\tau, \xi)$ such that $q_j(t) = \varepsilon u^{\mathrm{L}}(\varepsilon^3 t, \varepsilon(t+j))$. One now quickly derives that this two time-scale traveling wave Ansatz leads to the identity $u^{\mathrm{L}}_{\tau\xi} = \frac{\alpha}{4}u^{\mathrm{L}}_{\xi}u^{\mathrm{L}}_{\xi\xi} + \frac{1}{24}u^{\mathrm{L}}_{\xi\xi\xi\xi} + \mathrm{O}(\varepsilon^2)$. On setting $v^{\mathrm{L}} = u^{\mathrm{L}}_{\xi}$, this reduces to $v^{\mathrm{L}}_{\tau} = \frac{\alpha}{4}v^{\mathrm{L}}v^{\mathrm{L}}_{\xi} + \frac{1}{24}v^{\mathrm{L}}_{\xi\xi\xi} + \mathrm{O}(\varepsilon^2)$. This is the easiest way I know to formally obtain a KdV equation for the evolution of unidirectional waves—in this case traveling to the left.

By studying the KdV equation numerically, Kruskal and Zabusky discovered the stability of the interaction of its solitons. It was later proved by Gardner et al. [13] that the KdV equation has infinitely many integrals. In fact, Peter Lax realised that KdV is a member of a hierarchy of integrable equations that have a Lax-pair, and therefore a complete set of integrals. See [21] for a good overview of these results. We now know that the solutions of the KdV equation (and all other equations in the KdV hierarchy) are almost-periodic, with a dense set of quasi-periodic solutions, see [19]. This could explain, to some extent, the observation of quasi-periodicity in the FPU experiment, although the exact connection between FPU and KdV is not very clear from the above formal derivation.

In order to derive the KdV equation rigorously, one may proceed as follows. First, one writes an exact evolution equation for an interpolation function u: setting again $\varepsilon := \frac{1}{N} \ll 1$, and assuming that $q_j(t) = \varepsilon u(t, \varepsilon j)$ for a smooth function $u = u(t,x), \mathbb{R} \times \mathbb{R}/\mathbb{Z} \to \mathbb{R}$, it is clear that $q_j(t)$ satisfies the FPU equations of motion if u satisfies the evolution equation

$$u_{tt}(t,x) = \frac{1}{\varepsilon}W'\left(\varepsilon u(t,x+\varepsilon) - \varepsilon u(t,x)\right) - \frac{1}{\varepsilon}W'\left(\varepsilon u(t,x) - \varepsilon u(t,x-\varepsilon)\right)$$

One should think of this equation as a second-order ordinary differential equation on a space of smooth functions of x of period 1. One now proceeds by defining the discrete Riemann-invariants

$$U^{\mathrm{L}}(t,x) := \frac{1}{\varepsilon}\left(u_t(t,x) + u(t,x+\varepsilon/2) - u(t,x-\varepsilon/2)\right)$$
$$U^{\mathrm{R}}(t,x) := \frac{1}{\varepsilon}\left(u_t(t,x) - u(t,x+\varepsilon/2) + u(t,x-\varepsilon/2)\right)$$

and observing, by Taylor expanding $u(t, x \pm \varepsilon/2)$ and $u(t, x \pm \varepsilon)$ with respect to ε, that the corresponding evolution equations for U^{L} and U^{R} can be expressed as

$$U_t^{\mathrm{L}} = \varepsilon U_x^{\mathrm{L}} + \varepsilon^3\left(\frac{\alpha}{4}(U^{\mathrm{L}} - U^{\mathrm{R}})(U_x^{\mathrm{L}} - U_x^{\mathrm{R}}) + \frac{1}{24}U_{xxx}^{\mathrm{L}}\right) + \mathrm{O}(\varepsilon^5)$$
$$U_t^{\mathrm{R}} = -\varepsilon U_x^{\mathrm{R}} + \varepsilon^3\left(\frac{\alpha}{4}(U^{\mathrm{L}} - U^{\mathrm{R}})(U_x^{\mathrm{L}} - U_x^{\mathrm{R}}) - \frac{1}{24}U_{xxx}^{\mathrm{R}}\right) + \mathrm{O}(\varepsilon^5)$$

Quite remarkably, it turns out that it is possible to make a small transformation $(U^{\mathrm{L}}, U^{\mathrm{R}}) \mapsto (\tilde{U}^{\mathrm{L}}, \tilde{U}^{\mathrm{R}}) = (U^{\mathrm{L}}, U^{\mathrm{R}}) + \mathrm{O}(\varepsilon)$ in a suitable space of smooth periodic functions of x that removes all coupling terms from the above evolution equations. In other words, the evolution equations for $\tilde{U}^{\mathrm{L}}$ and $\tilde{U}^{\mathrm{R}}$ can be expressed as

$$\tilde{U}_t^{\mathrm{L}} = \varepsilon \tilde{U}_x^{\mathrm{L}} + \varepsilon^3\left(\frac{\alpha}{4}\tilde{U}^{\mathrm{L}}\tilde{U}_x^{\mathrm{L}} + \frac{1}{24}\tilde{U}_{xxx}^{\mathrm{L}}\right) + \mathrm{O}(\varepsilon^5)$$
$$\tilde{U}_t^{\mathrm{R}} = -\varepsilon \tilde{U}_x^{\mathrm{R}} + \varepsilon^3\left(\frac{\alpha}{4}\tilde{U}^{\mathrm{R}}\tilde{U}_x^{\mathrm{R}} - \frac{1}{24}\tilde{U}_{xxx}^{\mathrm{R}}\right) + \mathrm{O}(\varepsilon^5)$$

This means that $\tilde{u}^{\mathrm{L}}(t,x) := \tilde{U}^{\mathrm{L}}(t, x - \varepsilon t)$ and $\tilde{u}^{\mathrm{R}}(t,x) := \tilde{U}^{\mathrm{R}}(t, x + \varepsilon t)$ satisfy approximate KdV equations, arising after a coordinate transformation as a "resonant normal form". It is not very hard to prove the long (but finite) time validity of these KdV equations. A result of this kind was proved by Bambusi and Ponno in [3], where the above transformation is obtained by the so-called method of averaging and the above estimates are made precise. A similar result was obtained by Wayne and Schneider in [28], although these authors use a multiple scales method. I am at the moment not aware of any results stronger than the long time validity of the KdV equations. It seems to be completely unknown, for example, whether any of the quasi-periodic solutions of the KdV equations persist (as KAM tori) in the FPU lattice.

Persistence results for quasi-periodic tori are easier to obtain in the finite dimensional setting. In the remainder of this paper, we shall therefore view the FPU lattice as a finite dimensional dynamical system. As is well-known [2], periodic and quasi-periodic motion is typical in completely integrable finite dimensional Hamiltonian systems. Unfortunately, apart from the Toda lattice, the FPU lattice is not completely integrable. One possible explanation of the recurrent behavior of the lattice is therefore based on the famous Kolmogorov–Arnol'd–Moser (KAM) theorem [2, 4]. This theorem explains that large measure Cantor sets of quasi-periodic motions can also exist in classes of nonintegrable Hamiltonian systems, namely those that can be viewed as small perturbations of certain integrable Hamiltonian systems. The restrictive requirement is that the integrable system that we are perturbing satisfies a nondegeneracy condition, which requires that each quasi-periodic motion of the integrable system has a different frequency. Even though various—again heuristic—arguments advocate this approach, and I mention in particular [17], the big problem is that it is not at all a priori clear whether the finite dimensional FPU lattice can really be viewed as a perturbation of such a *nondegenerate* integrable Hamiltonian system. The only obvious integrable approximation to the FPU lattice is its linearization, which is highly degenerate as its frequency map is constant. Exactly this problem was pointed out for instance in the review paper by Ford [11] and the book by Weissert [29],

An interesting attempt to prove the applicability of the KAM theorem arises in a paper by Nishida [20], who in 1971 considered the FPU lattice with a finite number of particles, fixed endpoints and symmetric potential energy density function (the so-called β-lattice). Analogous to the normal form construction for the derivation of the two KdV equations, Nishida computes the so-called Birkhoff normal form for the finite dimensional FPU lattice. *Assuming* a rather strong nonresonance condition on the frequencies of this lattice, he shows that this normal form constitutes a nondegenerate integrable approximation to the original lattice Hamiltonian. In this way, he proves the applicability of the KAM theorem and the existence of a positive measure set of quasi-periodic motions in the nonlinear FPU lattice. But note that all of this is under the assumption of a nonresonance condition, which unfortunately is only satisfied in exceptional cases. The actual value of Nishida's computation therefore remains unclear.

This contribution is based on [25], which is devoted to a full proof of what Nishida intended to show. Let me summarize the main result of [25] as follows:

> *The Fermi–Pasta–Ulam lattice with fixed endpoints and an arbitrary finite number of moving particles possesses a completely integrable finite order Birkhoff normal form, which constitutes an integrable appoximation to the original Hamiltonian function. The integrals are the linear energies of the Fourier modes. When the potential energy density function of the lattice is an even function (β-lattice), this integrable approximation is nondegenerate in the sense of the KAM-theorem. This proves the existence of a large-measure set of quasi-periodic motions in the low-energy domain of the β-lattice.*

The key to proving this result lies in the fact that Nishida's nonresonance condition, which a priori seems highly necessary for computing the Birkhoff normal form, is actually obsolete. As in [23, 24], which treat the FPU lattice with periodic boundary conditions, discrete symmetries are the key to proving Nishida's "conjecture". The results of the present paper can be considered as an extension of [23] to the lattice with fixed endpoints with a considerably simpler proof which again uses discrete symmetry together with a simple algebraic trick.

I want to remark here that the results of this paper do not provide any explicit bounds on the domain of validity of the normal form approximation. In particular we have at this moment no estimates on the behavior of this domain when n grows to infinity. The principal interest of the result lies in the fact that, at least to my knowledge, it is the first complete proof of the very existence of quasi-periodic motion in the FPU lattice with fixed endpoints.

8.2 Discrete Symmetry

The Hamiltonian function (8.1) of the periodic FPU lattice (i.e. summation over $j \in \mathbb{Z}/N\mathbb{Z}$) has discrete symmetries of which we shall discuss some dynamical consequences. Two important symmetries of the periodic FPU lattice are the linear mappings $R, S : T^*\mathbb{R}^N \to T^*\mathbb{R}^N$ defined by

$$\begin{aligned} R :&(q_1, q_2, \ldots, q_{N-1}, q_N; p_1, p_2, \ldots, p_{N-1}, p_N) \mapsto \\ &(q_2, q_3, \ldots, q_N, q_1; p_2, p_3, \ldots, p_N, p_1) \\ S :&(q_1, q_2, \ldots, q_{N-1}, q_N; p_1, p_2, \ldots, p_{N-1}, p_N) \mapsto \\ &-(q_{N-1}, q_{N-2}, \ldots, q_1, q_N; p_{N-1}, p_{N-2}, \ldots, p_1, p_N) \end{aligned}$$

It is easily checked that R and S are canonical transformations that leave the periodic FPU Hamiltonian (8.1) invariant, i.e. $R^*(\mathrm{d}q \wedge \mathrm{d}p) = S^*(\mathrm{d}q \wedge \mathrm{d}p) = \mathrm{d}q \wedge \mathrm{d}p$ and $R^*H(= H \circ R) = S^*H(= H \circ S) = H$. This implies that $R^*X_{\mathrm{H}} = X_{R^*\mathrm{H}} = X_{\mathrm{H}}$ and $S^*X_{\mathrm{H}} = X_{S^*\mathrm{H}} = X_{\mathrm{H}}$, that is R and S conjugate the Hamiltonian vector field X_{H} to itself. This in turn implies that R and S commute with the time-t flows $\mathrm{e}^{tX_{\mathrm{H}}}$ of X_{H}. Canonical diffeomorphisms with this property are called symmetries of H and the group of symmetries of H is denoted G_{H}. The subgroup $\langle R, S\rangle = \{\mathrm{Id}, R, R^2, \ldots, R^{N-1}, S, SR, SR^2, \ldots, SR^{N-1}\} \subset G_{\mathrm{H}}$ is isomorphic to the Nth dihedral group, the symmetry group of the N-gon, as its elements satisfy the multiplication relations $R^N = S^2 = \mathrm{Id}$, $SR = R^{-1}S$. As R and S are linear mappings, the elements of $\langle R, S\rangle$ actually define a representation of D_N in $T^*\mathbb{R}^N$ by symplectic mappings.

For every subgroup $G \subset G_H$, we define the fixed point set

$$\mathrm{Fix}\ G = \{(q,p) \in T^*\mathbb{R}^N | P(q,p) = (q,p)\ \forall P \in G\} \tag{8.3}$$

Let $(q,p) \in \mathrm{Fix}\ G$ and $P \in G$. Then $P(\mathrm{e}^{tX_{\mathrm{H}}}(q,p)) = \mathrm{e}^{tX_{\mathrm{H}}}(P(q,p)) = \mathrm{e}^{tX_{\mathrm{H}}}(q,p)$, i.e. $\mathrm{e}^{tX_{\mathrm{H}}}(q,p) \in \mathrm{Fix}\ G$. Thus we see that Fix G is an invariant manifold for the flow of X_H. Classification of the fixed point subgroups of the different subgroups of G_H leads to a collection of invariant manifolds, listed for instance in [24]. Other authors, cf. [5], have baptized these invariant manifolds *bushes of normal modes.* The restriction of a Hamiltonian vector field to a fixed point set of a group is often easy to compute:

Proposition 8.1. *When G is compact and consists of linear symplectic isomorphisms of $T^*\mathbb{R}^n$, then* Fix G *is a symplectic manifold with the restriction* *to* Fix G *of* $\mathrm{d}q \wedge \mathrm{d}p$ *as symplectic form. This implies that whenever* X_{H} *is tangent to* Fix G*, in particular when* H *is* G*-symmetric,*

$$(X_{\mathrm{H}})|_{\mathrm{Fix}\,G} = X_{(\mathrm{H}|_{\mathrm{Fix}\,G})}$$

Proof. Clearly, $\mathrm{Fix}\ G = \cap_{P \in G} \ker(P - \mathrm{Id})$ is a subspace of $T^*\mathbb{R}^N$, and hence a submanifold. It remains to be proven that for every $(q,p) \in \mathrm{Fix}\ G$, the restriction of $\mathrm{d}q \wedge \mathrm{d}p$ to $T_{(q,p)}(\mathrm{Fix}\ G) \subset T_{(q,p)}(T^*\mathbb{R}^N)$ is nondegenerate. Let us first of all identify $T_{(q,p)}(T^*\mathbb{R}^N)$ by $T^*\mathbb{R}^N$ and $T_{(q,p)}(\mathrm{Fix}\ G)$ by Fix G and moreover note that since G is compact, it contains a unique left-invariant probability measure "dP," called the Haar-measure of G. We can therefore define the operator

$$\mathrm{av}_G : T^*\mathbb{R}^N \to \mathrm{Fix}\ G\ ,\ v \mapsto \int_G P(v)\mathrm{d}P$$

The operator av_G is a projection of $T^*\mathbb{R}^N$ onto Fix G, called "averaging over G." Now let $v \in \mathrm{Fix}\ G$ and $w \in T^*\mathbb{R}^N$. Then one easily computes that

$$(\mathrm{d}q \wedge \mathrm{d}p)(v, \mathrm{av}_G(w)) = (\mathrm{d}q \wedge \mathrm{d}p)(v, \int_G P(w)\mathrm{d}P) = \int_G (\mathrm{d}q \wedge \mathrm{d}p)(v, P(w))\mathrm{d}P$$
$$= \int_G (\mathrm{d}q \wedge \mathrm{d}p)(P(v), P(w))\mathrm{d}P = \int_G (\mathrm{d}q \wedge \mathrm{d}p)(v, w)\mathrm{d}P = (\mathrm{d}q \wedge \mathrm{d}p)(v, w)$$

where in the second equality we have used the linearity of $\mathrm{d}q \wedge \mathrm{d}p$ in its second argument, in the third equality the fact that $v \in \mathrm{Fix}\ G$, and in the fourth equality that every $P \in G$ is symplectic. We now observe that when $v \in \mathrm{Fix}\ G$ and $(\mathrm{d}q \wedge \mathrm{d}p)(v, w) = 0$ for every $w \in \mathrm{Fix}\ G$, then $(\mathrm{d}q \wedge \mathrm{d}p)(v, w) = 0$ even for every $w \in T^*\mathbb{R}^N$. Hence $\mathrm{d}q \wedge \mathrm{d}p$, when restricted to $\mathrm{Fix}\ G \cong T_{(q,p)}(\mathrm{Fix}\ G)$, is a nondegenerate anti-symmetric bilinear form. In other words, Fix G is a symplectic subspace of $T^*\mathbb{R}^N$. The final statement of this proposition follows trivially from this result.

Let us now look at the fixed point set of one particular subgroup of the symmetry group of the periodic FPU lattice with an even number $N = 2n+2$ of particles, namely the group $\langle S \rangle = \{\mathrm{Id}, S\}$:

$$\mathrm{Fix}\,\langle S \rangle = \{(q,p) \in T^*\mathbb{R}^N | q_j = -q_{2n+2-j}, p_j = -p_{2n+2-j}\ \forall j\}$$

Clearly, in $\mathrm{Fix}\,\langle S \rangle$, $q_0 = q_{n+1} = p_0 = p_{n+1} = 0$. Thus we see that $\mathrm{Fix}\,\langle S \rangle$ is filled with solutions $(q_1(t), \ldots, q_N(t); p_1(t), \ldots, p_N(t))$ for which the $(q_1(t), \ldots, q_n(t); p_1(t), \ldots, p_n(t))$ constitute the general solution curves of the FPU lattice with fixed endpoints and n moving particles. Hence, the FPU lattice with fixed endpoints and n particles is embedded in the periodic lattice with $2n+2$ particles. By Proposition 8.1, it can be described as a Hamiltonian system on $\mathrm{Fix}\,\langle S \rangle$, which has the restriction of $dq \wedge dp$ as symplectic form, and is determined by the Hamiltonian function $H|_{\mathrm{Fix}\,\langle S \rangle}$. As coordinates on Fix S one could choose $(q_1, \ldots, q_n; p_1, \ldots, p_n)$.

8.3 Quasi-particles

Of course, the representation of D_N on $T^*\mathbb{R}^N$ is the sum of irreducible representations. It is quite natural to choose coordinates on $T^*\mathbb{R}^N$ that are adapted to these irreducible representations. For the periodic lattice, we thus make the following real-valued Fourier transformation. For $1 \leq k < \frac{N}{2}$ define:

$$Q_k = \sqrt{\frac{2}{N}} \sum_{j \in \mathbb{Z}/_{N\mathbb{Z}}} \sin\left(\frac{2jk\pi}{N}\right) q_j \ , \ P_k = \sqrt{\frac{2}{N}} \sum_{j \in \mathbb{Z}/_{N\mathbb{Z}}} \sin\left(\frac{2jk\pi}{N}\right) p_j$$

$$Q_{N-k} = \sqrt{\frac{2}{N}} \sum_{j \in \mathbb{Z}/_{N\mathbb{Z}}} \cos\left(\frac{2jk\pi}{N}\right) q_j \ , \ P_{N-k} = \sqrt{\frac{2}{N}} \sum_{j \in \mathbb{Z}/_{N\mathbb{Z}}} \cos\left(\frac{2jk\pi}{N}\right) p_j$$

$$Q_N = \frac{1}{\sqrt{N}} \sum_{j \in \mathbb{Z}/_{N\mathbb{Z}}} q_j \ , \ P_N = \frac{1}{\sqrt{N}} \sum_{j \in \mathbb{Z}/_{N\mathbb{Z}}} p_j$$

and if N is even:

$$Q_{\frac{N}{2}} = \frac{1}{\sqrt{N}} \sum_{j \in \mathbb{Z}/_{N\mathbb{Z}}} (-1)^j q_j \ , \ P_{\frac{N}{2}} = \frac{1}{\sqrt{N}} \sum_{j \in \mathbb{Z}/_{N\mathbb{Z}}} (-1)^j p_j$$

The new coordinates (Q, P) are called *quasi-particles* or *phonons*. The transformation $(q, p) \mapsto (Q, P), T^*\mathbb{R}^N \to T^*\mathbb{R}^N$ is symplectic and one can express the Hamiltonian in terms of Q and P. If we write for (8.1)

$$H = H_2 + H_3 + H_4$$

where H_2 is a quadratic polynomial in (q, p) and H_3 and H_4 cubic and quartic polynomials in q, then we find that (see [18], [22] or [26])

$$H_2 = \sum_{k=1}^{N} \frac{1}{2}(P_k^2 + \omega_k^2 Q_k^2) \tag{8.4}$$

in which for $k = 1, \ldots, N$ the numbers ω_k are the well-known normal mode frequencies of the periodic FPU lattice:

$$\omega_k := 2 \ \sin\left(\frac{k\pi}{N}\right)$$

This means that written down in quasi-particles, the equations of motion of the harmonic lattice ($\alpha = \beta = 0$) are simply the equations for $N-1$ uncoupled harmonic oscillators and, as $\omega_N = 0$, one free particle. In fact, the Hamiltonian system is Liouville integrable in this situation. Integrals are for instance the linear energies

$$E_k := \frac{1}{2}(P_k^2 + \omega_k^2 Q_k^2)$$

The FPU model is of course much more interesting when the forces between the particles are nonlinear, i.e. when α or β is nonzero. The normal modes then interact in a complicated manner that is governed by the Hamiltonians H_r ($r = 3, 4$), which are of the form

$$H_r = \sum_{\theta:|\theta|=r} c_\theta \prod_{k=1}^{N-1} Q_k^{\theta_k} \tag{8.5}$$

Here the θ are multi-indices and the c_θ are real coefficients. Note that for every value of α and β, H is independent of $Q_N = \frac{1}{\sqrt{N}}\sum_j q_j$. Hence the total momentum $P_N = \frac{1}{\sqrt{N}}\sum_j p_j$ is a constant of motion and the equations for the remaining variables are completely independent of (Q_N, P_N). It is common to set the latter coordinates equal to zero, thus remaining with a system on $T^*\mathbb{R}^{N-1}$ with coordinates $(Q_1, \ldots, Q_{N-1}, P_1, \ldots, P_{N-1})$. As $\omega_1, \ldots, \omega_{N-1} > 0$, we can conclude by the Morse Lemma or Dirichlet's theorem [1], that the origin $(Q, P) = 0$ is a dynamically stable equilibrium of this reduced system.

Assume again that $N = 2n+2$. From the definition of the quasi-particles and the definition of S, we conclude that S acts as follows in Fourier coordinates

$$S: \ (Q_1, \ldots, Q_{N-1}; P_1, \ldots, P_{N-1}) \mapsto$$
$$(Q_1, \ldots, Q_n, -Q_{n+1}, \ldots, -Q_{N-1}; P_1, \ldots, P_n, -P_{n+1}, \ldots, -P_{N-1})$$

So that

$$\mathrm{Fix}\langle S\rangle = \{(Q, P) \in T^*\mathbb{R}^{N-1} \mid Q_k = P_k = 0 \ \forall \ n+1 \le k \le N-1 \ \}$$

which is a symplectic manifold isomorphic to $T^*\mathbb{R}^n$. Using the coordinates $(Q_1, \ldots, Q_n; P_1, \ldots, P_n)$ on $\mathrm{Fix}\langle S\rangle$, the restriction of the symplectic form $\sum_{j=1}^N \mathrm{d}Q_j \wedge \mathrm{d}P_j$ to $\mathrm{Fix}\,\langle S\rangle$ is simply $\sum_{j=1}^n \mathrm{d}Q_j \wedge \mathrm{d}P_j$. By Proposition 8.1, the Hamiltonian of the fixed endpoint lattice thus is simply the restriction of the periodic FPU Hamiltonian (8.4, 8.5) to $\mathrm{Fix}\langle S\rangle$, that is

$$\begin{aligned} H|_{\mathrm{Fix}\langle S\rangle} = & \sum_{k=1}^n \frac{1}{2}(P_k^2 + \Omega_k^2 Q_k^2) \\ & + \ H_3(Q_1, \ldots, Q_n, 0, \ldots, 0) + H_4(Q_1, \ldots, Q_n, 0, \ldots, 0) \end{aligned}$$

To distinguish we have used the notation $\Omega_k := \omega_k = 2\sin(\frac{k\pi}{2n+2})$ $(1 \le k \le n)$ for the linear frequencies of the fixed endpoint lattice.

8.4 The Birkhoff Normal Form

Nishida's idea was to study the Hamiltonian of the fixed endpoint lattice using Birkhoff normalisation, which is a way of constructing a symplectic near-identity transformation of the phase-space with the purpose of approximating the original Hamiltonian system by a simpler one. The study of this "Birkhoff normal form" can lead to important conclusions about the original system. For $r \geq 2$, let $\mathcal{F}_r$ be the finite-dimensional space of homogeneous rth degree polynomials in (Q, P) on $T^*\mathbb{R}^{N-1}$ and let $\mathcal{F} := \bigoplus_{r\geq 2} \mathcal{F}_r$. With the Poisson bracket $\{\cdot,\cdot\} : \mathcal{F} \times \mathcal{F} \to \mathcal{F}$ defined by

$$\{F, G\} := \sum_{k=1}^{N-1} \frac{\partial F}{\partial q_k}\frac{\partial G}{\partial p_k} - \frac{\partial F}{\partial p_k}\frac{\partial G}{\partial q_k}$$

$\mathcal{F}$ is a so-called graded Lie-algebra, which means that $\{\mathcal{F}_r, \mathcal{F}_s\} \subset \mathcal{F}_{r+s-2}$. With this definition, we have for each $F \in \mathcal{F}$, the "adjoint" linear operator

$$\mathrm{ad}_F : \mathcal{F} \to \mathcal{F} \ , \ G \mapsto \{F, G\}$$

We recall the following result, a complete proof of which can be found for instance in [6, 7, 12].

Theorem 8.1. (Birkhoff normal form theorem). *Let $H = H_2 + H_3 + \cdots \in \mathcal{F}$ be a Hamiltonian on $T^*\mathbb{R}^{N-1}$ such that $H_r \in \mathcal{F}_r$ for each r and*

$$\mathrm{ad}_{H_2} : G \mapsto \{H_2, G\} \ , \ \mathcal{F}_r \to \mathcal{F}_r$$

is semi-simple (i.e. complex-diagonalizable) for every r. Then for every finite $s \geq 3$ there is an open neighbourhood $0 \in U \subset T^*\mathbb{R}^{N-1}$ and a symplectic diffeomorphism $\Phi : U \to T^*\mathbb{R}^{N-1}$ with the properties that $\Phi(0) = 0$, $D\Phi(0) =$ Id and

$$\Phi^* H = H_2 + \overline{H}_3 + \cdots + \overline{H}_s + \mathrm{O}(||(Q, P)||^{s+1})$$

where

$$\mathrm{ad}_{H_2}(\overline{H}_r) = 0$$

for every $3 \leq r \leq s$. The transformed and truncated Hamiltonian $\overline{H} := H_2 + \overline{H}_3 + \cdots + \overline{H}_s$ is called a Birkhoff normal form of H of order s.

Idea of proof. For $H, F \in \mathcal{F}$, the curve $t \mapsto (\mathrm{e}^{tX_F})^* H = H \circ \mathrm{e}^{tX_F}$ in $\mathcal{F}$ satisfies the linear differential equation and initial condition

$$\frac{\mathrm{d}}{\mathrm{d}t}(\mathrm{e}^{tX_F})^* H = -\mathrm{ad}_F((\mathrm{e}^{tX_F})^* H) \ , \ (\mathrm{e}^{0X_F})^* H = H$$

This implies that

$$(\mathrm{e}^{-X_F})^* H = \mathrm{e}^{\mathrm{ad}_F}(H) = H + \{F, H\} + \frac{1}{2}\{F, \{F, H\}\} + \cdots$$

The transformation Φ is now constructed as the composition of a sequence of time-−1 flows $\mathrm{e}^{-X_{F_r}}$ $(3 \leq r \leq s)$ of Hamiltonian vector fields X_{F_r} with $F_r \in \mathcal{F}_r$. The idea is that we first choose $F_3 \in \mathcal{F}_3$, such that H is transformed into

$$(\mathrm{e}^{-X_{F_3}})^* H = \underbrace{H_2}_{\in \mathcal{F}_2} + \underbrace{H_3 + \{F_3, H_2\}}_{\in \mathcal{F}_3} + \underbrace{\cdots}_{\in \mathcal{F}_4 \oplus \mathcal{F}_5 \oplus \ldots}$$

When ad_{H_2} is semi-simple, then $\mathcal{F}_3 = \ker \mathrm{ad}_{H_2} \oplus \mathrm{im}\, \mathrm{ad}_{H_2}$ and we can decompose $H_3 = H_3^{\mathrm{ker}} + H_3^{\mathrm{im}}$ for uniquely determined $H_3^{\mathrm{ker}} \in \ker \mathrm{ad}_{H_2}$ and $H_3^{\mathrm{im}} \in \mathrm{im}\, \mathrm{ad}_{H_2}$. If we now choose F_3 such that $\mathrm{ad}_{H_2}(F_3) = H_3^{\mathrm{im}}$, which obviously is possible, then $(\mathrm{e}^{-X_{F_3}})^* H = H_2 + \overline{H}_3 + \cdots$ for $\overline{H}_3 = H_3 + \{F_3, H_2\} = H_3 - \mathrm{ad}_{H_2}(F_3) = H_3 - H_3^{\mathrm{im}} = H_3^{\mathrm{ker}} \in \ker \mathrm{ad}_{H_2}$, i.e. $\mathrm{ad}_{H_2}(\overline{H}_3) = 0$. We continue by choosing $F_4 \in \mathcal{F}_4$ such that $(\mathrm{e}^{-X_{F_4}})^*((\mathrm{e}^{-X_{F_3}})^* H) = H_2 + \overline{H}_3 + \overline{H}_4 + \cdots$ for which $\mathrm{ad}_{H_2}(\overline{H}_4) = 0$, etc. After $s - 2$ steps we obtain $\overline{H}$ with the desired properties. □

The normal form $\overline{H}$ is usually simpler than the original H because it Poisson commutes with the quadratic Hamiltonian H_2. This firstly means that H_2 is a constant of motion for $\overline{H}$ and secondly that the flow $t \mapsto \mathrm{e}^{tX_{H_2}}$ is a continuous symmetry of $\overline{H}$.

Also, H and $\overline{H}$ are symplectically equivalent modulo a small perturbation of order $\mathrm{O}(||(Q, P)||^{s+1})$. Studying $\overline{H}$ instead of H thus means neglecting this perturbation term. So we make an approximation error, but this error is very small in the low energy domain, that is for small $||(Q, P)||$. With Gronwall's lemma, precise error estimates can be made.

Finally, I would like to mention the ill-known bijective correspondence between the relative equilibria of the Birkhoff normal form and the bifurcation equations for periodic solutions obtained by Lyapunov–Schmidt reduction, as is explained in [8].

For Hamiltonian systems with symmetry, the following elegant and well-known result is often useful, see [6, 12]:

Theorem 8.2. *Let $H = H_2 + H_3 + \cdots \in \mathcal{F}$ and G be a group of linear symplectic symmetries of H. Then a normal form $\overline{H} = H_2 + \overline{H}_3 + \cdots + \overline{H}_s$ for H can be constructed such that also $\overline{H}$ is G-symmetric.*

This is obvious when one realizes that the "generating functions" F_r of the proof of Theorem 8.1 can be chosen G-symmetric as well.

We shall also use the following result on normal forms of symmetric subsystems, which trivially follows from Proposition 8.1 and the proof of Theorem 8.2, as the transformations $\mathrm{e}^{-X_{F_r}}$ induced by symmetric Hamiltonian functions F_r leave Fix G invariant.

Corollary 8.1. *Let H be a Hamiltonian function with compact symmetry group G consisting of linear symplectic mappings. Then the normal form of $H|_{\mathrm{Fix}\, G}$ is simply the restriction of the symmetric normal form $\overline{H}$ of H to* Fix G*, i.e.*

$$\overline{H|_{\mathrm{Fix}\, G}} = \overline{H}|_{\mathrm{Fix}\, G}$$

This corollary tells us that it is sufficient to compute the normal form of the full system to know the normal forms of its symmetric subsystems. In particular, to find the normal form of an FPU lattice with fixed endpoints, it suffices to know the normal form of the appropriate periodic lattice. Normal forms of periodic lattices have been studied elaborately in [23].

8.5 Nishida's Conjecture

In his 1971 paper, Nishida proved the following result:

Theorem 8.3. (Proven by Nishida in [20]). *Consider the FPU lattice with fixed endpoints, $\alpha = 0$, $\beta \neq 0$ and n arbitrary. Assume moreover the fourth order nonresonance condition on the $\Omega_k = 2\sin(\frac{k\pi}{2n+2})$ $(1 \le k \le n)$ requiring that*

$$\sum_{k=1}^{n}(l_k - m_k)\Omega_k \neq 0 \ \forall \ l, m \in \{0,1,2,\ldots\}^n \text{ with } \sum_{k=1}^{n}|l_k| + |m_k| = 4$$

$$\text{and } \sum_{k=1}^{n}|l_k - m_k| \neq 0$$

Then the quartic Birkhoff normal form $\overline{H} = H_2 + \overline{H}_4$ of the lattice is a function of the action variables $a_k := E_k/\Omega_k$ $(1 \le k \le n)$ only and is therefore integrable. Moreover it satisfies the Kolmogorov nondegeneracy condition

$$\det \frac{\partial^2 \overline{H}}{\partial a_k \partial a_{k'}} \neq 0$$

This implies that almost all low-energy solutions of the β-lattice with fixed endpoints are quasi-periodic and move on invariant tori. More precisely, the relative Lebesgue measure of all these tori lying inside the small ball $\{0 \le H \le \varepsilon\}$, goes to 1 as ε goes to 0.

As we shall see later, the numbers

$$\sum_{k=1}^{n}(l_k - m_k)\Omega_k \ , \ \text{for} \ \sum_{k=1}^{n}|l_k| + |m_k| = 4$$

are simply the eigenvalues of ad_{H_2} on $\mathcal{F}_4$. Nishida's requirement that they be nonzero except in the trivial case that $l_k = m_k$ for all k thus just means that the subspace $\ker \mathrm{ad}_{H_2} \in \mathcal{F}_4$ in which $\overline{H}_4$ must lie is very low-dimensional. It must therefore be remarked here that the integrability of the normal form follows almost trivially from Nishida's nonresonance assumption. Nishida's article consists mainly of the explicit computation of the normal form $\overline{H}$ of H under the nonresonance assumption in order to check its nondegeneracy.

But unfortunately, resonances do occur, implying that Nishida's nonresonance condition is often violated. We have for instance the relations

$$\sin(\pi/6) + \sin(3\pi/14) - \sin(\pi/14) - \sin(5\pi/14) = 0$$

$$\sin(\pi/6) + \sin(13\pi/30) - \sin(7\pi/30) - \sin(3\pi/10) = 0$$

$$\sin(\pi/2) + \sin(\pi/10) - \sin(\pi/6) - \sin(3\pi/10) = 0$$

which lead to a violation of Nishida's nonresonance condition if $n+1$ is a multiple of 21 or 15.

Nishida refers to an unpublished result of Izumi proving a much stronger nonresonance condition on the Ω_k in special cases. The result states that no $\mathbb{Z}$-linear relations between the Ω_k exist if $n+1$ is a prime number or a power of 2. I was not able to trace back Izumi's proof of this statement, but note that a more general result had already been obtained in 1959 by Hemmer [14], who actually derived an expression for the total number of independent $\mathbb{Z}$-linear relations between the Ω_k $(1 \leq k \leq n)$ in terms of Euler's phi-function. It turns out that no $\mathbb{Z}$-linear relations exist if and only if $n+1$ is a prime number or a power of 2.

Moreover, as the above examples illustrate, resonance relations between 4 eigenvalues exist for several n and Nishida's condition is therefore sometimes violated. In this paper we will nevertheless prove "Nishida's conjecture" that his theorem holds without having to impose any nonresonance condition.

8.6 Near-Integrability

Let us start with a review of some observation in [23] for the periodic FPU lattice. First of all we note that, as the symmetry R is symplectic,

$$(R^* \circ \mathrm{ad}_{H_2})(G) = R^*\{H_2, G\} = \{R^* H_2, R^* G\} = \{H_2, R^* G\} = (\mathrm{ad}_{H_2} \circ R^*)(G)$$

where we have used that H_2 is R-symmetric. From this result we read off that R^* and ad_{H_2} commute as linear operators $\mathcal{F}_r \to \mathcal{F}_r$. This means that they can be diagonalized simultaneously. In [23] this is done by introducing new canonical coordinates $(Q, P) \mapsto (z, \zeta)$ as follows. For $1 \leq k < \frac{N}{2}$, we define:

$$\begin{aligned} z_k &= \frac{1}{2}(P_{N-k} - \mathrm{i}P_k) + \frac{\omega_k}{2}(Q_k + \mathrm{i}Q_{N-k}) \\ z_{N-k} &= -\frac{1}{2}(P_{N-k} - \mathrm{i}P_k) + \frac{\omega_k}{2}(Q_k + \mathrm{i}Q_{N-k}) \\ \zeta_k &= \frac{1}{2\omega_k}(P_k - \mathrm{i}P_{N-k}) - \frac{1}{2}(Q_{N-k} + \mathrm{i}Q_k) \\ \zeta_{N-k} &= \frac{1}{2\omega_k}(P_k - \mathrm{i}P_{N-k}) + \frac{1}{2}(Q_{N-k} + \mathrm{i}Q_k) \end{aligned}$$

and if N is even:

$$z_{\frac{N}{2}} = \frac{1}{\sqrt{2}}(Q_{\frac{N}{2}} - \frac{\mathrm{i}}{2}P_{\frac{N}{2}}) \ , \ \zeta_{\frac{N}{2}} = \frac{1}{\sqrt{2}}(P_{\frac{N}{2}} - 2\mathrm{i}Q_{\frac{N}{2}})$$

It is then not hard to compute that

$$H_2 = \sum_{1 \le k < \frac{N}{2}} \mathrm{i}\omega_k (z_k \zeta_k - z_{N-k}\zeta_{N-k}) + \mathrm{i}\omega_{\frac{N}{2}} z_{\frac{N}{2}} \zeta_{\frac{N}{2}}$$

which implies that if $\Theta, \theta \in \{0, 1, 2, \ldots\}^{N-1}$ are multi-indices, then

$$\mathrm{ad}_{H_2} : \, z^{\Theta}\zeta^{\theta} \mapsto \nu(\Theta, \theta) z^{\Theta}\zeta^{\theta}$$

in which ν is defined as

$$\nu(\Theta, \theta) := \sum_{1 \le k < \frac{N}{2}} \mathrm{i}\omega_k (\theta_k - \theta_{N-k} - \Theta_k + \Theta_{N-k}) \; + \mathrm{i}\omega_{\frac{N}{2}} (\theta_{\frac{N}{2}} - \Theta_{\frac{N}{2}}) \quad (8.6)$$

In other words, ad_{H_2} is diagonal with respect to the basis of $\mathcal{F}_r$ consisting of the monomials $z^{\Theta}\zeta^{\theta}$ for which $|\Theta| + |\theta| := \sum_{j=1}^{N-1} |\Theta_j| + |\theta_j| = r$ and the corresponding eigenvalues are the $\nu(\Theta, \theta)$. In particular we observe that ad_{H_2} is semi-simple on every $\mathcal{F}_r$, so that Theorem 8.1 indeed applies. A $\mathbb{Z}$-linear relation in the frequencies ω_k is called a *resonance*. For this reason, the monomials $z^{\Theta}\zeta^{\theta}$ for which $\nu(\Theta, \theta) = 0$ are called resonant monomials. They are particularly important because they are exactly the ones that are not in im ad_{H_2} and thus, as is clear from the proof of Theorem 8.1, the ones that cannot be transformed away by Birkhoff normalisation. As $\Omega_k = \omega_k (1 \le k \le n)$, Nishida's nonresonance condition is a consequence of its analogon for periodic lattices, that can be formulated as follows:

When $|\Theta| + |\theta| = 4$ *and* $\nu(\Theta, \theta) = 0$ *then* $\theta_{\frac{N}{2}} - \Theta_{\frac{N}{2}} = 0$ *and* $\theta_k - \theta_{N-k} - \Theta_k + \Theta_{N-k} = 0$ *for each* $1 \le k < \frac{N}{2}$.

Of course, this condition is not valid either.

On the other hand, one may compute, see [23], that the operator $R^* : G \mapsto G \circ R$ acts as follows on the coordinate function $z_k, \; \zeta_k$:

$$\begin{aligned} R^* : \quad & z_k \mapsto \exp(2\pi \mathrm{i} k/N) z_k, \; \zeta_k \mapsto \exp -(2\pi \mathrm{i} k/N) \zeta_k, \\ & z_{N-k} \mapsto \exp(2\pi \mathrm{i} k/N) z_{N-k}, \; \zeta_{N-k} \mapsto \exp -(2\pi \mathrm{i} k/N) \zeta_{N-k} \;, \\ & z_{\frac{N}{2}} \mapsto -z_{\frac{N}{2}} \text{ and } \zeta_{\frac{N}{2}} \mapsto -\zeta_{\frac{N}{2}} \end{aligned}$$

And as a result we conclude that, as promised, R^* acts diagonally with respect to the monomials $z^{\Theta}\zeta^{\theta}$ as well:

$$R^* : \; z^{\Theta}\zeta^{\theta} \mapsto \exp(2\pi \mathrm{i} \mu(\Theta, \theta)/N) \; z^{\Theta}\zeta^{\theta}$$

in which μ is defined as:

$$\mu(\Theta, \theta) := \sum_{1 \le k < \frac{N}{2}} \mathrm{j}(\Theta_k + \Theta_{N-k} - \theta_k - \theta_{N-k}) + \frac{N}{2}(\Theta_{\frac{N}{2}} - \theta_{\frac{N}{2}}) \mod N \quad (8.7)$$

By Theorem 8.2 we now know that the normal form of the periodic FPU Hamiltonian must be a linear combination of monomials $z^{\Theta}\zeta^{\theta}$ that are both resonant and symmetric, i.e. for which $\nu(\Theta,\theta) = 0$ and $\mu(\Theta,\theta) = 0 \lfloor\ \rceil N$. The following theorem was proven in [23]. The proof below is considerably simpler though.

Theorem 8.4. *(i) The set of multi-indices $(\Theta,\theta) \in \mathbb{Z}_{\geq 0}^{N-1}$ for which $|\Theta| + |\theta| = 3$, $\mu(\Theta,\theta) = 0$ mod N and $\nu(\Theta,\theta) = 0$ is empty.*

(ii) The set of multi-indices $(\Theta,\theta) \in \mathbb{Z}_{\geq 0}^{N-1}$ for which $|\Theta| + |\theta| = 4$, $\mu(\Theta,\theta) = 0$ mod N and $\nu(\Theta,\theta) = 0$ is contained in the set defined by the relations $\theta_k - \theta_{N-k} - \Theta_k + \Theta_{N-k} = \theta_{\frac{N}{2}} - \Theta_{\frac{N}{2}} = 0$.

Proof. (i) Suppose that $|\Theta| + |\theta| = 3$ and $\mu(\Theta,\theta) = 0 \lfloor N \rceil$. Then we can conclude from looking closely at (8.7) and (8.6) that there must be integers $k, l, m \neq 0 \lfloor\ \rceil N$ with $k + l + m = 0 \lfloor\ \rceil N$ for which $\nu(\Theta,\theta) = 2\mathrm{i}\sin(\frac{k\pi}{N}) + 2\mathrm{i}\sin(\frac{l\pi}{N}) + 2i\sin(\frac{m\pi}{N}) = 2\mathrm{i}\sin(\frac{k\pi}{N}) + 2\mathrm{i}\sin(\frac{l\pi}{N}) - 2\mathrm{i}\sin(\frac{k\pi}{N} + \frac{l\pi}{N})$. Now I learnt the following trick from Frits Beukers: write $2\mathrm{i}\sin(\frac{k\pi}{N}) = x - 1/x$ and $2\mathrm{i}\sin(\frac{l\pi}{N}) = y - 1/y$ for some x, y on the complex unit circle. Then $\nu(\Theta,\theta) = x - 1/x + y - 1/y - xy + 1/xy = (1-x)(1-y)(1-xy)/xy$. This is zero only in the trivial cases that $x = 1$ $(k = 0 \lfloor\ \rceil N)$, $y = 1$ $(l = 0 \lfloor\ \rceil N)$ or $xy = 1$ $(m = 0 \lfloor\ \rceil N)$. But we already knew that $k, l, m \neq 0 \lfloor\ \rceil N$. The result also follows from the convexity of the sine function.

(ii) The proof of the second statement is similar but more remarkable, and based on the fact that $2\mathrm{i}\sin\alpha + 2\mathrm{i}\sin\beta + 2\mathrm{i}\sin\gamma - 2\mathrm{i}\sin(\alpha+\beta+\gamma) = x - 1/x + y - 1/y + z - 1/z - xyz + 1/xyz = (1-xy)(1-xz)(1-yz)/xyz$, which again is zero in trivial cases only. □

In spite of Theorem 8.4, resonances do exist, as was illustrated by the examples in Sect. 8.5. A full classification of third- and fourth-order resonance relations in the FPU eigenvalues is given in the appendix to [23]. Resonance relations lead to several nontrivial resonant monomials. But according to Theorem 8.4 we now know that these nontrivial resonant monomials are not R-symmetric and hence cannot occur in the normal form of the periodic FPU lattice. As a first result, we immediately see now that there are no nonzero elements of $\mathcal{F}_3$ that are both resonant and R-symmetric. As a result, $\overline{H}_3 = 0$ automatically.

To formulate a result for $\overline{H}_4$, we need to define the following *Hopf-variables*. For $1 \leq k < \frac{N}{2}$, let

$$a_k := \frac{1}{2\omega_k}(P_k^2 + P_{N-k}^2 + \omega_k^2 Q_k^2 + \omega_k^2 Q_{N-k}^2)\ ,\ b_k := Q_k P_{N-k} - Q_{N-k} P_k$$

$$c_k := \frac{1}{2\omega_k}(P_k^2 - P_{N-k}^2 + \omega_k^2 Q_k^2 - \omega_k^2 Q_{N-k}^2)\ ,\ d_k := \frac{1}{\omega_k}(P_k P_{N-k} + \omega_k^2 Q_k Q_{N-k})$$

and if N is even

$$a_{\frac{N}{2}} := \frac{1}{2\omega_{\frac{N}{2}}}(P_{\frac{N}{2}}^2 + \omega_{\frac{N}{2}}^2 Q_{\frac{N}{2}}^2)$$

Note that H_2 can be expressed as

$$H_2 = \sum_{1 \le k \le \frac{N}{2}} \omega_k a_k$$

We moreover observe that when $N = 2n+2$, the identities $a_{\frac{N}{2}} = b_k = d_k = 0$ and $c_k = a_k = E_k/\Omega_k$ $(1 \le k < \frac{N}{2})$ hold on Fix $\langle S \rangle$, so that our definitions agree with the definition of a_k in Theorem 8.3. The following result was proven in [23] for the periodic FPU lattice. The proof consists of a careful analysis of the subspace of resonant and $\langle R, S \rangle$-symmetric polynomials in $\mathcal{F}_4$ with the help of Theorem 8.4. It is not very deep and we do not repeat it here.

Theorem 8.5. *Let $H = H_2 + H_3 + H_4$ be the periodic FPU Hamiltonian (8.4, 8.5). Then there is a unique quartic Birkhoff normal form $\overline{H} = H_2 + \overline{H}_4$ of H which is $\langle R, S \rangle$-symmetric. For this normal form we have $\overline{H}_3 = 0$, whereas $\overline{H}_4$ is a linear combination of the quartic terms $a_k a_{k'}$, $b_k b_{k'}$ $(1 \le k, k' < \frac{N}{2})$ and if N is even also $a_{\frac{N}{2}} a_k$ $(1 \le k \le \frac{N}{2})$ and $d_k d_{\frac{N}{2}-k} - c_k c_{\frac{N}{2}-k}$ $(1 \le k \le \frac{n}{4})$.*

Corollary 8.2. (Conjectured by Nishida in [20]) *Independent of n, α and β, the quartic Birkhoff normal form $\overline{H} = H_2 + \overline{H}_4$ of the FPU lattice with fixed endpoints (8.6) is integrable with integrals E_k.*

Proof. By Corollary 8.1, the Birkhoff normal form of (8.6) is the restriction of the Birkhoff normal form of the periodic lattice with $N = 2n+2$ particles, to Fix $\langle S \rangle$. But on Fix $\langle S \rangle$, we have that $b_k = d_k = 0$ and $a_k = c_k = E_k/\Omega_k$. So according to Theorem 8.5, $\overline{H}_4$ is a quadratic function of the Poisson commuting E_k. So, clearly, is H_2.

Note that to prove the integrability of the normal form of the fixed endpoint lattice, we had to use the *hidden* symmetry of the periodic lattice in which it is embedded. It must also be remarked here that it is very exceptional for a high-dimensional resonant Hamiltonian system to have an integrable normal form.

Let us dwell a little longer on the dynamics of the normal form and consider the integral map $E : T^*\mathbb{R}^n \to \mathbb{R}^n$ that sends $(Q, P) \mapsto (E_1, \ldots, E_n)$. One checks that when $E_k > 0$ for every k, the derivatives $DE_k(Q, P)$ are all linearly independent. As the level sets of E are moreover compact, the theorem of Liouville–Arnol'd ensures that for each $e = (e_1, \ldots, e_n)$ with $e_k > 0$ for each k, the level set $E^{-1}(\{e\})$ is a smooth n-dimensional torus.

To compute the flow on these tori, we transform to action-angle coordinates $(Q, P) \mapsto (a, \varphi)$ as follows. Let arg : $\mathbb{R}^2 \backslash \{(0,0)\} \to \mathbb{R}/_{2\pi\mathbb{Z}}$ be the argument function, arg : $(r\cos\Phi, r\sin\Phi) \mapsto \Phi$ and define

$$\varphi_k = \arg(P_k, \Omega_k Q_k) \ , \ a_k = E_k/\Omega_k = \frac{1}{2\Omega_k}(P_k^2 + \Omega_k^2 Q_k^2) \ , \ 1 \le k \le n$$

With the formula $\mathrm{d}\arg(x,y) = \frac{x\mathrm{d}y - y\mathrm{d}x}{x^2+y^2}$, one can verify that (φ, a) are canonical coordinates: $\mathrm{d}Q \wedge \mathrm{d}P = \mathrm{d}\varphi \wedge \mathrm{d}a$. So in these coordinates the equations of motion read

$$\dot{a}_k = 0 \ , \ \dot{\varphi}_k = \Omega_k + \frac{\partial \overline{H}_4(a)}{\partial a_k}$$

This simply defines periodic or quasi-periodic motion. Remark: (φ, a) are sometimes called "symplectic polar coordinates".

8.7 Nondegeneracy

To verify that the normal form $\overline{H}$ is nondegenerate, we examine the frequency map Ω which assigns to each invariant torus the frequencies of the flow on it:

$$\Omega : a \ \mapsto \left(\Omega_1 + \frac{\partial \overline{H}_4(a)}{\partial a_1}, \ldots, \Omega_n + \frac{\partial \overline{H}_4(a)}{\partial a_n} \right)$$

The nondegeneracy condition of the KAM theorem requires that Ω be a local diffeomorphism, which is the case if and only if the constant derivative matrix $\frac{\partial^2 \overline{H}_4}{\partial a_k \partial a_{k'}}$ is invertible. To check this, we will unfortunately need to compute the Birkhoff normal form explicitly, where until now we had been able to avoid this. In the next Theorem we shall present the normal form of the FPU Hamiltonian in the case that $H_3 = 0$, i.e. $\alpha = 0$. This lattice, that has no cubic terms, is usually referred to as the β-lattice.

Theorem 8.6. (Conjectured by Nishida in [20]) *If $\alpha = 0$, then a quartic Birkhoff normal form of FPU lattice with fixed endpoints is given by $\overline{H} = H_2 + \overline{H}_4$, where*

$$\overline{H}_4 = \frac{\beta}{2n+2} \left(\sum_{1 \le k < l \le n} \frac{\Omega_k \Omega_l}{4} a_k a_l + \sum_{1 \le k \le n} \frac{3\Omega_k^2}{32} a_k^2 \right)$$

Proof. The computation of the normal form had already been performed by Nishida [20] who obtained exactly the above normal form, but under the assumption that resonant monomials are absent in the lattice Hamiltonian. We now know that these monomials can not occur in the Hamiltonian as they are not R-symmetric in the corresponding periodic lattice. Hence Nishida's computation gave the correct answer.

The reader can find similar computations in [15, 16, 23, 26, 27] of the normal form of the β-lattice with periodic boundary conditions. We can therefore obtain the result also by substituting $a_k = c_k = E_k/\Omega_k$, $b_k = d_k = 0$ on $\mathrm{Fix}\langle S \rangle$ in the normal form of the periodic lattice that was obtained for instance in theorem 10.1 in [23].

It is now an easy excercise to prove the invertibility of the matrix $\frac{\partial^2 \overline{H}_4}{\partial a_k \partial a_{k'}}$. Its nondegeneracy was also checked by Nishida himself by applying elementary row and column operations to compute the determinant that turns out to be nonzero. Thus we conclude:

Corollary 8.3. (Conjectured by Nishida in [20]) *If* $\alpha = 0$ *and* $\beta \neq 0$, *then the integrable quartic Birkhoff normal form* $\overline{H} = H_2 + \overline{H}_4$ *of the FPU lattice with fixed endpoints (8.6) satisfies the Kolmogorov nondegeneracy condition. Hence almost all low-energy solutions of the FPU lattice with fixed endpoints are quasi-periodic and move on invariant tori. In fact, the relative measure of all these tori lying inside the small ball* $\{0 \leq H \leq \varepsilon\}$, *goes to* 1 *as* ε *goes to* 0.

Nishida and we chose to compute normal form $H_2 + \overline{H}_4$ only for the β-lattice. This computation is already quite long, but it becomes even harder when $\alpha \neq 0$. It should nevertheless also be possible to write down an expression for the fixed endpoints normal form if $\alpha \neq 0$. For checking Kolmogorov's condition this will actually be necessary. We know a priori that the resulting normal form will be integrable and depends quadratically on the E_k. Preliminary results by Henrici and Kappeler, partially referred to in [15, 16] seem to prove exactly what one expects, namely that Kolmogorov's nondegeneracy condition is satisfied for "generic" α and β.

Acknowledgment

The author would like to address many thanks to F. Verhulst, J.J. Duistermaat, G. Gallavotti, F. Beukers and A. Henrici for many valuable remarks and comments. The EPSRC and MSRI Berkeley are kindly acknowledged for financial support.

References

1. R. Abraham and J.E. Marsden, *Foundations of mechanics.* Benjamin/ Cummings Publishing Co., Reading, Mass., 1978, Second edition.
2. V.I. Arnol'd and A. Avez, *Ergodic problems of classical mechanics.* W.A. Benjamin, New York-Amsterdam, 1968.
3. D. Bambusi and A. Ponno, *On metastability in FPU*, Comm. Math. Phys. **264**, 539–561 (2006).
4. H.W. Broer, G.B. Huitema and M.B. Sevryuk, *Quasi-periodic motions in families of dynamical systems.* Springer-Verlag, Berlin, 1996.
5. G.M. Chechin, N.V. Novikova and A.A. Abramenko, *Bushes of vibrational modes for Fermi–Pasta–Ulam chains*, Physica D **166**(3–4), 208–238 (2002).
6. R.C. Churchill, M. Kummer and D.L. Rod. *On averaging, reduction, and symmetry in Hamiltonian systems*, J. Diff. Eqs. **49**(3), 359–414 (1983).
7. R.H. Cushman and L.M. Bates, *Global aspects of classical integrable systems*, Birkhäuser Verlag, Basel, 1997.

8. J.J. Duistermaat, *Bifurcation of periodic solutions near equilibrium points of Hamiltonian systems*, Bifurcation theory and applications (Montecatini, 1983), LNM 1057. Springer-Verlag, Berlin, 1984, pp. 57–105.
9. E. Fermi, J. Pasta and S. Ulam, *Los Alamos report LA-1940*, E. Fermi, Collected Papers II (1955), University of Chicago Press, Chicago, 1965, pp. 977–988.
10. H. Flaschka, The Toda lattice. II. Existence of integrals, Phys. Rev. B **9**, 1924–1925 (1974).
11. J. Ford, The Fermi–Pasta–Ulam problem: paradox turns discovery, Phys. Rep. **213**(5), 271–310 (1992).
12. G. Gaeta, Poincaré normal and renormalized forms, Acta Appl. Math. **70**(1–3), 113–131, (2002), Symmetry and perturbation theory.
13. C.S. Gardner, J.M. Greene, M.D. Kruskal, and R.M. Miura, Method for solving the Korteweg-de Vries equation, Phys. Rev. Lett. **19**, 1095–1097 (1967).
14. P.C. Hemmer, *Dynamic and stochastic types of motion in the linear chain*, Thesis, University of Trondheim, Trondheim, 1959.
15. A. Henrici and T. Kappeler, *Birkhoff normal form for the periodic Toda lattice*, nlin.SI/0609045 (2006).
16. A. Henrici and T. Kappeler, *Normal form for odd periodic FPU chains*, nlin.SI/0611063 (2006).
17. F.M. Izrailev and B.V. Chirikov, Statistical properties of a nonlinear string, Sov. Phys. Dokl. **11**, 30–32 (1996).
18. E.A. Jackson, Nonlinearity and irreversibility in lattice dynamics, Rocky Mountain J. Math. **8**(1–2), 127–196 (1978).
19. T. Kappeler and J. Pöschel, *KAM and KdV*, Springer, 2003.
20. T. Nishida, A note on an existence of conditionally periodic oscillation in a one-dimensional anharmonic lattice, Mem. Fac. Engrg. Kyoto Univ. **33**, 27–34 (1971).
21. R.S. Palais, The symmetries of solitons, Bull. Am. Math. Soc. (N.S.) **34**(4), 339–403 (1997).
22. P. Poggi and S. Ruffo, Exact solutions in the FPU oscillator chain, Physica D **103**(1–4), 251–272 (1997).
23. B. Rink, Symmetry and resonance in periodic FPU chains, Comm. Math. Phys. **218**(3), 665–685 (2001).
24. B. Rink, Symmetric invariant manifolds in the Fermi–Pasta–Ulam lattice, Physica D **175**, 31–42 (2003).
25. B. Rink, Proof of Nishida's conjecture on anharmonic lattices, Comm. Math. Phys. **261**, 613–627 (2006).
26. B. Rink and F. Verhulst, Near-integrability of periodic FPU-chains, Physica A **285**, 467–482 (2000).
27. J.A. Sanders, *On the theory of nonlinear resonance*, Thesis, University of Utrecht, Utrecht, 1979.
28. G. Schneider and C.E. Wayne, *Counter-propagating waves on fluid surfaces and the continuum limit of the Fermi–Pasta–Ulam model*, International Conference on Differential Equations, Vol. 1 2 Berlin, (1999), World Science Publishing, River Edge, NJ, 2000, pp. 390–404.
29. T.P. Weissert, *The genesis of simulation in dynamics*, Springer-Verlag, New York, 1997.
30. N.J. Zabusky and M.D. Kruskal, Interaction of "Solitons" in a collisionless plasma and the recurrence of initial states, Phys. Rev. Lett. **15**, 240–243 (1965).

FSC
www.fsc.org
MIX
Papier aus verantwortungsvollen Quellen
Paper from responsible sources
FSC® C105338

If you have any concerns about our products,
you can contact us on
ProductSafety@springernature.com

In case Publisher is established outside the EU,
the EU authorized representative is:
Springer Nature Customer Service Center GmbH
Europaplatz 3, 69115 Heidelberg, Germany

Printed by Libri Plureos GmbH
in Hamburg, Germany